S0-BLA-251

TRACE METALS IN THE ENVIRONMENT

Volume 1–THALLIUM

TRACE METALS IN THE ENVIRONMENT

Volume 1–THALLIUM

by

Ivan C. Smith
Senior Advisor for Environmental Science

Bonnie L. Carson
Associate Chemist

Midwest Research Institute
Kansas City, Missouri

ANN ARBOR SCIENCE
PUBLISHERS INC
P.O. BOX 1425 • ANN ARBOR, MICH. 48106

TD
196
.T7
S57
vol. 1

ENGINEERING
& PHYSICS
LIBRARY

Published 1977 by Ann Arbor Science Publishers, Inc.
P.O. Box 1425, Ann Arbor, Michigan 48106

Library of Congress Catalog Card Number 77-088486
ISBN 0-250-40214-9
Manufactured in the United States of America

The work upon which this publication is based was performed pursuant to Contract No. N01-ES-2-2090 with the National Institute of Environmental Health Sciences, Department of Health, Education, and Welfare.

PREFACE

In 1972, the National Institute of Environmental Health Sciences initiated a program with Midwest Research Institute to assemble information on production, usage, natural environmental levels, anthropogenic sources, human and animal health effects, and environmental impacts of selected trace elements. This book on thallium is one of a series of comprehensive documents which has resulted from that program. Scientists in many disciplines should find this work to be of value.

Ms. Bonnie Carson compiled the information and was the major author of this book, Dr. Ivan C. Smith served as the program advisor and editor of the manuscript. The authors wish to express their appreciation to Dr. Ernest Angino of the University of Kansas and to our colleagues at Midwest Research Institute: Mr. Paul Gorman, Mr. Mark Lance, Dr. Edward W. Lawless, and Dr. Mark Marcus for their assistance with various portions of this book. Particular thanks are due Dr. Harold Orel and Mrs. Doris Nagel for editorial comments, to the MRI library staff, and to Miss Audene Cook and her staff for typing the manuscript.

We also wish to acknowledge the information provided by various industries, the U.S. Geological Survey, Dr. Robert Horton of the U.S. Environmental Protection Agency, Dr. P. Michael Terlecky, Jr., of Calspan Corporation, and the assistance from various scientists throughout the world who have offered constructive comments regarding scope and organization of these manuscripts and who have encouraged their publication.

Finally, we especially acknowledge the support, encouragement and patience of Dr. Warren T. Piver (project officer) and Dr. Hans Falk of the National Institute of Environmental Health Sciences.

TABLE OF CONTENTS

CONTENTS (concluded)

List of Figures

List of Tables

List of Tables (continued)

List of Tables (concluded)

SUMMARY

This report was prepared under the sponsorship of the National Institute of Environmental Health Sciences (NIEHS). It summarizes for thallium (Tl) and its compounds the available information on environmental occurrence, uses, chemistry, processes, and physiological effects. Sources of thallium losses to the environment have been identified. The probable magnitude of certain losses and their chemical forms have been assessed, as well as the associated health hazards to humans and other life forms.

Thallium is a rare toxic element for which there are very few present uses. Current consumption is about 0.5 ton/year and is not well defined. The major uses are apparently in electrical and electronic applications. If there is future expansion, it is likely to come from thallium's versatility as an intermediate or catalyst in organic synthesis. The former major use (perhaps 5 tons/year at its peak) of thallium compounds in rodenticides (chiefly western ground squirrel control) and insecticides was terminated in the United States in 1972 because of accidental or secondary poisoning of wild animals and birds (especially carnivores). Children were often severely poisoned by ingesting the attractive baits. The use of thallium acetate as a cosmetic depilatory (a 7% cream) around 1930, as well as its use for about 50 years as a therapeutic epilant (one-time dose of 8 mg/kg in children) in the treatment of fungal scalp infections, was often accompanied by severe poisonings and fatalities.

Thallium forms compounds in both the monovalent and trivalent state. In ionic radius and chemical behavior, Tl^+ resembles the alkali metal cations. TlOH is water soluble. However, thallium(I) forms sparingly soluble compounds: sulfides, iodides, chlorides, chromates, etc., analogous to the heavy metals of Group Ib, Cu^+, Ag^+ and Au^+, and to its nearest neighbors in the sixth period, Hg_2^{2+} and Pb^{2+}. The standard potential of the Tl^{3+}/Tl^+ couple apparently favors Tl^+. The Tl^{3+} is assumed attainable only through oxidation by very powerful oxidizing agents in very acid media. Recent arguments, however, propose that formation of Tl(III) chloride complexes in high chloride media would favor the trivalent state. Tl(III) inorganic compounds are generally more water soluble, but hydrolysis to colloidal Tl_2O_3 and $TlOH^{2+}$ is extensive. Tl(III) forms more stable organic compounds and has been methylated in vitro by methyl vitamin B_{12}.

In the natural environment, thallium(I) is widely dispersed as an isomorphous replacement for K^+ in potassium feldspars and micas (silicates). The crustal abundance of thallium (~ 1 ppm) is largely due to these minerals in clays, soils, and granites. Independent thallium minerals are rare, but thallium sulfide is found in widely varying amounts in sphalerite, galena, pyrite, and other sulfide mineral deposits. Some manganese oxide deposits are especially enriched (in the tenth of a percent range), and deep-sea manganese nodules usually contain hundreds of parts per million (ppm) of thallium. The usual levels reported for thallium in freshwater and seawater are from 0.01 to 14 parts per billion (ppb) and 0.01 ppb, respectively. The only air levels reported were in Chadron, Nebraska, and were 0.04 to 0.48 ng thallium per cubic meter. Most authorities believe thallium is transported in aqueous media as Tl^+. Under strong reducing conditions, thallium becomes enriched in the sediments. Under strongly oxidizing conditions, thallium(I) may be oxidized and precipitate and/or be adsorbed as Tl_2O_3 along with hydrated manganese and/or iron oxides. Thallium(I) is strongly fixed by soils.

The commerical recovery of thallium has always been from materials associated with sulfide ore processing; that is, dusts collected from roasting and smelting zinc, lead, copper, and iron sulfide concentrates, and from purification of zinc sulfate solutions in electrolytic zinc refining and lithopone production. The only identified U.S. producer of commercial thallium and thallium sulfate, at least since the 1930's, has been ASARCO Inc., at its cadmium refinery in Denver. Thallium is recovered along with cadmium by leaching smelter flue dusts with sulfuric acid and electrolyzing the solution in two stages. At the other U.S. cadmium refineries, thallium may be removed as an impurity, as Tl_2O_3 when iron is precipitated and as chromate and chloride precipitates. Only 30 ppm thallium is allowable in the product cadmium. Most of the thallium entering the zinc and cadmium refineries is not recovered as such but leaves in product zinc or is recycled, ultimately ending up in some undefined waste stream.

Levels of thallium exceeding 2 ppm in the ash of terrestrial plants are abnormal, and such levels usually occur only in highly mineralized regions. Levels are usually much less than 1 ppm in the plant ash. The concentration factor is usually less than 20, but may be as high as a few hundred for algae in thallium-contaminated waters. The concentration factor is the concentration in the plant ash divided by the concentration in the medium. Background levels in aquatic plants and in terrestrial mammal tissues, except hair and nails, have not been measured. Up to 2.93 ppm thallium was found in aquatic animals in Minamata Bay. A concentration factor for sea organisms has been calculated at greater than 700, but freshwater studies have generally indicated animal concentration factors of less than 20. Based on extensive tissue data for a 16-year-old German male, a human body burden of about 2 ppb (about 0.14 mg/70 kg man) has been calculated. Dietary intake and excretion of thallium is on the order of 2 μg/day. Absorption is apparently complete by any route of administration. Like potassium, thallium is distributed in the intracellular space of most tissues. Transplacental absorption of maternally toxic doses can kill fetuses. Most studies indicate that macro amounts, as opposed to radiotracer doses, of thallium salts are excreted predominantly in the urine rather than in the feces.

At low concentrations, Tl^+ has an affinity for certain enzymes and an activating ability 10 times that of K^+; its effects on Na^+, K^+-activated ATPase of various tissues are those that have been most thoroughly studied. Thallium appears to uncouple mitochondrial oxidative phosphorylation (the reaction of adenosine diphosphate with inorganic phosphate to produce ATP). Toxic doses adversely affect protein synthesis and cause disaggregation of ribosomes. Thallium salts inhibit several enzymes such as succinic acid dehydrogenase and alkaline phosphatase, which plays a major role in bone formation.

Thallium salts have marked antimitotic activity on mammalian, avian, and plant cells. Thallium(I) is the most active heavy metal ion found for breaking chromosomes of pea plants. Thallous acetate solutions retarded both mitosis and meiosis in mosquitoes when the larvae and nymphs were treated.

Selenium- and sulfur-containing compounds offer some protection against thallium toxicity, but the only antidote that has given uniformly favorable results in animal experiments and clinical trials is Prussian blue (potassium ferrihexacyanoferrate(II)). Thallium acetate aggravates the symptoms of dihydrotachysterol overdosage (including nephrocalcinosis). The salt also induces renal calcification when given with sodium acetate, sodium citrate, sodium dihydrogen phosphate, and disodium hydrogen phosphate. Thallium given together with zinc or barium decalcifies osseous tissue. Thallium induces riboflavin deficiency in rats, leading to optic nerve damage.

In mammals, toxic effects are usually delayed for 1 to 2 days even when a fatal dose is given. The systemic effects in humans of thallium poisoning include: gastrointestinal symptoms; limb pain and paralysis; polyneuritis; high blood pressure; optic nerve atrophy and blindness; psychic excitement; liver and kidney damage; lymphocytosis and other blood changes; and (at least 10 days after poisoning) hair fall. Without known association of the patient with possible sources the systemic effects are difficult to ascribe to thallium poisoning until alopecia occurs. Even with thorough investigation, numerous diagnoses have been listed on the death certificates of thallium-poisoned murder victims; but to recognize symptoms as being due to an environmentally caused thallium intoxication would be even more difficult. Symptoms in animals are similar. Weight loss is frequently noted. When occupational poisonings have occurred, despite protective clothing, the effects have usually been severe.

The U.S. threshold limit value for thallium compounds in workplace air is 0.1 mg/m^3, largely based on analogy with other highly toxic compounds. The USSR stipulates 0.01 mg/m^3 as a ceiling limit. Limits of 0.005 to 0.1 and 10 ppm thallium have been suggested for desired maximum levels in freshwater (aquatic life, wildlife, and public supply) and irrigation waters, respectively.

For humans, doses of 14 mg/kg and above are fatal. The minimum lethal dose in birds is about 15 to 50 mg/kg. In mammals, the LD_{50}'s of thallium(I) inorganic compounds range from 15.8 to 71 mg/kg; of thallium(III) inorganic compounds, 5.66 to 72 mg/kg. The least toxic doses were for the oxides by the intraperitoneal route. The LD_{Lo}'s for thallium(I) compounds range from 7 to 45 mg/kg; for thallium(III) compounds, 5 to 62 mg/kg. Lethal doses of dimethylthallium(III) bromide to mice are 5 to 6 mg/kg, and diorganothallium(III) salts are at least five times more toxic to a pathogen on sugar cane than either thallous or thallic chloride.

Few of the reported human and mammalian studies of thallium lend themselves to any conclusions about the dangers of very low chronic intakes. One equivocal human study implies that minor toxic effects, as evidenced by hair fall, arise at daily intakes of about 10 to 20 μg/day. Chronic levels as low as 30 ppm thallium in the diet increase mortality in rats; 7 ppm produced blindness and affected movement in aged rats; and although 4 ppm did not affect rats, rabbits showed behavioral changes, mental retardation, and rear limb paralysis.

Major histopathological lesions occur in the gastrointestinal tract, nervous system and eye, liver, and kidneys. Mitochondria are swollen and show degenerative changes. Thallium also damages the heart. Atrophy and hyperplasia of the skin and hair follicles have been used to explain the alopecia of thallium poisoning. In the 1920's, gastrointestinal proliferative lesions were reported, and a poorly documented study in 1958 reports female reproductive tract cancers in mice.

The deleterious effects of thallium compounds on reproductive organs have been judged to be no worse than those of any other general cellular poison; but fetal development is altered and fetal mortality is increased. Chicks exhibit achondroplasia, extreme leg-bone curvature and shortening, parrot beak, and other abnormalities. Rats show non-ossification of the phalanges and vertebral bodies when the mother is given thallium while fed a low potassium diet. Mucopolysaccharide synthesis is inhibited so that columnar cartilage of the long-bone is hypoplastic and the calcifying zone is defective. Sea urchin egg development is inhibited by 20 to 200 ppm thallous or thallic chloride, but 48 and 240 ppm do not affect tadpole development in the egg.

Thallium is toxic to both aquatic plants (reduced photosynthesis at levels of 1 to 2 ppm) and higher terrestrial plants at levels of about 7 ppm, but microorganisms usually are resistant to higher concentrations. The LD_{50} for Atlantic salmon is 30 ppb, which is about 1/1,300 of the lethal concentration for two other fish species tested. Tadpoles are killed by 400 ppb. Small crustaceans (Daphnia, Gammarus) are killed at 2 to 4 ppm thallium. Paramecia mating behavior is altered by 3 to 7 ppm thallous salts, concentrations that prove fatal within 10 to 14 days.

We conclude that at least 1,600 tons of thallium is processed annually in the United States as a small impurity in many raw materials. In addition, unknown amounts are being leached from sulfide-ore and coal mines and decades-old accumulations of solid wastes from smelters, refineries, and gold-ore mills. About 30% of the thallium identified probably remains as a few parts per million impurity in refinery end products (levels usually lower than in the initial concentrates). About 55% of the 1,600 tons represents new additions to tailings and other solid wastes described above, most of which are confined in industrial land or lagoon disposal sites. A final report by Calspan Corporation, issued about the same time as this report, indicates that sulfide-ore processors contribute significant pollution to water from thallium in leachate, runoff, and mud erosion. Where thallium levels are known to be harmful to organisms, organisms are scarce due to even more toxic levels of other heavy metals and low pH.

About 15% of the 1,600 tons thallium mobilized annually enters the atmosphere, most of it from 415 coal-burning power plants in highly populated regions. The estimated coal-burning emissions (about 180 tons) contain up to 76 ppm thallium in the finest particulates. In the immediate vicinity of the plants, ambient air levels could contribute 11 to 14 µg thallium per 24-hr period to human intake. Emissions from at least some of the 21 copper, zinc, and lead smelters could be significantly more concentrated. Other probable contributors to atmospheric thallium emissions are the iron and steel industry and the burning of fuel oil. No data were available on thallium emissions from production of ferromanganese and silicomanganese, but this source could conceivably contribute thallium emissions on a scale comparable to that of coal burning.

Potash-derived fertilizers could be a direct source of thallium to the food chain, but our estimate of 5 tons thallium annually is speculative. A few hundred parts per million thallium in formerly used arsenic-based pesticides is another source for speculation.

Environmental contamination from mining and use of mica and feldspar is not expected.

Biota in thallium-contaminated areas currently have thallium levels ($\leq$ 3 ppm) that could be high enough to cause toxic symptoms if they constituted the entire diet of a mammal. Direct toxicity to plants and soil organisms removed from the polluting source is not expected even if they are irrigated with contaminated water.

We recommend that thallium concentrations be determined in raw materials, products, and waste streams for the copper, lead, zinc, manganese ferroalloy, and iron and steel industries so that accurate assessments of their pollution potential can be made. Ambient air levels should be determined in highly populated regions exposed to emissions from these sources and fossil fuel combustion. Dietary and drinking water intakes should be determined. Ultra-low level, long-term animal experiments are required to assess the mutagenic, teratogenic, reproductive, carcinogenic, and systemic toxicity potential for environmental thallium exposure in humans.

I. INTRODUCTION

Thallium (Tl) is a rare but widely dispersed element. All forms of thallium are soluble enough to be toxic to living organisms. Organothallium(III) derivatives are among the most toxic forms, and the possibility of environmental methylation of Tl(III) compounds has been suggested. Many common thallium compounds are volatile at smelting and coal-burning temperatures. Commercial U.S. thallium production and use has probably never exceeded 5 tons annually at its peak (before its ban as a pesticide). Current use is estimated at 0.5 ton. Thus, at the onset of this study, it was recognized that intentional production and use would probably be minor sources of environmental thallium releases.

The aim of this study was to identify and quantify, if possible, potential and actual environmental releases of thallium from known and plausible sources and assess the associated hazards. Information was gathered from published literature and industrial and governmental contacts. The major mineral occurrences were first determined to identify those industrial processes most likely to release thallium from their raw materials. Data were not complete enough to insure accurate estimates of thallium amounts entering process streams, let alone industrial losses. At best, our estimates are correct to an order of magnitude; they should, however, be valuable in identifying and ranking the sources of environmentally hazardous thallium releases. We have attempted to identify the processing steps or the waste and product streams within a process that are most likely to concentrate or lose thallium.

Thallium concentrations in exposed and unexposed organisms were compared to assess the potential for bioaccumulation of thallium.

The findings of this study are presented in the following eight major sections: Uses of Thallium Compounds and Alloys; Chemistry; Occurrence in the Geochemical Environment and Geochemistry; Industrial Processes as Environmental Sources of Thallium; Occurrence in Living Organisms; Physiological Effects; Environmental Thallium Losses and Associated Health Hazards to Humans and Other Life Forms; and Conclusions and Recommendations. The Appendix is an extensive tabulation of geochemical occurrences of thallium and the reported concentrations. Each of the sections and the Appendix has its own bibliography. To assist workers in the health fields, the physiological effects section also includes an author index, which is expected to be more helpful than a bibliography arranged alphabetically by first author. In each section, the tables appear together after the narrative; but the figures are placed in the text where they are mentioned.

II. USES OF THALLIUM COMPOUNDS AND ALLOYS

Since the prohibition of thallium compounds in rodenticides and insecticides, no uses of comparable volume remain. In fact, the commodity experts at the Bureau of Mines have seldom clearly indicated which end uses were really commercial. Table II-1 gives the kinds of information provided by Bureau of Mines and similar publications on thallium uses. A recent reference[1/] estimates that 70% of thallium consumption is for electronics, 9% for pharmaceuticals, and about 21% for other uses with apparent consumption falling from 3,949 lb in 1971 to an estimated 1,000 lb in 1975. In 1970, Greenspoon[2/] estimated 60% used for consumer electronics and communication equipment, 15% in electronic computing equipment, and about 10% each for agricultural pesticide and an "other" category (total U.S. demand estimated at 6,500 lb).* More complete, but conflicting use "statistics" are given in Table V-1.

Munch (1933)[5/] reviewed the early uses of thallium compounds and alloys (103 references). Waggaman et al. (1950)[6/] of the Bureau of Mines prepared an extensive review (total 369 references) of thallium properties, sources, recovery, analysis, and uses (86 references).

A. Pesticides

1. Thallium pesticides in the United States: Thallium rodenticides were first used in Germany about 1920.** California was using thallium sulfate-treated bait for rodent control as early as 1923, and in large amounts by 1927. By the late 1920's, thallium sulfate in sugar syrups was being used as an insecticide, especially for ants. In 1944 and early 1945, there were heavy demands by the U.S. military for thallium rodenticides to control rats in the war theaters. By 1952, the sale of thallium pesticides to the U.S. public was apparently largely limited to exterminators, but it was not until 1965 that use of thallium products to control rodents and insects was limited to government personnel by the U.S. Department of Agriculture (USDA) under the Federal Insecticide, Fungicide, and Rodenticide Act. In 1972, all pesticide uses were banned.

---

* The data for thallium consumption in 1964 to 1970 in Minerals Facts and Problems 1975[3/] may have been generated from commodity statements for internal use after the Minerals Yearbooks appeared. The agriculture figures may be for pesticide production for export. All figures are probably estimates.[4/]

** Zelio (Celio) paste or granules containing about 2% thallium sulfate.[7/]

The program for California ground squirrel extermination was based not only on their being agricultural pests, but also on their being carriers of fleas infected with Bacillus pestis (cause of bubonic plague). Intensive rat eradication efforts after the San Francisco earthquake and fire had spread the fleas to other animals. The program began about 1923.[5/] In western United States ground squirrel control, 1% thallium sulfate on barley, oats, or wheat ("Thalgrain") was prepared under state or federal direction (Munch, 1968).[8/]

Crafts (1934)[9/] and (1936)[10/] assumed that the poisoned animals died in their burrows at various depths in the soil, not leaving any residual contamination on the surface. Waggaman et al. (1952)[6/] stated that the animals usually died away from their burrows while looking for water.

Thallium sulfate in syrups was used to exterminate small red ants (Monomorium pharaonis); pavement ants (Tetramorium cespitum); and fire ants (Solenopsis geminata), which severely damage young citrus trees in the lower Rio Grande Valley, Texas, Arizona, California, and Puerto Rico.

Two patents (1929 and 1931) for rendering textiles mold- and mildew-resistant, moth-proof, and/or insect-repellent with thallium composition may have been the sources for statements that thallium has been used in textile treatment.

By 1933, 64 human poisonings in the United States and Europe and 14 deaths were caused from accidental ingestion of thallium rodenticidal baits.[5/]

A 1956 National Center for Health Statistics survey of 152 human deaths due to pesticides attributed eight accidental deaths to thallium. A 1961 survey listed 2 of 111 deaths as due to thallium pesticides.[11/]

In 1960, the USDA limited the concentration in household pesticides to a maximum of 1.25% thallium sulfate (1.0% calculated as thallium). The baits were often sugar syrups or impregnated bakery crumbs or peanut butter. The 1960 regulation[12/] emphasized that such attractive baits should be placed where they were not accessible to children and pets. Less attractive fillers such as sawdust and addition of an emetic were suggested. To avoid secondary poisoning, it was recommended that "dead roaches should be swept up, and dead rats and mice picked up and carefully disposed of by burning or burying deeply. Uneaten baits should be picked up carefully and disposed of in a similar manner."

The registration of thallium products under the Federal Insecticide, Fungicide, and Rodenticide Act was finally cancelled in 1972 for all uses as an economic poison.[13/] Besides dermal absorption, accidental ingestion, and accumulation in the body, persistence in the environment was listed as a reason. Technically, the action did not legally prevent the use of thallium sulfate still on hand with Pest Control Operators, although the National Pest Control Association saw "no defense for the continued use of thallium sulfate."[14/] Presumably, values appearing in Bureau of Mines publications since 1972 are exported pesticides.[4/]

2. Pesticide use in other countries: In 1968, numerous ant and roach poison formulations, especially ant "traps" for household use, were on the market in Canada (Mathews and Anzarut).[15/] Between 1958 and 1964, about 13 cases of thallium poisoning a year were reported to the Canadian poison control centers.

Formulations containing 0.6% particulate thallium sulfate are sprayed by helicopters over Hokkaido forests (Japan) as a rodenticide. Kitayama and Saito (1972)[16/] did not detect any thallium sulfate in water samples at the locations of spraying 0.8 kg/ha or 0.4 kg/ha nor any at distances up to 3,950 m. Foxes, weasels, martens, and other predators suffer secondary poisoning.[17/]

By 1966, if not before, thallium use as a rodenticide was forbidden in France.[18/]

Thallium was still being used in marketed rodenticide formulations in 1971 in Switzerland. Of 95 animal poisoning incidents reported to the Swiss Poison Information Center between April 1966 and April 1970, thallium sulfate or acetate was responsible for eight of them.[19/]

Thallium sulfate has been used for rodent control in Denmark since the 1930's. The 1973 Danish Toxicological Board list of classified pesticides included 59 preparations containing thallium sulfate. About 1 ton of thallium sulfate is sold annually in Denmark. Unintentional thallium poisoning has been observed in partridges, pheasants, red foxes, badgers, and other members of the marten family.[20/]

In 1965/1966, about 2.3 lb thallium sulfate was used as a rodenticide in three United Kingdom flour mills; the amount fell to less than 0.1 lb (two mills) by 1972.[21/] As of 1974, thallium salts in Britain were not easily available commercially, but it could still be obtained under the regulations of the Poisons and Pharmacy Act.[22/]

B. Electrical and Electronic Uses

According to the Bureau of Mines publications, this use category is the major one.

1. Thalofide cell: A change in the electrical resistance occur in certain thallium compounds upon exposure to light. Thallium oxysulfide in the "Thalofide cell" is more sensitive to low-intensity, long-wavelength light than the selenium cell. Several 1919 and 1920 patents were issued for the Thalofide photoelectric cell to T. W. Case, who also had patented a thallium-bromine compound for use as an electrical resistance material in the Bellphotophone.[23] Munch (1933)[5] and Waggaman et al. (1950)[6] reviewed the early literature. The thallium photocell was important in World War II communication systems.

Thalofide photocells have been used in bolometers (semiconductors for recording in the visible part of the spectrum), photographic exposure meters, signal systems for measuring star radiation, and optical photometry to control and regulate temperature.[24]

2. Lamps: In 1964, Western Electric Corporation announced a new mercury lamp containing thallium, sodium, and iodine, whose light is yellow-white instead of the normal mercury lamp's blue-green. Between 1967 and 1972, about a dozen patents were abstracted in Chemical Abstracts for thallium-containing mercury vapor lamps [about 0.02 to 1.5 mg Tl (often as TlI) per cubic centimeter].

Suntan and other ultraviolet lamps can use thallium-activated alkali halides or alkaline earth silicates and phosphates as phosphors. For example, Sylvania Electric Products in 1947 patented a composition containing 25 to 50% KCl, 3% Tl_2SO_4, and the rest $Ca_3(PO_4)_2$.[25]

Thallium salts have also been used as a getter in tungsten lamps.[25]

3. Electronics: Between 1958 and 1971, at least eight foreign firms were granted patents on thallium-doped selenium rectifiers. Other electronic uses include two U.S. patents (1964)[26] and (1963)[27] for semiconductors containing thallium, e.g., $TlInS_2$ and Tl_2Te_3. Thallium-activate NaI or NaCl crystals have been used with photomultiplier tubes in certain portable scintillation counters to detect gamma radiation. Thallium-containing compositions are also used for potting electronics (see Glass below).

C. Catalysts

It has been remarked that the future expansion of thallium uses will probably be as an intermediate or catalyst in organic synthesis.[24] Table II-2 summarizes a 20-year review of Chemical Abstracts (for patented thallium use as an organic reaction catalyst) shown in Table II-3. Of the 94 patents covered, there were 12 U.S. patents (some foreign firms) and 11

U.S. firms involved (holding 16 patents, some foreign). Apparently, organic oxidations are the most likely application of commercial thallium catalysts. A few patent abstracts dealing with hydrochloric acid oxidation in the presence of thallium (e.g., TlCl) were also seen.

Section III.B. briefly reviews the uses of thallium compounds in organic synthesis. Actual commercial applications have not been identified, but current research along these lines probably accounts for a sizable fraction of present thallium sales.

D. Optical Uses

The use of mixed thallium bromide-iodide crystals in infrared spectrometers and sniper detection and optical lens coatings is mentioned in Table II-2. A review of Chemical Abstracts 1957 to 1976 found patents in the following areas: coatings on optical surfaces, 2; laser materials, 2; fiber optics, 7 (e.g., Tl_2O-SiO_2 fiber); multifocal lens, 1; and optical glass, 2.

The preparations of Tl_3VS_4, Tl_3NbS_4, and Tl_3PSe_4 single crystals for use in acoustooptical devices such as display devices, laser modulators, noncollinear optical filters, and acoustic delay lines are patented by Westinghouse Electric Corporation (1975).[28,29]

E. Glass

Added in small amounts to glass, thallium renders the glass dark, opaque brown or black; increases its density; and increases the refractive index. Thallium oxides produce greenish-yellow flint glass. Thallium glasses are better for production of artificial gems than corresponding lead glasses because of their higher refractive indices.[5] Patents for thallium-containing glass and ceramic compositions are frequent in Chemical Abstracts.

Several patents for use of thallium (usually its oxides) in semiconducting or electrically resistant glass have appeared in Chemical Abstracts since 1961. Most of the glasses were used to seal, or coat, or encapsulate semiconductor elements to protect them from atmospheric oxidation, contamination, and humidity. Pearson (1964)[30] of Bell Telephone Laboratories described the preparation and properties of low-melting arsenic-group chalcogenide glasses, some of which contained thallium, e.g., As-Se-Tl. The elements were fused in batch processes.

F. Alloys

Thallium forms binary alloys with antimony, arsenic, barium, bismuth (fusible alloys), cadmium, calcium, cerium, cobalt, copper, germanium, gold, indium, lanthanum, lead, lithium, magnesium, manganese, mercury, nickel, palladium, platinum, praseodymium, selenium, silver, strontium, sulfur (TlS, Tl_2S), tellurium, and tin.[25,31-34]

Among the ternary alloys of thallium are Tl-Pb-Bi, Tl-Al-Ag, Mg-Tl-Ag,[31,35] In-Hg-Tl (patented solder for semiconductors and ceramics),[36] Sn-Cd-Tl, Bi-Sn-Tl, and Bi-Cd-Tl.[25]

The quaternary thallium alloys include Tl-Sn-Cd-Bi,[31] Pb-Sn-Bi-Tl, and Pb-Cd-Bi-Tl.[25]

Howe and Smith in 1950[25] reviewed suggested uses for thallium alloys, none of which appeared to be commercial. The Hg-Tl system was apparently being used in 1950 for low temperature applications, but no reference is cited. Petar in her 1931 review[23] first mentioned such a use, but no reference was cited.

Table II-1 notes reported uses of thallium alloys, typically bearings and fusible alloys (solders and fuse metal); it is uncertain if any of these were commercial uses. A few patents for solders and fuse metals containing thallium have appeared in Chemical Abstracts since 1958.

G. Medical and Cosmetic Uses

Thallium compounds have been used internally to treat gonorrhea, syphilis, dysentery, and "night sweats" of tuberculosis, and externally in depilatories (also called "epilants"--agents that loosen hair at the roots) in the treatment of Trichophyton (favus, tinea capitis) and microsporon (ringworm) infections of the scalp mostly in prepubescent children.[5,37]

Thallium compounds were widely used from 1912 to 1930 as epilants to treat ringworm of the scalp in children. An ointment (petrolatum-lanolin base) containing 1% thallium acetate was commonly used. Thallium epilants fell into disrepute because of their toxicity.[28]

Munch (1933)[5] found records for 8,006 patients treated with thallium acetate depilatory; 5.5% (447) patients showed some degree of intoxication. About one child per 1,000 treated died. Among 33 pubescent children or adults who showed poisoning symptoms, 22 died.

Koremlu cream, a depilatory containing 7.18% thallium acetate 5.6% thallium), was widely advertised in the United States in the late 920's and early 1930's. Polyneuritis occurred after 4 months' use. Symp-oms resembling those from intracranial tumors were suspected in three atients until their illness was traced to the prolonged use of the cream.[39,40] ltogether more than 51 cases of poisoning from use of Koremlu cream were eported in the literature.[7] In many cases, post-optic hypertrophy and lindness resulted. Munch (1968)[8] relates that "This was one of the 'hor-ible examples' considered by Congress in amending the Federal Food and rugs Act in 1938."

Surprisingly enough, thallium treatment has been suggested in zech medical literature for stimulating hair growth. Sinka and Antosovsky 1970)[41] treated 83 patients suffering from alopecia areata (patchy bald-ess of the scalp) aged 13 to 57 years with an orally administered thallium reparation for 2 to 6 months. The therapy (comprising 10 drops three imes a day of the following mixture: 0.09 mg thallium acetate, 0.03 mg chinacea purpurea, and 120 ml 45% ethanol) was well tolerated and was as ffective as other current methods. At 2 ml of solution (1.2 µg Tl) per ay for 6 months, a patient would have received less than 0.5 mg.

Intravenous injection of ^{201}Tl (half-life = 12.2 days) has been uggested for myocardial imaging to evaluate the distribution of regional yocardial perfusion.[42,43] Unlike presently used agents for kidney studies, hich concentrate in the cortex, thallium concentrates in the renal medulla. adiothallium may be useful for melanoma diagnosis. It has been shown that l^{+} is taken up more by tissues of pigmented rabbits than by those of albinos.[44]

. Miscellaneous

1. Photography and xerography: Chemical Abstracts in the past 5 years has abstracted at least 11 patents involving thallium compounds in hotographic formulations (in the roles of sensitizer, photopolymerization atalysts, halide layer, developer addition, and antifoggant). Thallium as also been patented as a selenium dopant in xerography (two patents) and n electroless oxide coating (one patent).

Munch (1933)[5] reviewed the early uses of thallium in photography, nd reported that it was used to modify the rate of development. TlCl forms he insoluble, dark complex $TlCl_3 \cdot 3TlCl$ on light exposure, sometimes as fast s the formation of silver chloride.

2. Plating: Since 1962, Chemical Abstracts has included a dozen or so patents involving thallium compounds in chemical or electrolytic plat-ing. ASARCO has recently sold some thallium for use in gold plating.[45]

Sel-Rex Corporation holds an unexamined German patent[46] for the use of 15 to 140 mg Tl/liter in a gold-plating bath. A recent Japanese Kokai includes 0.3 to 1 g/liter thallium in a high-speed gold cyanide electroplating bath.[47]

3. Antiknock compounds for gasolines: Use of thallium compounds as antiknock additives for gasoline for internal combustion engines has been mentioned since a 1927 British patent and a 1930 U.S. patent were issued to the same inventor. Among the suitable compounds were thallium benzoate and thallium alcoholates (about 2 g compound per liter of fuel).[5] An antiknock composition containing 10% thallous formate in aqueous methano was patented in 1957 by California Research Corporation.[48] A patent applied for by Ethyl Corporation in 1952 was finally granted in 1967.[49] Cyc pentadienylthallium (0.03 to 8 g Tl/gal) was used in an example as the Grou III metal antiknock compound.

4. Other uses: Munch (1933)[5] mentions the preparation of thal lium chromate pigments (orange-yellow, -red, and -brown). He assumed they had some commercial value. Waggaman et al. (1950)[6] state that thallium chromate pigments were of "considerable commercial value" for artists' pain

TABLE II-1

HISTORICAL REVIEW OF THALLIUM USES

Year	Medicines	Electrical and Electronics	Optical Uses	Alloys	Pesticides	Miscellaneous	Reference
~ 1896	X						2
1910		Thalofide cell patented					2
1922		Photoelec. device similar to Se cell termed "a recent use"					50
1923					Calif. rodent control beginning		5
1924	Salts (mainly acetate)	Photoelec. cell	Optical glass				51
1925	X	Elec. lamp industry photoelec. cell	Optical glass		X		52
1926						Heavy liquids to determine specific gravity of minerals. Antiknock additive for gasoline	53, 25
1927					Rodent and ant control most important use		54
1929					Tl_2SO_4 in sugar syrup for ant control		55
1930	X	Photoelec. cell			X (major use)	Photography	56
1937	X		Optical glass	Quite small amounts for special alloys. Pb-Tl	Insecticide and fungicide. Mothproofing. Fabric finishing	Antiknock additive. Depilatory use discouraged	M.Y.[a]
1942				3-65% Tl-Pb alloys proposed as a bearing metal			M.Y.

TABLE II-1 (continued)

Year	Medicines	Electrical and Electronics	Optical Uses	Alloys	Pesticides	Miscellaneous	Reference
1943					Rodent and ant control still major use		M.Y.
1944					Rat control by military in war theaters		M.Y.
1945				Patents issued on Cu-Tl, Cu-Pb-Tl, and Ag-Tl bearing materials	Rodenticides in Europe until mid-summer. Sale to public barred		M.Y.
1946			Tl bromoiodide crystals in sniper detection and signaling devices	Low temperature 8.5% Tl-Hg alloy			M.Y., 1946, 1948, 1952
1947		Phosphor activator for ultraviolet fluorescent suntan lamps (Sylvania patent)					25
1948		Photoelec. cells. Se rectifiers	Tl bromoiodide crystals as lenses, plates, and prisms in infrared spectrometers (research use at the time). Special glasses	Ag and Pb alloys	Mold- and insect-proofing	High-density liquids	25 M.Y.
1949			Predicted future bulk use of Tl bromoiodide crystals in military infrared signal transmission and research infrared devices. (Harshaw Chemical Company produced the crystals commercially)		Rodenticide still major use		M.Y.

TABLE II-1 (continued)

Year	Medicines	Electrical and Electronics	Optical Uses	Alloys	Pesticides	Miscellaneous	Reference
1950				No alloys commercially important			25
1951			Lens components for microscope objectives for use in infrared optical instruments				M.Y.
1952		Photoelec. cells. Minor use in Se rectifiers and scintillation counters. Phosphor activator for cathode ray tubes			Sale to public generally forbidden		M.Y.
1953		Tl-activated NaCl or NaI crystals in scintillation counters for U and Th prospecting		Tarnish-resistant silverware of Ag-Tl		Pigments, glass, colorizers, signals, and rockets	M.Y.
1954				Bi-Pb-Sn-Tl alloys considered by AEC as a coolant for atomic piles			M.Y.
1955						Green flares	M.Y.
1956		Expected military electronic and phosphor use expansion	Use of Tl bromo-iodide crystals small but important	Hg-Tl low-temperature switches for military. Expected military alloy use expansion	Tl_2CO_3 termed a good fungicide. Protection of textiles, leather, etc., against weather, fungi, and insects	Photog. emulsions	57
1957					Fungicide is first use listed		M.Y.

TABLE II-1 (concluded)

Year	Medicines	Electrical and Electronics	Optical Uses	Alloys	Pesticides	Miscellaneous	Reference
1958		Electronics uses small but important, e.g., portable scintillation counters					M.Y.
1961		Feasibility of use in Hg vapor lights studied					M.Y.
1962		Glass seals for protecting electronic components					M.Y.
1964		New Hg lamp contg. Tl (Westinghouse Elec. Corp., Lamp Div.)					M.Y.
1965					Private use of insecticides and rodenticides banned		M.Y.
1969						High-purity Tl metal used chiefly for metallurgical research	M.Y.
1970		Electronics area of future expansion, if any. As-Se-Tl semiconductor glasses studied	Optical instrument glass	Ag-Tl in elec. contacts. Pb-Tl in solder and fuse metal. Sn-Pb-Tl anodes for electrowinning Cu. V-Tl catalyst for H_2SO_4 manuf. Small amounts of Tl alloy used in telephones		Minerals sepn. by thallous silver nitrate and thallous formate in water. Color glass brown or black. Synthetic gems	2
1972					Pesticide use banned for all applications		1
1977		Source of Tl in electronics may be imports		Special alloys (ASARCO sales)		Specialty gold plating (ASARCO sales)	45

a/ M.Y. = Minerals Yearbooks of the year cited.

TABLE II-2

SUMMARY OF ORGANIC REACTIONS CATALYZED BY THALLIUM, 1957-1976

Reaction Class	Tl Critical	Tl Optional	U.S. Patents	U.S. Firms
Epoxidation	5	1	1	1 (Union Carbide)
Oxidation of hydrocarbons, olefins, unsatd. aldehydes	18	17	3	2 (Dow) (Union Oil Company of California)
Ammoxidation (prepn. of nitriles from olefins)	5		2	2 (Du Pont)
Olefin polymerization, oligomerization, or dimerization	5	17	2	7 (Hercules Powder Company) (2-Phillips Petroleum Company) (Celanese) (2-Monsanto) (GAF)
Condensation polymerization (polysulfides, polyethers, polyesters, polysiloxanes)	4	3		2 (Hooker Chemical Corporation) (Union Carbide)
Dehydrogenation	2	2	1	
Hydrogenation		2		
Isomerization (reforming)		4		
Degradation		3	1	
Miscellaneous	2	4	2	2 (Ethyl Corporation)
Total	41	53	12	16 (11 firms)

TABLE II-3

ORGANIC REACTION CATALYZED BY THALLIUM [PATENTS ABSTRACTED IN CHEMICAL ABSTRACTS (CA), 1957-1976]

Reaction	Form if Tl Critical	Tl Optional	Country of Patent	Firm or Nationality	CA Reference
Epoxidation					
Propylene to propylene oxide		X	Ger. Offen.	Russian	80:83867p
Propylene to propylene oxide	Tl oxide prepared from a Tl(III) salt and alkali		Japan	Asahi Chem. Industries Co., Ltd.	77:115046b
Prepn. olefin oxides, e.g., propylene to propylene oxide	CuO-Tl_2O_3; molar ratio ~ 100-0.01%		Japan	Asahi Chem. Ind. Co., Ltd.	83:44004x
Ethylene oxide, catalyst for	Ag-Ba-Tl-K-Cs-O with Sn and/or Sb		Ger. Offen.	Nippon Shokubai Kagaku Kogyo Co., Ltd.	83:59781x
Prepn. of ethylene, propylene, and butylene oxides	Tl oxide/C (5% Tl)		Japan. Kokai	Asahi Chem. Ind. Co., Ltd.	83:179900y
Prepn. of ethylene oxide and propylene oxide	Tl_2O_3/SiO_2 (Tl_2O reoxidized by O_2)		U.S. 3,436,409	Union Carbide	71:38313h
Oxidation					
Oxidn. of acetylenic hydrocarbons	$TlCl_3$? (abstr. says Th)		Japan		72:100040t
Succinic acid or anhydride or maleic anhydride to γ-butyrolactone	Mo-Ni-Ba or Mo-Ni-Tl/SiO_2-Al_2O_3		Ger. Offen.	Mitsubishi Petrochem. Co., Ltd.	78:158962k
Oxidn. of furfural to maleic anhydride		Ag, Cd, or Tl promoter	USSR	Soviet	72:21345t
Naphthalene to phthalic anhydride		Tl_2O	Japan	Japan. Catalytic Chem. Ind. Co., Ltd.	75(7):48727t
Prepn. phthalic anhydride		Tl oxide	Japan	Japan Mining Co., Ltd.	76:99367d
Oxidn. of aromatic hydrocarbons (prepn. of a catalyst for)		Nine metals claimed	Czech.		66:57608w
Pseudocumene to trimellitic acid or anhydride		Several metal oxides	Japan. Kokai	Japan Catalytic Chem. Ind. Co., Ltd.	79:136815h
Anthracene to anthraquinone		TiO_2-V_2O_5-MO-P_2O_5-Nb_2O_5 where M = 2 or more alkali metals and Tl	Ger. Offen.	Japan Catalytic Chem. Ind. Co., Ltd.	79:115369a

TABLE II-3 (continued)

Reaction	Form if Tl Critical	Tl Optional	Country of Patent	Firm or Nationality	CA Reference
Anthraquinone manufacture		X	Ger. Offen.	Badische Anilin- und Soda-Fabrik A.-G.	79:66098a
Anthraquinone manufacture	Th?		Ger. Offen.	Badische Anilin- und Soda-Fabrik A.-G.	76:72313w
Anthracene to anthraquinone	Tl salt of vanadic acid on a support		Switzerland	CIBA, Ltd.	66(7):28581b
Olefins to glycols	Tl(III)		Japan	Teijin, Ltd.	73:47165e
Olefins to glycols	TlCl (with $CuCl_2$ and HCl)		Japan	Teijin, Ltd.	74:111549h
Olefins to glycols	Any sol. Tl(III) salt, preferably the sulfate or nitrate		U.S. 3,048,636	Dow Chem. Co.	57:14938g
Olefins to halohydrins	$Tl(OH)_3$		Japan	Teijin, Ltd.	80:47445r
Olefins to unsaturated aldehydes:					
Propylene to acrolein, CH_2:CHCHO			Japan. Kokai	Sumitomo Chem. Co., Ltd.	79:136506h
Propylene to acrolein, CH_2:CHCHO	Oxide mixt. of Mo, Bi, Fe, Ni, Tl, P, etc.		Ger. Offen.	Sumitomo Chem. Co., Ltd.	78:71435s
Isobutylene to methacrolein			Japan. Kokai	Sumitomo Chem. Co., Ltd.	79:136505g
Isobutylene to methacrolein			Japan. Kokai	Sumitomo Chem. Co., Ltd.	79:125848v
Isobutylene to methacrolein			Ger. Offen.	Sumitomo Chem. Co., Ltd.	76:72023b
Isobutylene to methacrolein	Mo, Bi, Ni, Cr, Tl, and optionally Sn and/or P		Japan. Kokai	Japan Syn. Rubber Co., Ltd.	83:29534b
Acrolein manufacture			Ger. Offen.	Sumitomo Chem. Co., Ltd.	76:58981f
Acrolein and methacrolein manufacture	Contains Si and Tl oxides (oxide mixt. similar to that of Sumitomo patents)		Ger. Offen.	Japan Catalytic Chem. Ind. Co., Ltd.	78:813825g
Oxidn. of aldehydes to acids		Mo, Ni and/or Co, V, Tl and/or In	Japan	Nippon Zeon Co., Ltd.	80:95271c
Oxidn. of aldehydes to acids (regeneration of catalysts for)		P-Mo-X where X = Tl or 11 other elements	Japan. Kokai	Nippon Zeon Co., Ltd.	78:102517b
Methacrolein to methacrylic acid		Mo-P-(W)-(B, Nb, or Ta)-(Cs, K, or Tl)	Ger. Offen.	Sumitomo Chem. Co., Ltd.	84:122537k
Acrolein to acrylic acid and methacrolein to methacrylic acid	Tl and Mo oxides		Japan	Asahi Chem. Ind. Co.	77:33955h

TABLE II-3 (continued)

Reaction	Form if Tl Critical	Tl Optional	Country of Patent	Firm or Nationality	CA Reference
Acrolein to acrylic acid and methacrolein to methacrylic acid	$MoTl_{a}X_{b}O_{c}$		Ger. Offen.	Asahi Chem. Ind. Co.	76:104381k
Prepn. unsatd. carboxylic acids from unsatd. aldehydes		Mo-P with Cu, Zn, Ni, Ba, Tl, and/or Cs	Japan. Kokai	Asahi Chem. Ind. Co.	83:194019v
Unsatd. aldehydes to acids	Mo-Ni-V-Tl-W oxide catalyst		Japan. Kokai	Japan Synthetic Rubber Co., Ltd.	81:119991m
Unsatd. aldehydes to acids, e.g., methacrolein to methacrylic acid	Tl(III) carboxylate		Japan. Kokai	Asahi Chem. Ind. Co.	84:43315h
Unsatd. aldehydes to unsatd. esters, e.g., methacrolein to methyl methacrylate	Tl(III) carboxylate		Japan. Kokai	Asahi Chem. Ind. Co.	84:16779f
Isobutyraldehyde to methacrolein and methacrylic acid	Mo, P, Tl, O, and optionally V		Japan. Kokai	Mitsubishi Chem. Ind. Co., Ltd.	83:59769z
Oxidn. of alkylbenzenes such as p-xylene to aromatic acids	Thallous acetate, p-toluate, hydroxide, or propionate		U.S. 2,887,511	Union Oil Co. of Calif.	53:18916g
Oxidn. of alkylbenzenes, e.g., o-xylene to phthalic anhydride		Vanadyl vanadates of Tl, Ag, K, Na, Li, or Rb	U.S. 3,012,043	Imperial Chem. Ind., Ltd.	56:4144a
Ammoxidation					
Propylene, NH_3, and O_2 to acrylonitrile	Tl-P-Mo-Fe-Bi-X-Y-O			Sumitomo Chem. Co.	80:14559c
Propylene, NH_3, and O_2 to acrylonitrile			Japan	Sumitomo Chem. Co.	81:92194q
Prepn. HCN, acrylonitrile, or methacrylonitrile	Tl in a complex catalyst		Ger. Offen.	Sumitomo Chem. Co.	81:153048u
Isobutylene to methacrylonitrile	Oxides of Tl, P, Mo, Fe, Bi, etc.		Ger. Offen.	Sumitomo Chem. Co.	77:4983k
Compds. contg. an alkyl-C:C linkage plus NO	Tl oxide or silicate or compounded with an Fe group, noble metal, B, or Pb (0.5% Tl if supported; $\geq$ 90% Tl if unsupported Tl oxide(s) are used)		U.S. 3,157,688; Nov. 17, 1964	Du Pont	62:2714c

TABLE II-3 (continued)

Reaction	Form if Tl Critical	Tl Optional	Country of Patent	Firm or Nationality	CA Reference
Propylene plus NO to acrylonitrile	Ag-Cs-Tl nitrates (1:0.5:0.5 g-atom ratio)		U.S. 3,125,538	Du Pont	60:13921e
Olefin polymerization, oligomerization, or dimerization		X			63:15006e
Olefin polymerization, oligomerization, or dimerization		Alkylmetal halides	French		62:16404c
Olefin polymerization, oligomerization, or dimerization		Tl hydride or complex with alkali metal halide	Brit.	Phillips Petroleum Co.	54:21857g
Olefin polymerization, oligomerization, or dimerization			Brit.	Phillips Petroleum Co.	52:21242a
Ethylene polymerization		Alkylaluminum plus TlCl, $GaCl_3$, $TlCl_3$	U.S. 2,914,520	Hercules Powder Co.	54:5168d
Prepn. of low-temp., low-pressure polyethylene	?	?(Tl not in abstr; $AlCl_3$-$TiCl_4$ claimed)	Brit.	Wacker-Chemie G.m.b.H.	61:5804h
Crystalline polypropylene prepn.		(o-aminobenzenethiolato)thallium	Japan	Ube Industries, Ltd.	76:100406z
Vinyl chloride polymn.		MR_n where M = Tl or seven other metals		SOLVIC Soc. Anon.	58:14217f
Vinyl chloride polymn.		iso-Pr_3Ta?	Ital.	Solvic-Industria Delle Materie Plastiche S.p.A.	57:6146f
Vinyl chloride polymn.		$(iso\text{-}Pr)_3Tl$		SOLVIC Soc. Anon.	53:11892c
Styrene polymn. to cryst. polystyrene		Alkyl deriv. with V compd.	Brit.	Imperial Chem. Industries, Ltd.	54:21850f
Styrene polymn. to cryst. polystyrene		Alkyl deriv. with V compd.	Brit.	Imperial Chem. Industries, Ltd.	53:15649b
Ethylene polymerization		Organomercury or -thallium salt such as Ph_2TlOAc, Ph_2TlOPh, or Et_2TlBr	French	Badische Aniline- und Soda-Fabrik A.-G.	66:29363u

TABLE II-3 (continued)

Reaction	Form if Tl Critical	Tl Optional	Country of Patent	Firm or Nationality	CA Reference
Ethylene and/or propylene polymn.		R_xAl_y + $TiCl_3$ activated by Be, K, B, In, or Tl halide	Japan	Mitsui Toatsu Chemicals	82:31782d
1,3-Dienes polymd. to 1,3-polydienes		Cation exchangers having complexed or adsorbed transition metal ions, e.g., 1.5% Tl(III)	Ger.	Badische Anilin- und Soda-Fabrik A.-G.	68:40804a
Prepn. of cis-1,4-polybutadiene from 1,3-butadiene		Group III organometal hydrides	Brit.	Phillips Petroleum Co.	61:4572g
$Me_2C:CH_2$ to $CH_2:C(CH_3)CH_2CH_2C(CH_3):CH_2$ (2,5-dimethyl-1,5-hexadiene)	Tl_2O_3/Al_2O_3		Japan. Kokai	Sumitomo Chem. Co.	78:147316b
$Me_2C:CH_2$ to 2,5-dimethyl-1,5-hexadiene	Tl_2O_3/Al_2O_3 in the presence of O_2 to keep Tl_2O_3 oxidized				78:147320y
Oligomerization and halogenation of vinyl compds. to dihaloalkanes		Tl^+ one of $\geq$ 8 metal cations	U.S.	Celanese	78:158886p
Acrylonitrile dimerization	1% Tl-Pb electrodes (more corrosion resistant than usual Ag-Pb electrodes)		French	Monsanto Co.	75:58073r
Acrylonitrile dimerization	0.5-2% Tl-Pb electrodes		Ger. Offen.	Monsanto Co.	73:83281y
Photopolymerization of vinylidene monomers (for photog. plates)	Tl compds. plus various sensitizers		Brit.	General Aniline and Film Corp.	57:3628g
Vulcanization of liq. rubber, e.g., the polysulfide Thiokol LP-32	Tl alkoxide or organothallium compd.		Japan. Kokai	Denki Kagaku Kogyo K.K.	85:79435g
Condensation polymerization					
Prepn. of polyether esters	Tl compd. such as TlCl		Japan	Unitaka Co., Ltd.	80:4077e
Prepn. of polyethylene terephthalate	0.01-1.0% TlOAc, Tl_2O, Tl, isopropoxide, or TlOMe		Japan	Mitsubishi Rayon Co. Ltd.	78:97345n
Prepn. of bis(β-hydroxyethyl)-iso- or terephthalate oligomers	Tl oxide or salt added to prevent glycol ether formation		Japan. Kokai	Toyobo Co. Ltd.	81:26295x

TABLE II-3 (continued)

Reaction	Form if Tl Critical	Tl Optional	Country of Patent	Firm or Nationality	CA Reference
Prepn. of terephthalate polyesters		Ca, Sr, Ba, Tl, or Pb salt	Japan	Toyobo Co. Ltd.	81:78529w
Nitril condensation to give triazines, nitrile polymers, and crosslinked polymers or copolymers contg. a metal		X	French	Hooker Chem. Corp.	71:125482f
Prepn. of polysiloxanes	?	Hg, Th?, or Pb carboxylate	Brit.	Union Carbide Corp.	58:9143a
Pyrolysis of 1,2-dichloroethane to manuf. vinyl chloride	Nb and/or Tl added to decrease butadiene formation		Japan	Kureka Chem. Ind. Co., Ltd.	70:3245q
Combustion modifiers in solid propellants		Eight organometallics	U.S.	Ethyl Corp.	67:101571a
Prepn. of dealkylated aromatics (e.g., benzene) and H_2	Ni, Th, Tl, and K on alumina		Ger. Offen.	Mitsubishi Petrochem. Co.	79:81421u
Hydrocortisone acetate to prednisolone acetate	$Tl(OAc)_3$		U.S. 2,927,121	Farbwerke Hoechst Akt.-Ges. vorm. Meister Lucius und Brüning	54:13178c
Amine dehydrogenation to nitriles		Molten Zn, Ga, Tl, or In	Japan. Kokai		79:67045t
Dehydrogenation catalysts		X	Ger. Offen.	French inventor	80:5535q
Dehydrogenation of isobutyric acid to methacrylic acid	Mo-P-Tl-O		Japan. Kokai	Mitsubishi Chem. Industries Co., Ltd.	83:44007a
Hydration of unsatd. nitriles		X	Japan. Kokai	Mitsubishi Petrochem. Co.	80:71332x
Metalation of organoboranes		Ag, Cu, Pg, Hg, and Tl salts			80:R96060v
Prepn. of metal carbonyls useful as antiknock agents in gasoline		Ph_3Tl	U.S. 2,952,517 U.S. 2,952,523	Ethyl Corp.	55:4903f 55:4903i
Hydrogenation		5-10% In or Tl promoters for Pd catalyst	USSR		81:96794h
Reduction of acyclic hydroperoxides		Poisoned Pd catalyst	Belg.	British Petroleum Co.	60:7917c

TABLE II-3 (concluded)

Reaction	Form if Tl Critical	Tl Optional	Country of Patent	Firm or Nationality	CA Reference
Isomerization					
Reforming hydrocarbons		Pt, Pb, and "another metal" catalyst	Ger. Offen.	Asahi Chem. Ind. Ltd.	76:74539y
Reforming		0.2% Pt, 0.05% Ir, 1-1.6% Cl, and 0.03-3.5% In or Tl on Al_2O_3 support	Ger. Offen.	French inventors	78:113636r
Pentane and hexane isomerization		?(Tl not in abstr.) $HSbF_6$ complex with metal fluorides	French		79:136452n
Alkene isomerization, e.g., 2-methyl-1-pentene to 2-methyl-2-pentene		Numerous metals as poisons for Pt-group-metal catalyst	Belg.	British Petroleum Co.	60:8690a
Degradation					
Thermal degradation of polyalkenes to lower mol. wt. products useful in coating applications		Tl_2O_3	Neth. Appl.	Allied Chem. Corp.	64:6854d
Unsatd. fatty acid salts to less unsatd. fatty acids by heating with alkali			Ger.	Henkel und Cie G.m.b.H.	57:16776h
Degradation of unsatd. fatty acid such as soybean or oleic acid		Cd, Pb, Bi, Tl, their alloys, nitrates, or chlorides	U.S.	Henkel und Cie G.m.b.H.	54:9329i

BIBLIOGRAPHY. SECTION II.

1. DeFilippo, R. J., "Thallium" in Commodity Data Summaries 1976, Bureau of Mines, U.S. Department of the Interior, 1976, p. 172.

2. Greenspoon, G. N., "Thallium" in Minerals Facts and Problems, Bureau of Mines Bulletin 650, U.S. Department of the Interior, U.S. Government Printing Office, Washington, D.C., 1970, pp. 749-758.

3. Babitzke, H. R., "Thallium" in Minerals Facts and Problems, 1975 ed., Bureau of Mines Bulletin 667, U.S. Department of the Interior, Washington, D.C., 1976, pp. 1109-1114.

4. Hague, J. M., Bureau of Mines, Thallium commodity specialist, U.S. Bureau of Mines, Washington, D.C., personal communication (telephone), March 1977.

5. Munch, J. C., "Recent Developments in the Preparation and Uses of Thallium," Foote-Prints, **6**, 1-21 (1933).

6. Waggaman, W. H., G. G. Heffner, and E. A. Gee, Thallium. Properties, Sources, Recovery, and Uses of the Element and Its Compounds, Bureau of Mines Information Circular 7553, U.S. Department of the Interior, Washington, D.C., 1950.

7. Heyroth, F. F., "Thallium. A Review and Summary of Medical Literature," Public Health Rep. (U.S.), Suppl., No. 197, 1-23 (1947).

8. Munch, J. C., "Development of Thallium as a Pesticide in the United States," Ind. Med. Surg., **37**(7), 543 (1968).

9. Crafts, A. S., "The Effects of Thallium Sulfate on Soils," Science, **79**, 62 (1934).

10. Crafts, A. S., "Some Effects of Thallium Sulfate Upon Soils," Hilgardia, **10**, 377-398 (1936).

11. Hayes, W. J., Jr., and C. I. Pirkle, "Mortality from Pesticides in 1961," Arch. Environ. Health, **12**, 43-55 (1966).

12. Anonymous, "Interpretation with Respect to Pesticides (Economic Poisons) Containing Thallium Compounds Intended for Household Use," Federal Register, **25**, 6115-6116 (June 30, 1960).

13. Anonymous, "Suspension of Registration for Certain Products Containing Sodium Fluoroacetate (1080), Strychnine and Sodium Cyanide," Federal Register, 37(54), 5718-5720 (1972).

14. Anonymous, "Federal Registration of Thallium Sulfate Suspended," Natl. Pest Control Assoc. Tech. Release, No. 6-72, 1-2 (March 24, 1972).

15. Mathews, J., and A. Anzarut, "Thallium Poisoning," Can. Med. Assoc. J., 99(2), 72-75 (1968).

16. Kitayama, M., and M. Saito, "Studies on the Prevention of Poisoning by Agricultural Chemicals (Part 17). Water Pollution by Rodenticide (Thallium Sulfate) Strewn on Forest Region from Helicopter," Hokkaidoritsu Eisei Kenkyusho-ho (Rep. Hok), 22, 92-95 (1972); abstract in TOXLINE thallium search.

17. Utagawa, T., Jr., "Effects of Environmental Pollution of Birds and Animals," Kogai to Taisaku, 7(1), 35-39 (1971).

18. Fradi, "I Wish to Ask Professor Roche to Give Us Several Examples of Poisoning with Agricultural Products," Lyon Med., 51, 1512 (1966).

19. Wangenheim, M., and J. E. Pasi, "Rodenticide Poisonings of Animals in Switzerland," Schweiz. Arch. Tierheilk., 113, 350-360 (1971).

20. Munch, B., B. Clausen, and O. Karlog, "Thallium Poisoning in Red Foxes (Vulpes vulpes) and Badgers (Meles meles) in Denmark," Nordisk Veterinarmedicin, 26(5), 323-338 (1974).

21. Papworth, D. S., and J. K. Taylor, "Survey of Pesticides Used in Flour Mills," J. Flour Animal Feed Milling, 157(6), 21-22 (1975).

22. Cavanagh, J. B., N. H. Fuller, H. R. M. Johnson, and P. Rudge, "The Effects of Thallium Salts, with Particular Reference to the Nervous System Changes," Quart. J. Med., 43, 293-319 (1974).

23. Petar, A. V., Thallium, Bureau of Mines Information Circular 6453, Bureau of Mines, U.S. Department of Commerce, Washington, D.C., 1931, 6 pp.

24. Wade, K., and A. J. Banister, The Chemistry of Aluminum, Gallium, Indium and Thallium, Vol. 12 Pergamon Texts in Inorganic Chemistry, Pergamon Press, New York, N.Y., 1973.

25. Howe, H. E., and A. A. Smith, Jr., "Properties and Uses of Thallium," J. Electrochem. Soc., 97, 167C-170C (1950).

26. Offergeld, G. R. (to Union Carbide Corporation), "Semiconductive Materials Containing Thallium," U.S. Patent 3,110,685, Nov. 12, 1963, 2 pp.; Chem. Abstr., 60, 6325b (1964).

27. Rabenau, A. T. (to North American Philips Company), "Semiconductor Tl_2Te_3," U.S. Patent 3,096,151, July 2, 1963, 7 pp.; Chem. Abstr., 59, 8246e (1963).

28. Isaacs, T. J., M. S. Gottlieb, J. D. Feichtner, and A. A. Price (Westinghouse Electric Corporation), "Trithallium Vanadium Sulfide (Tl_3VS_4) and Trithallium Niobium Sulfide (Tl_3NbS_4) Crystals and Acoustooptical Devices," U.S. Patent 3,929,976, Dec. 30, 1975, 5 pp.; Chem. Abstr., 84, 114705x (1976).

29. Isaacs, T. J., M. S. Gottlied, J. D. Feichtner, and A. A. Price (Westinghouse Electric Corporation), "Trithallium Phosphorus Selenide (Tl_3PSe_4) Single Crystals," U.S. Patent 3,929,970, December 30, 1975, 5 pages; Chem. Abstr., 84, 114705y (1976).

30. Pearson, D. A., "Preparation of Sulfide, Selenide, and Telluride Glasses," The Glass Industry, 45(12), 666-669, 712-713 (1964).

31. Hampel, C. A., Ed., Rare Metals Handbook, 2nd ed., Reinhold Publishing Corporation, London, 1961.

32. Hansen, M., and K. Anderko, Constitution of Binary Alloys, 2nd ed., McGraw-Hill Book Co., Inc., New York, N.Y., 1958.

33. Krohn, A., and C. W. Bohn, "Electrodeposition of Alloys: Present State of the Art," Electrodeposition and Surface Treatment, 1(3), 199-211 (1973).

34. Lee, A. G., The Chemistry of Thallium, Elsevier Publishing Company, New York, N.Y., 1971.

35. Addicks, L., Ed., Silver in Industry, Reinhold Publishing Co., New York, N.Y., 1940.

36. King, V. J., "Indium-Mercury-Thallium Alloys for Soldering Semiconductors and Ceramics," U.S. Patent 3,374,093, March 19, 1968, 3 pp.; Chem. Abstr., 68, 98239m (1968).

37. Bank, W. J., D. E. Pleasure, K. Suzuki, M. Nigro, and R. Katz, "Thallium Poisoning," Arch. Neurol., 26(5), 456-463 (1972).

38. Barry, R. H., "Chapter 18. Depilatories" in Cosmetics: Science and Technology, Vol. 2, 2nd ed., M. S. Balsam, E. Sagarin, S. D. Gershon, S. J. Strianse, and M. M. Rieger, Eds., Wiley-Interscience, a division of John Wiley and Sons, Inc., New York, N.Y., 1972, pp. 39-72.

39. Polson, C. J., and R. N. Tattersall, Clinical Toxicology, 2nd ed., J. B. Lippincott Company, Philadelphia, Pa., 1969.

40. Prick, J. J. G., W. G. S. Smitt, and L. Muller, Thallium Poisoning, Elsevier Publishing Co., Amsterdam, 1955.

41. Sinka, L., and J. Antosovsky, "Contribution to the General Treatment of Alopecia Areata," Cesk. Dermatol., 45(1) 16-19 (1970).

42. Strauss, H. W., K. Harrison, J. K. Langan, E. Lebowitz, and B. Pitt, "Thallium-201 for Myocardial Imaging. Relation of Thallium-201 to Regional Myocardial Perfusion," Circulation, 51(4), 641-645 (1975); Chem. Abstr., 82, 151493g (1975).

43. Bradley-Moore, P. R., E. Lebowitz, M. W. Greene, H. L. Atkins, and A. N. Ansari, "Thallium-201 for Medical Use. II: Biologic Behavior," J. Nucl. Med., 16(2), 156-160 (1975).

44. Lebowitz, E., M. W. Greene, R. Fairchild, P. R. Bradley-Moore, H. Atkins, A. N. Ansari, P. Richards, and E. Belgrave, "Thallium 201 for Medical Use. I," J. Nucl. Med., 16(2), 151-155 (1975).

45. DeFilippo, R. J., Thallium commodity specialist, U.S. Bureau of Mines, Washington, D.C., personal communications (telephone), January 1977.

46. Fletcher, A., and P. T. Smith, "Gold Electroplating Bath Containing Thallium," Ger. Offen. 2,131,815, Jan. 5, 1972, 12 pp.; Chem. Abstr., 76, 67459u (1972).

47. Serizawa, S., "High-Speed Gold Electroplating Bath," Japan. Kokai 75 90,538, July 19, 1975, 3 pp.; Chem. Abstr., 84, 81724y (1976).

48. Denison, G. H., Jr., and M. R. Barusch (to California Research Corporation), "Antiknock Compositions for Spark-Ignition Engines," U.S. Patent 2,776,262, Jan. 1, 1957; Chem. Abstr., 51, 5405g (1957).

49. Shapiro, H., E. G. DeWitt, and J. E. Brown (to Ethyl Corporation), "Antiknock Compounds for Gasolines," U.S. Patent 3,328,440, June 27, 1967, 5 pp.; Chem. Abstr., 67, 83674e (1967).

50. Meyer, H. C., "Uncommon Ores and Metals," Eng. Mining J. - Press, 115(3), 105-106 (1923).

51. Meyer, H. C., "Uncommon Ores and Metals," Eng. Mining J. - Press, 119(3), 93-94 (1925).

52. Meyer, H. C., "Uncommon Ores and Metals," Eng. Mining J. - Press, 121(3), 94-95 (1926).

53. Meyer, H. C., "Uncommon Ores and Metals," Eng. Mining J., 123(4), 136-137 (1927).

54. Meyer, H. C., "Uncommon Ores and Metals," Eng. Mining J., 125(3), 96-97 (1928).

55. Anonymous, "Minor Metals. Thallium," Mineral Industry 1929, 38, 700 (1930).

56. Anonymous, "Minor Metals. Thallium," Mineral Industry 1930, 39, 660-661 (1931).

57. Sargent, J. D., "Thallium" in Mineral Facts and Problems 1956, Bulletin 556, Bureau of Mines, U.S. Department of the Interior, Washington, D.C., 1956, pp. 871-875.

III. CHEMISTRY

This section summarizes the inorganic, organic, and analytical chemistry of thallium pertinent to the aims of this study. Extensive reviews of these topics are available. Geochemistry is treated in Section IV.

A. Inorganic Chemistry

Recent reviews of thallium inorganic and organic chemistry include *The Chemistry of Thallium* (1971)[1/] by A. G. Lee (336 pages); "The Coordination Chemistry of Thallium(I)" (1972)[2/] by A. G. Lee (59 pages); and *The Chemistry of Aluminum, Gallium, Indium, and Thallium* (1973)[3/] by Wade and Banister (53 pages on thallium). Two short reviews (six pages and about 50 references each on thallium) appeared in the 1972 and 1975 (Series 1 and 2) of *MTP International Reviews of Science: Inorganic Chemistry* by J. B. Farmer and K. Wade[4/] and by E. A. Forman and K. Wade.[5/] Siegel (1968)[6/] reviewed Group IIIa oxyhalides. Zitko (1975)[7/] briefly reviewed all aspects of thallium chemistry (as well as its toxicity and pollution potential).

The electronic configuration of thallium is [Xe] $4f^{14}$ $5d^{10}$ $6s^2$ $6p$, placing it in Group IIIa* of the periodic classification of elements with boron, aluminum, gallium, and indium. In this group, the univalent state is important only in thallium chemistry. The resistance of the pair of 6*s* electrons of thallium to becoming lost or to participate in covalent bond formation is often given as the reason thallium usually preferentially forms Tl^{I} rather than Tl^{III} inorganic compounds.[8/]

Thallium is a gray-white, soft, ductile metal similar to lead. Freshly cut surfaces have a metallic luster which dulls in air; but unlike lead, a heavy oxide crust forms after a few days. The gray film of thallous oxide (including some Tl_2O_3) formed at normal temperatures becomes brown Tl_2O_3 at 100°C. Thallium metal is rapidly corroded by water containing dissolved air or oxygen. It does not dissolve in solutions of alkalies or in liquid ammonia. Thallium alloys with most metals,** dissolves in mercury, and reacts with certain metal salts and organic compounds. Numerous physical properties of thallium metal were tabulated by Wade and Banister (1973).[3/] It melts at 302.4°C and boils at 1453°C.

---

* Some authors refer to these same elements as Group IIIb.

** Exceptions: Al, As, B, Fe, Mn, Ni, and Zn.[3/]

Following are descriptions of inorganic thallium compounds and complexes. Table III-1 and Table III-2 summarize information regarding the volatility and solubility of certain thallium compounds.

Thallous carbonate is prepared by treating the aqueous hydroxide with carbon dioxide. Thallium(III) does not form a carbonate or bicarbonate. Tl_2CO_3 solutions are strongly alkaline because of hydrolysis and will dissolve acidic oxides (even precipitated silica when hot).[3/]

TlCN (solubility 16.8 g/100 g H_2O at 28°) can be precipitated from solutions of KCN and thallous acetate. Unlike AgCN, TlCN does not complex with an additional cyanide ion (but Tl_2CN_2 is known). Tl^+ forms salts of the complex cyanides of other metals, or of Tl^{+3}; e.g., $TlAg(CN)_2$, $Tl_2Zn(CN)_4$, $Tl_3Cu(CN)_4$, $Tl^I[Tl^{III}(CN)_4]$, $Tl_2Ni(CN)_4$, and water-insoluble $Tl_4Fe(CN)_6 \cdot 2H_2O$.[3/]

Thallium(I) halides are air stable, predominantly ionic, and sparingly water soluble except for TlF. In water solutions, thallous halides are incompletely dissociated. At high halide ion concentrations, the anionic species TlX_2^-, TlX_3^{2-} and TlX_4^{3-} are formed. The thallium(III) halides are relatively unstable, lose halogen on heating, and are readily hydrolyzed. Complexes with halide ions, and ligands with nitrogen or oxygen donors are known. Very water-soluble (86.2 g/100 g solution), $TlCl_3 \cdot 4H_2O$ can be prepared from a boiling aqueous suspension of TlCl and chlorine gas followed by evaporation.[3/] It loses chlorine at about 40°C to give TlCl.[8/]

Reacting thallous chloride with chlorine monoxide at 3.8°C for prolonged periods gives chlorine and thallium(III) oxychloride (TlOCl) mixed with thallous chloride.[6/]

Oxides of thallium include the well-defined Tl_2O and Tl_2O_3, ill-defined intermediate oxides, the peroxide Tl_2O_4, and the nonstoichiometric oxides $TlO_{1.5}$-$TlO_{1.75}$. Black Tl_2O forms when thallium is oxidized in air (Tl_2O_3 at dull redness) or when TlOH is heated to 100°C or Tl_2CO_3 to 700°C out of contact with air. Tl_2O attains a vapor pressure of $\sim$ 1 mm Hg at 580°C. It is very hygroscopic, forming TlOH with water. Tl_2O_3 is more conveniently prepared by treating an alkaline Tl(I) solution with H_2O_2 or by treating aqueous $TlNO_3$ with chlorine or bromine followed by OH^- than by reacting thallium with oxygen. Solid Tl_2O_3 vaporizes by forming gaseous Tl_2O and O_2.[1,3/]

Tl^{3+} precipitates from basic solutions. Solid hydrates or hydroxides of Tl(III) may not exist, but $[Tl(OH)(H_2O)_5]^{2+}$ and $[Tl(OH)_2(H_2O)_4]^+$ may exist in solution. TlOH is slightly less water soluble than the alkali hydroxides, but TlOH is also a strong base.[1,3/]

Tl_2O_3 is insoluble in water but reacts with mineral acids, acetic acid, and oxalic acid to give thallic salts.[3/]

In water solution, Group III elements probably form octahedral $[M(H_2O)_6]^{3+}$ ions, which are acidic, giving for example, $[M(H_2O)_5(OH)]^{2+}$ + H^+. Group III salts of weak acids cannot exist in contact with water because they are completely hydrolyzed. The thallic ion is hydrolyzed to $TlOH^{2+}$ and the colloidal Tl_2O_3 even at pH 1-2.5.[8/]

Thallous nitrate, prepared from thallium or Tl_2CO_3 in dilute nitric acid or from Tl_2SO_4 and $Ba(NO_3)_2$, is water soluble and useful for preparing other thallous salts by metathetical precipitation reactions. Thallic nitrate is a strong oxidizing agent. It is prepared by treating Tl_2O_3 with warm concentrated nitric acid.[3/]

Thallous perchlorate, $TlClO_4$, is much more thermally stable than the chlorite $TlClO_2$, which detonates on shock, and the chlorate $TlClO_3$. Evaporating $TlNO_3$ in excess $HClO_4$ and evaporating Tl_2O_3 in $HClO_4$ give the Tl^I and Tl^{III} perchlorates, respectively. Anodic oxidation of $TlClO_4$ also gives $Tl(ClO_4)_3$.[3/]

Thallium metasilicate, Tl_2SiO_3, can be prepared by treating TlOH with dilute Na_2SiO_3 in dilute $NaNO_3$. The orthosilicate, Tl_4SiO_4, is prepared from TlOH and Na_2SiO_3. Both are water insoluble. Cooling a molten thallium silicate gives a glass of high refractive index and dispersivity. Heating kaolin and Tl_2SO_4 gives thallium aluminosilicate, $TlAlSiO_4$.[3/]

Pure thallium(I) sulfate is prepared electrolytically or by dissolution of TlOH or $TlNO_3$ in dilute sulfuric acid. Thallium(I) salts of all the common sulfur oxyacids are known; thallium(III) forms sulfates, basic sulfates, double sulfates, and the mixed sulfate $Tl^ITl^{III}(SO_4)_2$. Sulfatothallium(III) complexes are more stable than sulfatothallium(I) complexes.[3/]

Adding hydrogen sulfide to alkaline solutions of thallous salts or thallic salts gives thallium(I) sulfide, which is soluble in dilute mineral acids but not in ammonium sulfide. It is readily air-oxidized to thallous oxide and thallium thiosulfate. Wade and Banister[3/] also describe the properties of the selenides and tellurides, some of which are used industrially as semiconductors. The binary compounds are prepared by heating the elements in a sealed tube at about 400°C.

Stability constants for thallium(I) complexes may be found in the handbook by Sillen and Martell (1964).[9/] Many are also given by Lee (1972).[2/] Complexes between most ligands and Tl(I) are relatively weak.

The logarithms of the formation constants with most organic ligands are in the range of 0.5 to 2.0. Tl(I) is not complexed by humic acid. Thallium(III) forms much stronger complexes, generally of the type TlX_4^- and TlX_6^{3-}.[7]

Thallium(I) is not able to form strong complexes due to the two σ-antibonding electrons in the outer (6) s orbital. Thallium(I) derivatives adopt distorted octahedral or cubic structures in the solid state. The greater molecular polarizability of thallium(I) leads to differences between Tl^+ and alkali cations.[2]

In aqueous solution, 4 M in perchlorate, weak thallium(I) halide complexes form. At high halide ion concentrations, the absorption spectrum of the thallium(I) ion in solution resembles that of the K(Tl)Cl phosphors, indicating an assembly of ions in which Tl(I) is symmetrically surrounded by six halide ions. However, anionic halide complexes had not been isolated as of 1972.[2]

The Tl^+ ion in aqueous solution is more stable than Tl^{3+} and not very sensitive to pH. For the half-reaction $Tl^{3+} + 2e \rightleftharpoons Tl^+$, the standard electrode potential is +1.25 v; in 1 M hydrochloric acid, the electrode potential is +0.77 v; and in 1 M perchloric acid, +1.26 v. Even at pH 1.0-2.5, however, Tl^{3+} is extensively hydrolyzed to $TlOH^{2+}$ and colloidal Tl_2O_3 so that the reduction potential is dependent on pH and on the presence of complexing ions. As can be seen by the reduction potential in hydrochloric acid compared with that in perchloric acid, the reduction potential falls due to strong complexing of Tl^{3+} by chloride ions.

Few thallium(I) complexes are known, and coordination numbers in solution are seldom greater than 4. The complexes are generally neutral or anionic, the ligand atoms being halide, nitrogen, oxygen, or sulfur. The stability constants of many thallium(I) and thallium(III) inorganic complexes are given in Tables III-3 and III-4. For stability constants of certain other species not likely to be found in the environment, see Sillen and Martell (1964)[9] or the 1971 supplement. The best known thallium(I) halo complexes are the crystal phosphors comprising alkali halide crystals activated with Tl^I that are used in scintillation radiation detectors.[3]

The most stable metal chloride complexes known include $TlCl^{2+}$, $TlCl_2^+$ and $TlCl_4^-$, recognized even in aqueous solutions of $TlCl_3$. Wade and Banister (1973)[3] review the 4-, 5-, and 6-coordinate thallium(III) halo complexes. Hexachlorothallate(III) complexes may not all contain the $TlCl_6^{3-}$ ion.

Batley and Florence (1975)[10] present thermodynamic arguments that predict, and experiments that are said to confirm the prediction, that thallium in both freshwater (pH 6.5) and seawater (pH 8.1) predominates in equilibrium with atmospheric oxygen in the trivalent state. Their calculations considered complex formation with only chloride, bromide, sulfate, and hydroxyl ions, which are present in significant concentrations. Stability constant data were available for these complexes but not for phosphate, borate, carbonate, and organic complexes, which were all expected to stabilize Tl(III). At pH 8.1 in seawater, Batley and Florence conclude that the predominant Tl(I) species is TlCl and Tl(III) present as $Tl(OH)_2^+$ rather than $Tl(OH)^{2+}$.

On the other hand, Lee[1] states that addition of Cl^- to aqueous solutions of Tl^{3+} (probably firmly bound to only two water molecules) produces the linear $TlCl_2^+$ ion, possibly aquated weakly by four water molecules, and the tetrahedral $TlCl_4^-$ ion.* Lee further states that solutions "must have a very high chloride-ion concentration before any higher species, possibly $TlCl_6^{3-}$, is formed." He also states, "...Tl(I) salts are oxidised only by very powerful oxidizing agents such as MnO_4- and Cl_2 in very acid media."

Lanford (1969)[11] found that treating a solution containing 2.4×10^{-4} mole/liter thallium (plus 50 mg/liter each of 19 other metals) with a stoichiometric amount of lime did not reduce the thallium concentration. This would indicate that Tl^+ was the predominant species since Tl_2O_3 would precipitate if it were Tl^{3+}. When stoichiometric amounts of both lime and sulfide were added, the residual thallium concentration in solution was 8.7×10^{-7} moles/liter (0.178 mg/liter), indicating 99.6% removal from solution.

B. Organic Chemistry

In recent years, the usefulness of thallium compounds in organic synthesis has been widely studied, and numerous reviews have appeared. Nesmeyanov and Sokolik (1967)[12] described known preparations of organothallium compounds. Annual thallium surveys by J. P. Oliver (1971, 1972) or Kurosawa and Okawara (1967, 1968, 1973, 1974, 1975)[13] have appeared in the Journal of Organometallic Chemistry. Other extensive reviews include those by Yasuda and Okawara (1967), Henry (1968),[14] Taylor and McKillop

---

* Organic solvents always extract $HTlCl_4$ hydrate from aqueous hydrochloric acid solutions of Tl(III). Even evaporating a solution of $TlCl_3$ in concentrated hydrochloric acid gives $HTlCl_4 \cdot 3H_2O$ crystals!

(1970),[15] A. G. Lee (1970) and (1971),[1] McKillop and Taylor (1973),[16,17] Banerji et al. (1974),[18] and McKillop (1975).[19] Maher (1973)[20] and (1975)[21] has produced some shorter annual reviews. Recently McKillop and Taylor (1976)[22] reviewed oxythallation limitations and described two new experimental procedures to overcome them.

Whereas inorganic thallium(I) compounds are more stable, covalent organothallium compounds are stable only in the +3 covalent valence state.[17]

Nesmeyanov and Sokolik (1967)[12] describe known preparations of $RTlX_2$, R_2TlX, R_3Tl, and RTl organothallium compounds. Cyclopentadienylthallium and aryl- and alkylcyclopentadienylthalliums are the only representatives of monovalent RTl. $RTlX_2$ compounds include alkyl-, alkenyl-, and arylthallium dihalides and carboxylates. Diorganothallium halides (aliphatic, alicyclic, aromatic, and heterocyclic derivatives are known) are the most stable organothalliums. Derivatives with other anions are also known. Aliphatic and aromatic triorganothalliums exist. Only the lower trialkythalliums are liquid; the other organothalliums are generally crystalline.

Organothallium(III) compounds have been prepared via organolithium, organosodium, organomagnesium (Grignard reagents), organozinc, organoboron, organoaluminum, organomercury, organotin, organolead, and organobismuth compounds as well as aryldiazonium salts and diphenylthalonium salts. Aromatic thallium(III) compounds can also be prepared by "thallation," a reaction analogous to mercuration:

$$ArH + Tl(OOCR)_3 \longrightarrow ArTl(OOCR)_2 + RCOOH$$

Cyclopentadienylthallium(I), first prepared in 1956, is rather stable. Little decomposition occurs on storing for months in sealed bottles; it is not attacked by water or dilute alkali and is oxidized only slowly in air at room temperature. However, it is rapidly decomposed by acid.

Cyclopentadienylthallium and methylcyclopentadienylthallium (oxidizes vigorously in air) are prepared by treated TlOH or thallous salts with the cyclopentadiene in aqueous media at 0-20°C. Other alkylcyclopentadienyl thalliums require the presence of base, an inert gas atmosphere, and temperatures below 100°C.[12] m- and p-Fluorophenylcyclopentadienylthalliums(I) were prepared in 1974.[13]

Cyclopentadienylthallium(I) is a mild reagent for preparing transition-metal and other cyclopentadienyl compounds.[20]

The trialkyl- and triarylthalliums are decomposed by water to give stable dialkylthallium hydroxides. The R_3Tl compounds are oxidized by elemental oxygen; $(CH_3)_3Tl$ ignites spontaneously in air. The R_2TlX compounds are stable in water, air, or acids.[12] The monoalkylthallium-(III) compounds are unstable; the monovinyl- and monoarylthallium compounds are more stable and can be isolated readily.[16]

Bajpai and Bajpai (1976)[23] produced water-soluble, thermally stable compounds of the type C_6H_5TlXCl where X = Br, I, CN, NCS, and NO_3 by treating $(C_6H_5)_2Tl(III)$ derivatives with the appropriate mercury(II) salt in refluxing dry alcohol and/or acetone.

For example:

$$(C_6H_5)_2TlX + HgX'_2 \longrightarrow C_6H_5HgX' + C_6H_5TlXX'$$

In the hard acid-soft base classification of R. G. Pearson, thallium is the only Group IIIa element classified as a soft acid.* A softer acid than Tl(I), Tl(III) resembles mercury(II) and lead(IV) rather than aluminum(III), which is a hard acid. In its characteristic properties Tl(I) resembles both potassium(I), a hard acid, and silver(I), a soft acid.[15]

The salts of only these few metals undergo metallation or oxymetallation reactions: mercury(II), thallium(III), lead(IV), palladium(II), gold(III), and platinum(II).[16]

As a relatively soft acid, Tl+ tricoordinates with bidentate ligands and compounds containing acidic hydrogens to give complexes that react regiospecifically with organic electrophiles. Since the electrophilicity of TlX_3 derivatives can be controlled by varying the nature of X, specific thallium(III) salts can be designed for purposes such as Friedel-Crafts reactions, electrophilic aromatic thallation, or oxythallation. The very weak C-Tl bonds (25-30 kcal/mole) cleave under heterolytic or homolytic conditions. Conversion of $RTlX_2$ compounds into organic products and TlX occurs readily because of the thermodynamic ease of reducing Tl(III) to Tl(I). Thallium(III) is an efficient two-electron oxidizing agent with a redox potential intermediate between those of Hg(II) and Pb(IV).[17]

Interest in the applications of thallium compounds to organic synthesis intensified about a decade ago at Princeton (Taylor) and the University of East Anglia (McKillop). Metallic thallium itself selectively reduces nitro compounds to azoxy compounds in refluxing ethanol. Taylor and McKillop (1970)[15] reviewed the use of thallium(I) salts in alkylations, acylations, and/or tosylation of 1,3-dicarbonyl compounds,

---

* A soft acid is characterized by a low charge, large size, and often _d_ electrons in its outer shell.[16]

phenols, and heterocylic amides; alkylation of purines; and synthesis of anhydrides, primary alkyl bromides, and biphenyls. Thallium(III) had been used to brominate and iodinate aromatic compounds selectively; catalyze Friedel-Crafts reactions; thallate aromatic compounds; synthesize aromatic nitriles, thiophenols, biphenyls, and phenols; and selectively deuterate aromatic compounds. Thallium(III) trifluoroacetate (TTFA) converts parasubstituted phenols to p-quinones and promotes oxidative coupling of aromatic compounds in the presence of boron trifluoride etherate. TTFA in ether or thallium(III) nitrate in methanol causes oxidative ring contraction of cyclic olefins to aldehydes. Thallium(III) nitrate is also good for controlled oxidation of acyclic olefins. By 1973, more than 50 specific synthetic organic transformations had been described.

Henry (1968)[14/] reviewed the similarities and differences in oxidizing olefins by Pd(II) and Tl(III) compounds. Oxymetallation is the rate controlling step. Tl(III) gives carbonyl compounds (epoxides)[16/] and glycols, whereas Pd(II) gives only carbonyl compounds. In Tl(III) oxidations, π-complex formation is not observed. The effects of olefin structure on rate and salt effects are different.

In 1975, Maher[21/] noted a "slight decrease" in the amount being published on organothallium compounds. He noted that little work has been done on Ziegler-Natta olefin polymerization catalysts involving gallium, indium, or thallium as the Group III element.

$ArTlX_2$ compounds can be attacked by either electrophilic or nucleophilic reagents which enter the ring where thallium was originally attached. Since the position of the initial aromatic thallation can be well controlled, organothallium compounds can "play a unique and indispensable role" in organic synthesis.[16/] The regiospecificity observed in organothallium reactions is remarkable. McKillop and Taylor (1973)[17/] proclaimed, "Thallium and its salts must now be regarded as essential reagents for modern organic synthesis."

The 1974 McKillop review[19/] specifically covered applications of thallium(III) nitrate(TTN) as an oxidant in organic synthesis. Oxidations and oxidative rearrangements of olefins, acetylenes, ketones, and compounds with -C:N- bonds are described. TTN is highly ionic and reacts as a powerful electrophile almost instantaneously with olefins. McKillop listed several reasons for its being the "reagent of choice for oxidation of many organic functional groups":

1. It is prepared easily, rapidly, and almost quantitatively by dissolving Tl_2O_3 in nitric acid.

2. It can be stored for months in sealed bottles as a stable crystalline solid.

3. It is readily dissolved by both inorganic and organic solvents; whereas the reduction product, $TlNO_3$, is insoluble in most common solvents, precipitating from solution.

4. Because nitrate is a poor nucleophile, it is not incorporated into reaction products.

Another advantage of thallium in organic synthesis is that, unlike sodium ethoxide, thallium(I) ethoxide is soluble in most organic solvents and can be used for homogeneous base-catalyzed reactions in non-polar solvents.[3/]

Table II-3 is an extensive tabulation of patented organic reactions systems catalyzed* by thallium compounds, but these would not necessarily indicate the potential industrial applicability of certain reactions in which organothallium intermediates participate. With so many organic reactions in which thallium compounds appear to be the reagent of choice, the chances of thallium's future expansion in industrial organic synthesis are good.

Stability constants for complexes of thallium ions with organic ligands with donor oxygen or nitrogen atoms are given in Table III-4. (See also Table III-3.)

Is the chemistry of organothallium(III) compounds really applicable to environmental and physiological systems? Thayer,[25/] citing his own unpublished research, states that "teratological" effects of $(CH_3)_2Tl^+$ in water solution on sprouting of cucumber seeds begins at the concentration 2.3 mg/liter compared with 0.13 mg/liter for $(C_2H_5)_3Pb^+$ or 0.98 mg/liter for CH_3Hg^+. The organometallic derivatives of platinum, gold, mercury, thallium, tin, lead, phosphorus and arsenic apparently suppress chlorophyll formation at lower concentrations and damage cell membrane walls at higher concentrations. He further discusses the fungicidal and bactericidal properties of the organothallium compounds R_2TlX studied by Srivastava et al.[26/] and the finding by Agnes et al. (1971)[27/] that methyl vitamin B_{12} (MeB_{12}) can methylate Tl(III) by an acid-base reaction over a wide pH range but not Tl(I). (The reaction is first order in Tl(III) and in MeB_{12}.) He concludes that the "ability of thallium to undergo biological methylation may well depend on conditions used, especially as to whether chelating groups may be present."

---

* The table includes reactions in which thallium appears to be a catalyst. In addition, Syntex Corporation (1971)[24/] has patented substituted prostaglandin derivatives that are prepared from cyclopentadienylthallium(I) as an intermediate in a series of steps.

C. Analytical Chemistry

An excellent review with an extensive bibliography was made by Korenman in 1960.[28] This review covered the early analytical chemistry of thallium by gravimetric, titrimetric, and electrochemical methods. A more recent review by Zitko in 1975 covered the determination of thallium at trace levels.[7] The latter review has an extensive section on the chromatography of thallium to separate it from other materials and to concentrate environmental samples before chemical analysis. Recent analysis procedures are also included in this review.

It is not the intent of this discussion to summarize these review articles but rather to focus on analytical problems associated with the analysis of thallium in biological fluids and tissue and environmental samples. Emphasis is placed on procedures and precautions that should be followed during the development of new analytical methods for the analysis of thallium.

1. Measurement: The analysis of thallium in environmental and biological samples at trace levels is not limited by instrumental capabilities but the availability of standard reference materials. The accuracy of any given method of analysis is established by the analysis of a material of known composition in the same sample matrix. In the absence of standard reference material, standard addition and round-robin analysis of common samples are ways of approaching accuracy levels for unknown elemental concentrations.

The National Bureau of Standards is actively preparing standard reference materials (SRM) for environmental and biological matrix samples. To date, coal (SRM 1632), fly ash (SRM 1633), bovine liver (SRM 1577), orchard leaves (SRM 1571), spinach (SRM 1570), tomato leaves (SRM 1573), and pine needles (SRM 1575) are available. The thallium content of coal (SRM 1632) has been certified. Thallium contents in fly ash (SRM 1633) and bovine liver (SRM 1577) are listed but not certified. The thallium levels in spinach (SRM 1570) and pine needles (SRM 1575) are given, but due to the recent release of these materials, we do not know if they are certified.[29] The National Bureau of Standards should be commended for the service they have provided by preparing these standard reference materials, and continued encouragement and support should be given to them in developing additional materials and in the certification of increased numbers of elements in these standards.

The analyst should select the SRM of the same or closest matrix type to be included during sample preparation and analysis. Even if the SRM is not certified for thallium, it should be included and the results reported. If at a later time the SRM is certified for thallium, the accuracy of those results can be determined.

If no suitable reference material is available for a sample matrix of interest, a careful standard addition experiment should be performed. It should be realized that the chemical form and oxidation state of the added thallium may be different from those of the sample. Thus, the results of these experiments need to be carefully interpreted.

As analytical procedures become more sensitive (in the parts per billion range) all reagents used in the analysis method must be carefully checked for thallium. Reagent blank analyses become critical as lower levels are analyzed.

2. Detection: The methods for the detection of thallium in biological and environmental materials are based on single- and multi-element techniques. The single-element methods include atomic absorption and electrochemical methods. The multi-element techniques include emission, X-ray, neutron activation, and mass spectrometric. This subsection discusses examples of applications of these techniques with attention given to those papers that carefully evaluate the method for analysis of thallium in environmental sample matrices.

a. Atomic absorption: The flame atomic absorption detection limit at 3 ppm is not sensitive enough for most environmental samples. Flame methods cannot be used without sample preconcentration. These procedures are summarized by Zitko (1975)[7] and by Van Ormer (1975).[30]

Flameless atomic absorption has the necessary sensitivity for thallium analysis in environmental samples. The detection limit at about 10 ppb has allowed determinations of thallium in many types of samples. J. Y. Marks et al. (1977)[31] evaluate the performance for three commercial heated graphite atomization systems for thallium analyses. A method for determining thallium in urine was developed by Kubasik and Volosin (1973).[32] Satisfactory recovery was obtained at 30 µg Tl/liter. This paper is an example of the careful and systematic procedure which must be followed in the development of a flameless method. Blood analysis for thallium by the flameless method was reported by Machata and Binder (1973).[33] The blood iron levels caused interferences with the thallium determination. These problems were overcome by changing digestion procedures.

The interference problems with flameless procedures should be emphasized. The analysis of thallium by flameless techniques can be invalid if careful attention is not paid to species that are present during analysis. The nitrate of thallium(I) decomposes at 450°C and will be lost during the ashing step. The thallium(III) nitrate is also converted to thallium(I) nitrate at 145°C. Because of this problem, the nitrate form should be avoided. This is an important consideration because nitric acid is a popular and useful sample matrix for most other flameless work.

A recent paper by Skogerboe et al. (1975)[34] suggests an interesting development for the atomic absorption determination of thallium. The volatile thallium(I) chloride is produced from the aqueous sample and swept into an atomic absorption flame. This procedure is analogous to hydride production of arsenic, antimony, and selenium. The advantage of this kind of system is the ability to analyze a large volume of sample in a short time, resulting in lower detection limits. The thallium detection limit by this technique is 0.050 ppb.

b. Electrochemical methods: Polarographic and anodic stripping methods have been applied to thallium determination in biological samples. Reviews of these methods are given in References 7 and 28. The well-behaved redox nature of thallium makes it amenable to electrochemical study and allows for the determination of complexed and uncomplexed material. Pulse polarography was used to analyze tissue samples for thallium.[35] Thallium could not be detected in normal tissue at a detection level of 0.001 ppb wet weight but could be detected in tissue when thallium exposure was known to have occurred. The high sensitivity and increased resolution of pulse polarography make it more suitable to complex matrices than conventional polarography.

Anodic stripping voltammetry was used to determine thallium in urine at 50 ppb (~ 50 μg/liter) with 2.5% precision.[36] Lead and organic interferences were evaluated. Lead at elevated concentrations causes problems, but the presence of organic material does not affect the analysis Another anodic stripping voltammetry paper[37] noted the interference of cadmium during analysis for thallium at low parts per billion levels in urine. A mercury-film electrode was used to increase resolution. The cadmium and lead were complexed with EDTA to shift the oxidation wave away from thallium. Standard addition and time-stability studies were made to demonstrate the technique. A comparison of hanging mercury-drop electrode and mercury-plated wax-impregnated graphite electrode was made using differential pulse anodic stripping voltammetry in urine.[38]

The use of anodic stripping voltammetry to determine the chemical nature of thallium in natural water demonstrates another application of this technique.[39] Such studies provide much needed information on speciation of metals in environmental samples.

c. Emission spectrography: Relatively few studies of thallium by emission spectroscopy have been reported. Zitko[7] reviewed spectro graphic studies. The limited use for environmental and biological samples results from insensitivity (20 ppm) and spectral interference problems. Kartha et al. (1967)[40] used thallium's 2767.87A line for thallium determination in sodium chloride. This line is bracketed by iron emission lines at 2767.52 and 2768.11. If much iron is present in the sample (as is the case in most environmental samples), this can create a spectral interference problem.

The low levels of thallium in environmental samples severely limit the use of emission spectrographic analysis. The poor sensitivity and precision for thallium have precluded many determinations by this method.

d. Inductively coupled plasma emission spectrometry (ICPES): The use of ICPES on environmental and biological samples is becoming very popular because of the multi-element capability, decreased interference problems, and generally better sensitivity than flame atomic absorption. Thallium has received little attention by this technique because of its poor sensitivity. Fassel and Kniseley (1974)[41] gave thallium's detection limit as 0.2 ppm for aqueous solutions. This poor sensitivity plus the inability of plasma systems to handle high dissolved solids results in detection limits above those required for most environmental samples. Skogerboe et al.[34] suggested the introduction of gaseous thallium chloride into the plasma source. This procedure would greatly lower the detection limit for thallium and is worthy of consideration for environmental and biological samples.

e. X-ray techniques: Marcie (1967)[42] reported thallium analysis in water at 5 ppb by X-ray fluorescence. This detection limit was achieved by extraction of water samples and evaporation onto filter paper. Goldman et al. (1966)[43] reported a detection limit of 5 ppm by X-ray emission spectrography for thallium in dried biological tissue without additional sample concentration. (The method was to be used to determine thallium in poisoned birds.) The biological matrix is well suited for heavy metal determination by X-ray methods because it is transparent to most of the matrix elements present. The use of X-ray methods for air particulate analysis is increasing; the problems of sample and standard preparations were addressed in two recent papers,[44,45] but thallium was not among the elements considered.

f. Neutron activation analysis: The analysis of thallium by neutron activation has a detection limit of 1 ppm if no sample concentration or separation is applied. This limits the utility of the techniques for environmental and biological samples. Specht reported thallium analysis of human hair in a case of chronic thallium intoxication.[46] Thallium in geological samples has been analyzed by this method.[47,48] A review of activation analysis of environmental samples was published by Filby and Shah (1974).[49] A round-robin test of National Bureau of Standards coal and fly ash by neutron activation analysis did not include thallium.[50] This is probably because of sensitivity problems.

g. Mass spectroscopy: Spark source mass spectroscopy (SSMS) has received recent attention for multi-element analysis of environmental and biological samples. The 0.1 ppm detection limit for thallium is attractive for this kind of work. Fitchett determined thallium concentrations in biological media.[51] This study demonstrates the utility of SSMS with minimum sample pretreatment. The accuracy of SSMS does not match that of other analytical techniques, but SSMS is an excellent method for surveying metal concentration of environmental samples.

Isotope dilution mass spectrometry and isotope dilution spark source mass spectrometry were used by the National Bureau of Standards to certify thallium content of Standard Reference Material 1632 (coal). Although these techniques are beyond the cost and skill of the normal analyst, their use in preparing reference materials is very valuable

TABLE III-1

VAPOR PRESSURES OF THALLIUM AND SOME OF ITS COMPOUNDS

Compound	Heat of Sublimation, (kcal/mole)	Temperature	Vapor Pressure mm Hg	Vapor Pressure log p, mm
Tl		500°C	7.4×10^{-4} [3]	
Tl_2SO_4		700°C	19.5 [52]	
TlSe	30.845 [53]	615-689°K		12.433-6,744.2/T [53]
Tl_2Se	26.9051 [53]	615-689°K		9.8052-5,880.9/T [53]
	26.960 [54]	615-689°K		9.8052-5,800.9/T [54]
Tl_2S_3	33.9721 [53]	615-689°K		9.2481-7,425.5/T [53]
Tl_2O	30.21 [55]	580°C	~ 1 [1]	11.5-6,612/T [55]
Tl_2S	20.45 [55]			7.354-4,484/T [55]

TABLE III-2

SOLUBILITY AND SOLUBILITY PRODUCTS OF THALLIUM COMPOUNDS[28]/

Compound	Formula	Solubility, g/ℓ (20°C)	Solubility Product
Thallium azide	TlN_3	3.5 (25°C)	1.5×10^{-4}
Thallium bromate	$TlBrO_3$	3.46	8.9×10^{-5}
Thallium bromide	$TlBr$	0.48	2.6×10^{-6}
Thallium carbonate	Tl_2CO_3	52.3 (18°C)	
Thallium chlorate	$TlClO_3$	39	
Thallous chloride	$TlCl$	3.4	1.5×10^{-4}
Thallic chloride	$TlCl_3 \cdot 4H_2O$	826 (17°C)	
Thallium chromate	Tl_2CrO_4	0.042	2.0×10^{-12}
Thallium cyanide	$TlCN$	168 (28.5°C)	
Thallium ferrocyanide	$Tl_4[Fe(CN)_6] \cdot 2H_2O$	3.7 (18°C)	3.8×10^{-14}
Thallium fluoride	TlF	786 (15°C)	
Thallous hydroxide	$TlOH$	350	
Thallium iodide	TlI	0.063	3.6×10^{-8}
Thallium nitrate	$TlNO_3$	86.7	
Thallium nitrite	$TlNO_2$	270	

TABLE III-2 (concluded)

Compound	Formula	Solubility, g/ℓ (20°C)	Solubility Product
Thallium oxalate	$Tl_2C_2O_4$	15.8	
Thallium phosphate	Tl_3PO_4	5.0	6.6×10^{-9}
Thallium selenate	Tl_2SeO_4	28	
Thallium selenide	Tl_2Se	5.8×10^{-7}	6.9×10^{-27}
Thallium sulfate	Tl_2SO_4	46.4	
Thallium sulfide	Tl_2S	Various values: 0.695 g/ℓ to 2.78×10^{-6} g/ℓ	10^{-24} to 7×10^{-20}
Thallium sulfite	Tl_2SO_3	32.4 (15°C)	
Thallium telluride	Tl_2Te	1.5×10^{-6}	9.5×10^{-26}
Thallium thiocyanate	TlSCN	3.16	1.1×10^{-4}
Thallium thiosulfate	$Tl_2S_2O_3$	1.92 (25°C)	1.08×10^{-7}

TABLE III-3

STABILITY CONSTANTS OF THALLIUM IONS WITH INORGANIC LIGANDS[9]

Form	Inorganic Ligand	Log of Equilibrium Constant[a]
Tl^{3+}	e^-	$Tl^{3+} + e^- \rightleftharpoons Tl^+$ ~ 41-44 in H_2SO_4, HNO_3, $HClO_4$ 26.0 (0.5 - 1.0 M HCl)
Tl^{3+}	e^-	$1/2\ Tl_2O_3(s) + 3/2\ H_2O + 2e^- \rightleftharpoons Tl^+ + 3OH^-$ 0.7
Tl^+	e^-	$Tl^+ + e^- \rightleftharpoons Tl(s)$ -5.595 to -5.819
Tl^+	e^-	$TlCl(s) + e^- \rightleftharpoons Tl(s) + Cl^-$ -9.42
Tl^+	OH^- hydroxide	$K_1 = 0.42 - 0.85$
Tl^{3+}	OH^- hydroxide	$K_1 = -0.2$
Tl^+	$Fe(CN)_6^{4-}$ cyanoferrate(II)	$K_1 = 3.00 - 3.22$
Tl^+	$Fe(CN)_6^{3-}$ cyanoferrate(III)	$K_1 = 1.83$
Tl^+	CN^- cyanide	No evident complex in 4 mole/ℓ $NaClO_4$

TABLE III-3 (continued)

Form	Inorganic Ligand	Log of Equilibrium Constant[a]
Tl^{3+}	CN^- cyanide	$Tl^{3+} + 4CN^- \rightleftharpoons Tl(CN)_4^-$ β_4 = 35 (various media)
Tl^+	SCN^- thiocyanate	K_1 = 0.94 (0°C), 0.80 (25°C), 0.64 (40°C)
Tl^+	NH_3 ammonia	$K_1 \cong -0.9$ (2 mole/ℓ NH_4NO_3)
Tl^{3+}	NH_3 ammonia	β_4 = 17 (doubtful)
Tl^+	N_3^- azide	K_1 = 0.39 (25°C)
Tl^+	NO_2^- nitrite	K_1 = 0.80 to 0.85
Tl^+	NO_3^- nitrate	K_1 = 0.31 to 0.44
Tl^{3+}	NO_3^- nitrate	K_1 = 0.41 (10°C), 0.18 (18°C) (various media)
Tl^+	S^{2-} sulfide	(log of solubility constant ~ -20)
Tl^+	$S_2O_3^{2-}$ thiosulfate	K_1 = 1.91
Tl^{3+}	$S_2O_3^{2-}$ thiosulfate	β_4 = 41 (various media)
Tl^{3+}	SO_3^{2-} sulfite	$\beta_4 \cong 34$ (various media)
Tl^+	SO_4^{2-} sulfate	K_1 = 1 to 2

TABLE III-3 (concluded)

Form	Inorganic Ligand	Log of Equilibrium Constant[a]
Tl^{3+}	SO_4^{2-} sulfate	$K_1 \cong 1$ (various media)
Tl^{3+}	SeO_3^{2-} selenite	(log of solubility constant -38.7)
Tl^{+}	Cl^{-} chloride	$K_1 = 0.47 - 0.78$
Tl^{3+}	Cl^{-} chloride	$K_1 = 8.1$, $K_2 = 5.5$, $K_3 = 2.2$, $K_4 \cong 2.2$ (at 1.2 mole/ℓ $NaClO_4$, 1 mole/ℓ H+, or 0.4 mole/ℓ Na, $HClO_4$: $K_1 = 6.25 - 7.50$, $K_2 = 4.50 - 5.15$, $K_3 = 2.75 - 3.10$, $K_4 = 2.25 - 2.5$, $K_5 = 1.95 - 2.15$, $K_6 = 1.75 - 1.80$)
Tl^{+}	HOP_4^{2-}	$K_1 = 0.73 \pm 0.1$[b]
Tl^{+}	Ribose phosphate^{2-}	$K_1 = 0.87 \pm 0.15$[b]
Tl^{+}	ADP^{3-}	$K_1 = 1.32 \pm 0.1$[b]
Tl^{+}	ATP^{4-}	$K_1 = 1.99 \pm 0.1$[b]
Tl^{+}	$HP_2O_7^{3-}$	$K_1 = 2.34 \pm 0.1$[b]
Tl^{+}	PO_4^{3-}	$K_1 = 2.41 \pm 0.1$[b]
Tl^{+}	$P_2O_7^{4-}$	$K_1 = 3.05 \pm 0.04$[b]

a/ Extrapolated to zero ionic strength unless stated otherwise; temperature usually ~ 25°C.

TABLE III-4

STABILITY CONSTANTS OF COMPLEXES OF THALLIUM IONS WITH ORGANIC LIGANDS[9/]

Form	Organic Ligand	Log of Equilibrium Constant
Tl^+	$C_2H_2O_4$ oxalic acid	$K_1 = 2.03$
Tl^+	$C_2H_4O_2$ acetic acid	$K_1 = -0.11$
Tl^{3+}	$C_2H_4O_2$ acetic acid	$\beta_4 = \sim 15.4$
Tl^+	$C_2H_8N_2$ ethylenediamine	$K_1 = 0.3\text{-}0.4$
Tl^+	$C_6H_8O_7$ citric acid	$K(Tl^+ + HL^{3-} \rightleftharpoons TlHL^{2-}) =$ 0.65 - 1.04 (ionic strength 0.1 - 0.5 mole/ℓ)
Tl^+	$C_6H_9O_6N$ nitrilotriacetic acid	$K_1 = 3.44$, 4.74 (ionic strength 1 mole/ℓ and 0.1 mole/ℓ, respectively)
Tl^{3+}	$C_{10}H_{16}O_8N_2$ ethylenediaminetetraacetic acid	$K_1 = 22.5$
Tl^+		$K_1 = 6.55$[1/] (ionic strength 0.1 mole/ℓ), 6.47[2/]
Tl^+	$C_4H_6O_6$ tartaric acid	$K_1 = 1.39$
Tl^+	$C_3H_7NO_2$ -alanine	$K_1 = 1.49$
Tl^+	$C_3H_7NO_2$ -alanine	$K_1 = 1.04$

BIBLIOGRAPHY. SECTION III.

1. Lee, A. G., The Chemistry of Thallium, Elsevier Publishing Company, New York, N.Y., 1971.

2. Lee, A. G., "The Coordination Chemistry of Thallium(I)," Coord. Chem. Rev., 8, 289-349 (1972).

3. Wade, K., and A. J. Banister, The Chemistry of Aluminum, Gallium, Indium and Thallium, Vol. 12 Pergamon Texts in Inorganic Chemistry, Pergamon Press, New York, N.Y., 1973.

4. Farmer, J. B., and K. Wade, "Aluminum, Gallium, Indium, and Thallium" in MTP (Med. Tech. Publ. Co.) Int. Rev. Sci.: Inorg. Chem., Ser. One, No 4, Butterworths, London, England, 1972, pp. 105-140.

5. Forman, E. A., and K. Wade, "Aluminum, Gallium, Indium, and Thallium" in MTP Int. Rev. Sci.: Inorg. Chem., Ser. Two, 4, B. J. Aylett, Ed., Butterworths, London, England, 1975, pp. 119-156.

6. Siegel, B., "Oxyhalides of Boron, Aluminum, Gallium, Indium, and Thallium," Inorg. Chim. Acta, Rev., 2, 137-146 (1968).

7. Zitko, V., "Chemistry, Applications, Toxicity, and Pollution Potential of Thallium," Tech. Rep.--Fish. Mar. Serv. (Can.), 518, 1-41 (1975).

8. Cotton, F. A., and G. Wilkinson, Advanced Inorganic Chemistry. A Comprehensive Text, Interscience Publishers, a Division of John Wiley and Sons, Inc., U.S.A., 1962.

9. Sillen, L. G., and A. E. Martell, Stability Constants of Metal-Ion Complexes, Special Publication No. 17, The Chemical Society, London, England, 1964.

10. Batley, G. E., and T. M. Florence, "Determination of Thallium in Natural Waters by Anodic Stripping Voltammetry," J. Electroanal. Chem. Interfacial Electrochem., 61(2), 205-211 (1975).

11. Lanford, C., "Effect of Trace Metals on Stream Ecology," paper presented at the Cooling Tower Institute meeting, January 20, 1969.

12. Nesmeyanov, A. N., and R. A. Sokolik, Methods of Elemento-Organic Chemistry. Vol. 1, The Organic Compounds of Boron, Aluminum, Gallium, Indium, and Thallium, North Holland Publishing Co., Amsterdam, 1967.

13. Kurosawa, H., and R. Okawara, "Thallium. Annual Survey Covering the Year 1974," J. Organomet. Chem., 98, 467-493 (1975).

14. Henry, P. M., "Oxidizing Olefins by Pd(II) and Tl(III)" in Homogeneous Catalysis. Industrial Applications and Implications, Advances in Chemistry Series No. 70, American Chemical Society, Washington, D.C., 1968, pp. 126-154.

15. Taylor, E. C., and A. McKillop, "Thallium in Organic Synthesis," Accounts of Chem. Res., 3(10), 338-346 (1970).

16. McKillop, A., and E. C. Taylor, "Organothallium Chemistry," Advan. Organometal. Chem., 11, 147-206 (1973).

17. McKillop, A., and E. C. Taylor, "Thallium in Organic Synthesis," Chem. Brit., 9(1), 4-11 (1973).

18. Banerji, A., J. Banerji, and R. Das, "Use of Thallium in Organic Reactions," J. Sci. Ind. Res., 33(10), 510-532 (1974).

19. McKillop, A., "Applications of Thallium(III) Nitrate (TTN) to Organic Synthesis," Pure Appl. Chem., 43(3-4), 463-479 (1975).

20. Maher, J. P., [Organometallic Compounds of] "Group III. Aluminum, Gallium, and Thallium," Organometal. Chem., 2, 89-136 (1973).

21. Maher, J. P., "Group III: Aluminum, Gallium, Indium, and Thallium," Organomet. Chem., 4, 68-101 (1975).

22. McKillop, A., and E. C. Taylor, "Thallium(III) Nitrate in Organic Synthesis: New Experimental Approaches to Oxythallation," Endeavour, 35(125), 88-93 (1976).

23. Bajpai, K. K., and B. Bajpai, "Interaction Products of Diphenylthallium(III) Compounds With Mercury(II) Salts," Inorg. Chim. Acta, 19, L64-L66 (1976).

24. Crabbe, P., and A. Guzman, "Substituted Prostaglandin Derivatives," Fr. Demande 2,162,212, August 1973, 75 pp.; Chem. Abstr., 80, 82218r (1974).

25. Thayer, J. S., "Organometallic Compounds and Living Organisms," J. Organometal. Chem., 76, 265-295 (1974).

26. Srivastava, T. N., K. K. Bajpai, and K. Singh, "Antimicrobial Activities of Diaryl Gallium, Indium, and Thallium Compounds," Indian J. Agric. Sci., 43(1), 88-93 (1973).

27. Agnes, G., S. Bendle, H. A. O. Hill, F. R. Williams, and R. J. P. Williams, "Methylation by Methyl Vitamin B_{12}," J. Chem. Soc. D (Chem. Communications), No. 15, 850-851 (1971).

28. Korenman, I. M., Analytical Chemistry of Thallium, Israel Program for Scientific Translations, Jerusalem, 1963, translation of Analiticheskaya Khimiya Talliya, Izdatel'stvo Akademii Nauk SSSR, Moscow, 1960.

29. Reese, K. M., "NBS Develops Standard Spinach for FDA Program," Chem. Eng. News, 55(11), 48 (1977).

30. Van Ormer, D. G., "Atomic Absorption Analysis of Some Trace Metals of Toxicological Interest," J. Forensic Sci., 20(4), 595-623 (1975).

31. Marks, J. Y., G. G. Welcher, and R. J. Spellman, "Atomic Absorption Determination of Lead, Bismuth, Selenium, Tellurium, Thallium, and Tin in Complex Alloys Using Direct Atomization From Metal Chips in the Graphite Furnace," Appl. Spectr., 31(1), 9-11 (1977).

32. Kubasik, N. P., and M. T. Volosin, "A Simplified Determination of Urinary Cadmium, Lead and Thallium, With Use of Carbon Rod Atomization and Atomic Absorption Spectrophotometry," Clin. Chem., 19(9), 954-958 (1973).

33. Machata, G., and R. Binder, "Determination of Lead, Thallium, Zinc, and Cadmium Trace Elements in Biological Material by Flameless Atomic Absorption," Zeitschrift fuer Rechtsmedizin, Berlin, 73(1):29-34 (1973).

34. Skogerboe, R. K., D. L. Dick, D. A. Pavlica, and F. E. Licte, "Injection of Samples Into Flames and Plasmas by Production of Volatile Chlorides, Anal. Chem., 47, 568-570 (1975).

35. Reinhardt, G., and P. Zink, "Analytische Probleme bei der polarographisch Thalliumbestimmung in kleinen Organproben" ["Analytic Problems in the Polarographic Determination of Thallium in Small Organ Samples"], Beitr gerichtl. Med., 30, 371-375 (1973); abstract in TOXLINE thallium search

36. Vassos, B. H., "Averaging Polarograph With Digital Output," Anal. Chem., 45(7), 1292-1295 (1973).

37. Curtis, A. R., "Determination of Trace Amounts of Thallium in Urine, Using the Mercury-Film Electrode and Differential Pulse Anodic Stripping Voltammetry," J. Assoc. Offic. Agric. Chemists, 57(6), 1366-1372 (1974).

38. Franke, J. P., P. M. J. Coenegracht, and R. A. De Zeeuw, "A Rapid and Specific Method for the Determination of Traces of Thallium Directly in Urine by Differential Pulse Anodic Stripping Voltammetry," Arch. Toxicol., 34(2), 137-143 (1975).

39. O'Shea, T. A., and K. H. Mancy, "Characterization of Trace Metal-Organic Interactions by Anodic Stripping Voltammetry," Anal. Chem., 48, 1603-1607 (1976).

40. Kartha, V. B., T. S. Krishnan, N. Patel, and S. Gopal, "A Spectrographic Method for the Analysis of Impurities in Sodium Chloride," BARC-284, U.S. At. Energy Comm., 1967.

41. Fassel, V. A., and R. N. Kniseley, "Inductively Coupled Plasma Optical Emission Spectroscopy," Anal. Chem., 46, 1110A-1111A (1974).

42. Marcie, F. J., "X-Ray Fluorescence Determination of Trace Toxic Elements in Water," Environ. Sci. Technol., 1(2), 164-166 (1967).

43. Goldman, M., R. P. Anderson, J. P. Henry, and S. A. Peoples, "X-Ray Emission Spectrographic Determination of Thallium in Biologic Materials," J. Agric. Food Chem., 14(4), 367-369 (1966).

44. Camp, D. C., J. A. Cooper, and J. R. Rhodes, "X-Ray Fluorescence Analysis. Results of a First Round Intercomparison Study," X-Ray Spectrometry, 3, 47-50 (1974).

45. Camp, D. C., A. L. Van Lehn, J. R. Rhodes, and A. H. Pradzynski, "Intercomparison of Trace Element Determinations in Simulated and Real Air Particulate Samples," X-Ray Spectrometry, 4, 123-137 (1975).

46. Specht, W., and D. Rohner, "Bestimmung von Thalliumspuren in menschlichen Haaren mittels Neutronen-Aktivierungsanalyse" ["Determination of Traces of Thallium in Human Hair by Neutron-Activation Analysis"], Arch. Toxikol., 18, 359-367 (1960).

47. Morris, D. F. C., and R. A. Killick, "The Determination of Silver and Thallium in Rocks by Neutron-Activation Analysis," Talanta, 4, 51-60 (1960).

48. Morris, D. F. C., and R. A. Killick, "Silver and Thallium Content of Rocks," Geochim. Cosmochim. Acta, 19, 139-140 (1960).

49. Filby, R. H., and K. R. Shah, "Activation Analysis and Applications to Environmental Research," Toxicol. Environ. Chem. Rev., 2, 1-44 (1974).

50. Ondov, J. M., W. H. Zoller, I. Olmez, N. K. Aras, G. E. Gordon, L. A. Rancitelli, K. H. Abel, R. H. Filby, K. R. Shah, and R. C. Ragaini "Elemental Concentrations in the National Bureau of Standards Environmental Coal and Fly Ash Standard Reference Materials," Anal. Chem. 47, 1102-1109 (1975).

51. Fitchett, A. W., R. P. Buck, and P. Mushak, "Direct Determination of Heavy Elements in Biological Media by Spark Source Mass Spectrometry," Anal. Chem., 46(6), 710-713 (1974).

52. Favorskaya, L. V., and E. I. Ponomareva, "The Pressure of Thallium Sulfate Vapor," Izvest. Akad. Nauk Kazakh. S.S.R., Ser. Met. Obogashchen. i Ogneuporov, No. 2, 30-33 (1960); Chem. Abstr., 54, 1114g (1961).

53. Shakhtakhtinskii, M. G., and A. A. Kuliev, "Investigation of Saturated Vapor Pressures of Compounds in the System Tl-Se With the Use of Radioactive Isotopes," Doklady Akad. Nauk Azerbaidzhan. S.S.R., 15, 891-895 (1959); Chem. Abstr., 54, 9410f (1960).

54. Shakhtakhtinskii, M. G., and A. A. Kuliev, "Saturated Vapor Pressure of Thallium Selenides," Fiz. Metal. i Metalloved., Akad. Nauk S.S.S.R., 9, 202-204 (1960); Chem. Abstr., 54, 18049h (1960).

55. Shakhtakhtinskii, M. G., and A. A. Kuliev, "The Vapor Pressure of Some Thallium Compounds," Doklady Akad. Nauk S.S.S.R., 123, 1071-1073 (1958) Chem. Abstr., 55, 6086b (1961).

IV. OCCURRENCE IN THE GEOCHEMICAL ENVIRONMENT AND GEOCHEMISTRY

This section includes discussions of the background occurrence of thallium in the natural environment (supplemented by an extensive tabulation in the Appendix) and a description of the geochemistry of thallium, including its transport.

A. Occurrence

For many years, only five independent thallium minerals were known--the very rare lorandite, vrbaite, hutchinsonite, crookesite, and avicennite. (These and other thallium minerals are in Table IV-1.) They form in arsenic-antimony or lead-zinc deposits in the very last stages of the hydrothermal* process. They are also associated with arsenic and antimony minerals or in limonite-calcite veins formed under supergene** conditions (avicennite).[1/] Reports of new thallium minerals have appeared recently.[2/]

Most of the ores containing high concentrations of thallium occur in Switzerland, Yugoslavia, the Caucasus, and Soviet Central Asia.

Thallium behaves as an accessory of more common elements during natural processes. It is geochemically similar in some instances to the alkali metals potassium, rubidium, and cesium (principal mineral carriers of thallium are micas and potash feldspars); in other instances, to manganese (oxide minerals and rhodochrosite, which form during the late stages of the hydrothermal process), lead (galena, sulfate salts), silver, zinc (sphalerite), and iron (sulfides, especially colloidal pyrite and marcasite, and jarosites).[1/] The concentrations of thallium in these minerals are quite variable (see the table in the Appendix). There are apparently few "rules of thumb" for predicting which U.S. deposits of these minerals are most likely to be enriched in thallium. The geochemists and geologists with whom we have discussed the problem have conflicting views. Angino[4/] suggests that each deposit would have to be examined individually to make any judgments. Table V-6 summarizes published data on the occurrence of thallium in U.S. sulfide ores.

Table IV-2 shows the generalizations made by Ivanov regarding thallium and cadmium in sphalerites. One of the generalizations apparently borne out by observations in the United States is that sphalerite from high temperature paragenesis contains less than 2 ppm thallium. Lowest temperature zinc deposits are expected to be most enriched. A

---

* Hydrothermal--pertaining to the action of hot water in dissolving, redepositing, and otherwise changing crustal minerals.

** Supergene--concentrated by weathering action at or near the surface.

less pronounced dependence of a similar type is seen for thallium in chalcopyrite ($CuFeS_2$). Various observations imply that both the U.S. high- (Balmat, New York) and low-temperature zinc deposits (e.g., Missouri; Kansas; Tennessee; Rossie, New York; and Colorado)[5/] are not especially enriched in thallium.

The sporadic occurrence of high concentrations of thallium in other low-temperature sulfides (particularly colloidal ones) such as realgar, orpiment, antimonite, and cinnabar may be due to mechanical contamination by independent minerals.[1/]

Even the use of the word "enriched" is rather precarious. Estimates cited in the Appendix for the crustal abundance range from 0.3 ppm to 3 ppm. About 1 ppm is a generally accepted value.

Ahrens suggested and Shaw concurred[6/] that nearly all of the thallium in the earth's crust is carried in potassium minerals.

For igneous rocks in general, the concentrations of thallium increase with increasing acidity (SiO_2 content) of rock type. See Figure IV-1.

B. Geochemistry

An excellent review (unattributed, but probably by V. V. Ivanov) of thallium geochemistry appears in the book edited by Vlasov (1964).[3/] Other reviews of the topic have been prepared by Shaw (1952)[6/] and (1957),[7/] Wedepohl (1972),[8/] and Velikii et al. (1968).[9/]

Thallium associates geochemically with potassium, rubidium, and cesium in high-temperature magmatic, metasomatic, and pegmatitic formations; with lead, iron, zinc, copper, arsenic, and antimony in mesothermal and epithermal deposits; and with manganese in supergene formations.[3/]

1. Potassium minerals: Tl^+ has a coordination number of at least eight in silicates and is thus associated with the alkali metals of large ionic radius.[6/] Tl^+ (ionic radius 1.49 Å) fits more readily into K^+ (1.33 Å) sites in minerals where potassium occurs in high coordination (octahedral or dioctahedral) rather than in minerals of low coordination (tetrahedral).[3/] Clays and micas have sheet-like, sieve-like horizontal structures that are more likely to hold thallium irreversibly than sylvite and other evaporite minerals found in potash deposits.[4/] Biotite is the best thallium concentrator, with "magmatic" muscovite a close second; but the thallium content of the latter varies widely. More closely packed structures are less favorable for potassium-thallium isomorphism, especially those of plagioclases and hornblendes. The more potassium in an aluminosilicate, the greater the tendency for thallium enrichment.[3/]

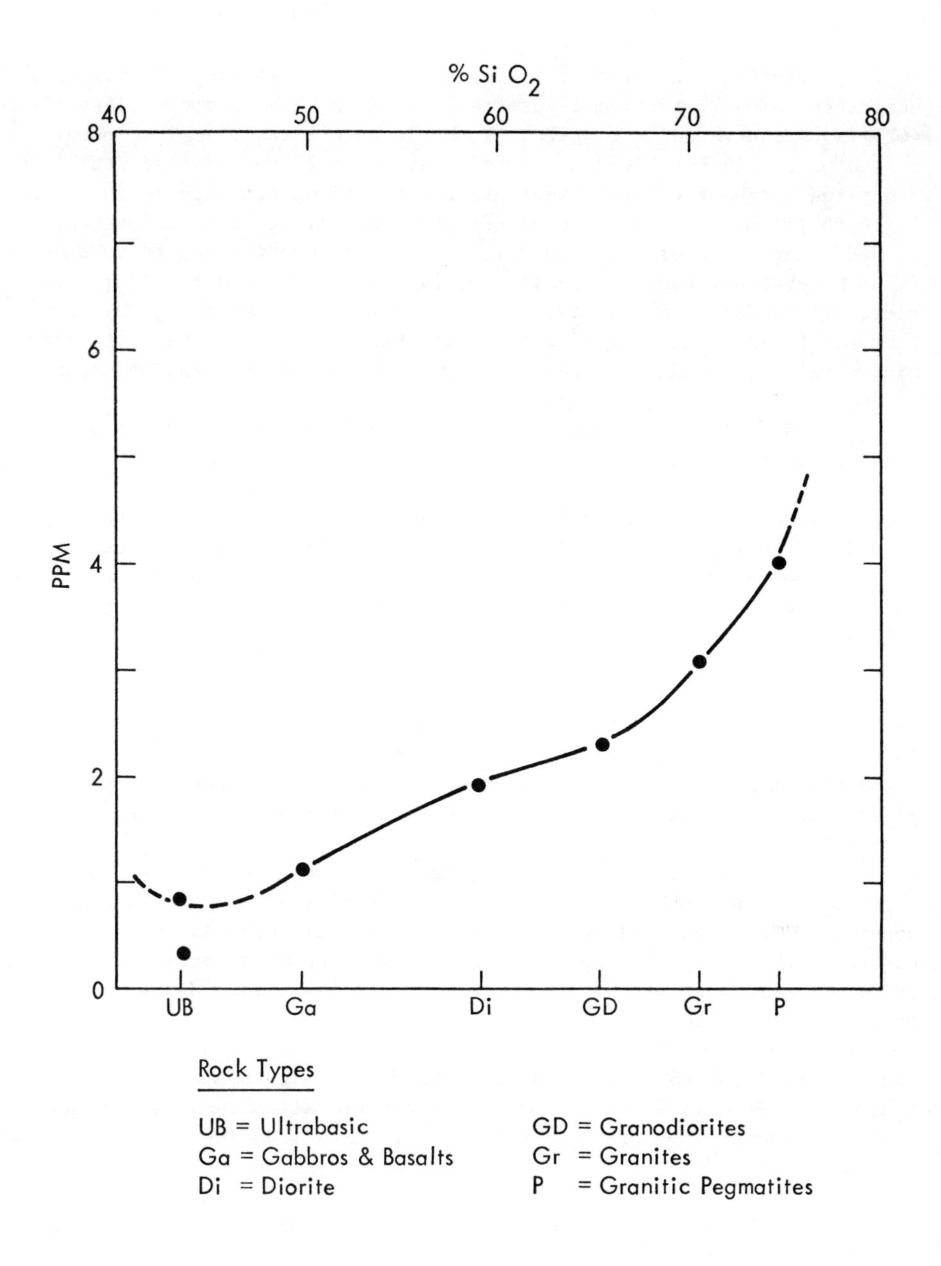

Figure IV-1 - Content of Thallium in Igneous Rocks

During the magmatic process, thallium is mostly (60-95%) concentrated in the potassium feldspars of acid and alkali rocks; less (1-40%) is concentrated in biotite. In basic rocks, thallium is concentrated mostly in hornblendes and feldspars. Thallium was preferably concentrated during the late stages of pegmatization, at least in the USSR. It is thallium's chalcophilic nature that usually governs its presence in hydrothermal deposits. However, in high-temperature quartz-rare metal (tin, molybdenum, tungsten [sulfide]) deposits, the main thallium carriers are feldspars and micas. In silicate-sulfide deposits, thallium's behavior is more lithophilic than chalcophilic so that in silicate-lead-zinc deposits, thallium probably accumulates in sericite and feldspar.[3/]

Shaw (1957)[7/] suggests minerals in which Tl^{3+} (0.95 Å) might substitute for Ca^{2+}.

2. Sulfide minerals: Velikii et al. (1966)[9/] state that thallium is found in all types of hydrothermal sulfide deposits, occasionally as thiosalts and selenides, but usually in the disseminated state in low temperature lead-zinc, antimony-mercury, and other deposits. Deposits high in arsenic are often also high in thallium.

Factors promoting increased thallium in hydrothermal formations are given as: presence of galena, sphalerite, pyrite, or lead sulfosalts in the ore complex; low temperature and small depth of formation of the deposits; increased antimony and arsenic in the low-temperature ore complex; colloidal origin; and relatively recent formation.[3/]

Although cadmium, gallium, and thallium have similar geochemical characteristics and behave similarly in hydrothermal processes, cadmium appears not to have followed gallium and thallium entirely during the formation of various types of lead-zinc deposits. The Ga/Tl ratio is 0.6 to 10 in low-temperature carbonate (barite, fluorite)--sulfide deposits in carbonate and silicate rocks. The ratio is 6 to 15 in medium-temperature deposits and effusive rocks, granites, sandstones, and carbonate rocks. Gallium and thallium concentrations increased from high-temperature to low-temperature deposits, while the cadmium concentration remained unchanged. Thallium concentrated predominately in ores localized in carbonate rocks (Ivanov, 1967).[10/]

High thallium concentrations occur in epithermal deposits in very different rocks. Thus, the surrounding rock composition does not govern the concentration or dispersion of thallium in the rocks.

3. Oxides, hydroxides, etc.: Thallium that has dissolved during oxidation of sulfide deposits is fixed in situ by gels (e.g., manganese hydroxides, alkali-metal and lead sulfates). Iron gossans* contain less thallium than the primary ore. During the endogene process, thallium that is removed by water from the oxidation zone is either dispersed in the general groundwater cycle or adsorbed by clay minerals.[3] The high redox potential of the thallous-thallic couple (+1.25 v at pH 0 and +0.82 v at pH 7) indicates that thallous ion will be the only species present under normal weathering and sedimentation processes.[6] However, under certain highly oxidizing aqueous conditions, thallium(I) can apparently be oxidized, precipitated, and dehydrated; the resultant Tl_2O_3 may be adsorbed by MnO_2 or may penetrate the Mn_2O_3 lattice.[3,11] Thus, manganese nodules in the oceans and certain USSR and U.S. manganese oxide deposits are enriched in thallium. (We have not found data for thallium concentrations in manganese oxide ores from other countries.)

Razenkova and Galaktionova (1963),[12] showed in laboratory experiments that thallium is adsorbed by iron hydroxide.** Most thallium adsorbed by limonite is desorbed easily by solutions present in the oxidation zone of sulfide deposits. In highly carbonaceous medium, $Tl(OH)_3$ forms. Tl^{3+} is not substituted for Fe^{3+} in the hematite lattice, but apparently thallium enters the crystal lattice during formation of gel-like manganese compounds and jarosite.

4. Transport: The following discussion of thallium transport is largely adapted from Shaw (1952)[3] and Wedepohl (1972).[8]

The close relations between thallium and rubidium and potassium, the adsorption of thallium in clays, and its enrichment in sedimentary rocks formed in reducing conditions led Wedepohl to conclude that in most natural environments, thallium is present in univalent compounds.

Thallium is largely retained during rock weathering and may be enriched in the clay, silt, and loam from weathering granodiorite, andesite, and shale. Ionic adsorption on clay sediments would explain why eroded thallium, like the large alkali ions, is improverished in the oceans.

Shaw indicated the unlikelihood of thallium chloride's being precipitated from river water despite its sparing solubility (0.33 g/100 g water at 20°C).

---

* Principally hydrated iron oxide from the oxidation and removal of sulfur, copper, etc., superficially covering or filling pyrite masses or mineral veins.

** 75-100% at pH $\sim$ 4-10.

Both Shaw and Wedepohl believed that thallium is probably transported in solution as Tl^+. Thallium becomes enriched in sediments in strong reducing environments where organic matter is accumulating under quiet, anaerobic conditions. In strongly oxidizing conditions, Tl^+ will be removed from solution as Tl^{3+} by precipitation with manganese and/or iron. Adsorption is also important in the concentration of thallium in colloform ores and perhaps in galena. Both the Calspan report (1977)[13/] and Batley and Florence (1975)[14/] suggest thallium solution transport is as Tl(III) species, and that chloride ions control solution transport by forming chloride complexes, which are either stable with respect to reduction to Tl(I), readily adsorbed (by anion exchange), and/or precipitated. The arguments for Tl(III) speices predominating at natural pH values in seawater or fresh water are not entirely convincing. When the seawater is acidified to pH 1 with hydrochloric acid, up to 80% of the thallium is adsorbed on an anion-exchange resin in the absence of an oxidizing agent (bromine). The remaining 20% was thought to be cationic or neutral thallium(III) complexes or thallium(I) species.

Crafts (1934)[15/] and (1936)[16/] found that Tl_2SO_4 was strongly fixed in the top 10 cm by four California soils, regardless of soil type or water-holding capacity, and resisted leaching. Yolo clay loam had a saturation capacity of $\sim$ 10,000 ppm Tl_2SO_4 (dry weight basis). In further experiments, the strong localization of thallium was shown when thallium-treated barley grains had to be within 0.25 cm of germinating or growing potted oats to affect them adversely.

McCook, 1933,[17/] concluded that soils remove thallium from solution by base (anion) exchange. After 25 g samples of sandy loam and silt loam were shaken with 100 ml, 5.5% Tl_2SO_4 solution (4.5 g Tl), let stand 24 hr, shaken again, and filtered, the sandy loam had removed 0.362 g Tl from solution and the silt loam, 0.44 g. (Thus, the saturation capacities for thallium of these soils were 14,500 and 18,000 ppm, respectively.) The thallium treatment increased the aluminum content of the sandy loam soil extract from less than 1 to 20 ppm and that of the silt loam from less than 1 to 15 ppm. At the same time, 0.047 g calcium was displaced from the sandy loam and 0.064 g from the silt loam.

When sandy loam and silt loam soils were treated with $\leq$ 2% Tl_2SO_4 solution (1 ml/g soil); leached with $\leq$ 36 in. water; and planted with corn, rye grass, or wheat, the extent of plant injury was not reduced significantly compared with experiments in unleached soils.

TABLE IV-1

THALLIUM MINERALS[8]

Mineral	Chemical Composition	Paragenesis	Comments
Lorandite	$TlAsS_2$	Realgar, orpiment, colloform pyrite	Found in Macedonia
Picotpaulite	$TlFe_2S_3$		
Chalkothallite	Cu_3TlS_2		
Vrbaite	$Hg_3Tl_4As_8Sb_2S_{20}$	Realgar, orpiment, colloform pyrite	Macedonia, 29-32% Tl
Hutchinsonite	$(Pb,Tl)_2(Cu,Ag)As_5S_{10}$	Sulfides, etc., of As, Zn, Pb, Fe	13-35% Tl. Occurs in white dolomite of the Lengenbach Quarry, Binnental, Switz.
Bukovite	$Cu_{3+a}Tl_2FeS_{4-a}$		
Wallisite	$PbTlCuAs_2S_5$		
Hatchite	$PbTlAgAs_2S_5$		
Crookesite	$(Cu,Tl,Ag)_2Se$	Selenides	Found in Skrikerum, Sweden[18]
Avicennite	Tl_2O_3	Hematite, calcite	

TABLE IV-2

CONTENT OF CADMIUM AND THALLIUM (g/ton) IN SPHALERITES ACCORDING TO IVANOV, 1966 [cited by Kabanova][19]

Type of Deposit	Cd, g/ton	Tl, g/ton
• Pb-ore high- and medium-temp. silicate-sulfide	3,300	1
• Pb-Zn high-temp. deposits in highly metamorphosized silicate rocks	2,000	-
• Pb-Zn high- and medium-temp. deposits in skarns	5,000	-
• Cu-Pb-Zn chalcopyrite medium-temp. deposits in effusive-sedimentary rocks	2,500	8
• Pb-Zn medium-temp. deposits in granitoids, sandstones and schists	3,000	1
• Pb-Zn medium-temp. deposits in carbonate rocks	2,000	2
• Pb-Zn low-temp. deposits in carbonate rocks	3,000	80
• Pb-Zn low-temp. deposits in silicate rocsk	3,000	10

BIBLIOGRAPHY. SECTION IV.

1. Ivanov, V. V., and V. Y. Volgin, "Thallium Minerals" in Mineralogy of Rare Elements, Vol. 2 of Geochemistry and Mineralogy of Rare Elements and Genetic Types of Their Deposits, K. A. Vlasov, Ed., Russian ed., Publishing House "Nauka," Moscow, 1964; Z. Lerman, translator, Israel Program for Scientific Translations, Jerusalem, 1966 [Distributed by Daniel Davey and Company, Inc., New York, N.Y.], pp. 590-605.

2. Fleischer, M., U.S. Geological Survey, Reston, Virginia, personal communication (telephone), January 1977.

3. Anonymous [probably V. V. Ivanov], "Thallium" in Geochemistry of Rare Elements, Vol. 1 of Geochemistry and Mineralogy of Rare Elements and Genetic Types of Their Deposits, K. A. Vlasov, Ed., Russian ed., Publishing House "Nauka," Moscow, 1964; Z. Lerman, translator, Israel Program for Scientific Translations, Jerusalem, 1966 [Distributed by Daniel Davey and Co., Inc., New York, N.Y.], pp. 491-524.

4. Angino, E. E., Consulting geochemist, personal communications, 1976, 1977.

5. Stoiber, R. E., "Minor Elements in Sphalerite," Econ. Geol., **35**, 501-519 (1940).

6. Shaw, D. M., "The Geochemistry of Thallium," Geochim. Cosmochim. Acta, **2**, 118-154 (1952).

7. Shaw, D. M., "The Geochemistry of Gallium, Indium, and Thallium--A Review," Phys. Chem. Earth, **2**, 164-211 (1957).

8. Wedepohl, K. H., Ed., Handbook of Geochemistry, Element 81, Vol. 2 No. 3, Springer-Verlag, Berlin Heidelberg, 1972, pp. 81-B-1 to 81-0-1.

9. Velikii, A. S., V. Yu. Volgin, and V. V. Ivanov, "Thallium Deposits," Geokhim. Mineral. Genet. Tipy Mestorozhd. Redk. Elem. Akad. Nauk S.S.S.R., Gos. Geol. Kom., S.S.S.R., Inst. Mineral. Geokhim. Kristallokhim. Redk. Elem., **3**, 376-386, 834-835 (1966); Chem. Abstr., **68**, 4786n (1968).

10. Ivanov, V. V., A. A. Garmash, G. M. Meituv, and N. V. Nechelyustoe, "Kadmii, Tallii i Gallii v Svintsovo-Tsinkovykh Mestorozhdeniyakh" ["Cadmium, Thallium, and Gallium in Lead-Zinc Deposits"] in Formy Nakhozhdeniya i Osobennosti Raspredeleniya Redkikh Elementov v Nekotorykh Tipakh Gidrotermal'nykh Mestorozhdenii [Forms of Occurrence and Characteristics of the Distribution of Rare Elements in Some Types of Hydrothermal Deposits], V. V. Ivanov, Ed., Publisher "Nauka," Moscow, 1967, pp. 111-179.

11. Voskresenskaya, N. T., and T. D. Usevich, "The Occurrence of Thallium in Manganese Minerals," Geochemistry (USSR), 710-721 (1957).

12. Razenkova, N. I., and G. F. Galaktionova, "Eksperimental'noe Izuchenie Formy Nakhozhdeniya Talliya v Zone Okisleniya Sul'fidnykh Mestorozhden ["Forms of Thallium Occurrence in the Oxidation Zone of Sulfide Deposits"], Tr. Inst. Mineralog, Geokhim. i Kristallokhim. Redkikh Elemento Akad. Nauk S.S.S.R., No. 18, 5-19 (1963).

13. Calspan Corporation, Heavy Metal Pollution From Spillage at Ore Smelters and Mills, preliminary copy of final version (as of March 7, 1977), Calspan Report No. ND-5189-M-1, prepared for the U.S. Environmental Protection Agency, Cincinnati, Ohio (in press).

14. Batley, G. E., and T. M. Florence, "Determination of Thallium in Natural Waters by Anodic Stripping Voltammetry," J. Electroanal. Chem. Interfacial Electrochem., **61**(2), 205-211 (1975).

15. Crafts, A. S., "The Effects of Thallium Sulfate on Soils," Science, **79**, 62 (1934).

16. Crafts, A. S., "Some Effects of Thallium Sulfate Upon Soils," Hilgardia, **10**, 377-398 (1936).

17. McCool, M. M., "Effect of Thallium Sulfate on the Growth of Several Plants and on the Nitrification in Soils," Contrib. Boyce Thompson Inst., **5**(3), 289-296 (1933).

18. Petar, A. V., Thallium, Information Circular 6453, Bureau of Mines, U.S. Department of Commerce, Washington, D.C., 1931, 6 pp.

19. Kabanova, E. S., "Elementy-Primesi v Sfalerite" ["Impurity Elements in Sphalerite"], Itogi Nauki: Geokhim., Mineral., Petrogr.: 1966, 53-99 (1968).

V. INDUSTRIAL PROCESSES AS ENVIRONMENTAL SOURCES OF THALLIUM

Thallium was discovered spectrographically by Crookes in 1861 in the selenium-bearing deposit of a German sulfuric acid factory and soon after by Lamy in the lead chamber deposit of a French sulfuric acid plant where pyrites were roasted.[1/] Since then, thallium has been produced commercially not only from sulfuric acid plant materials (especially the dusts collected to purify the pyritic roasting gases used as the source of sulfur dioxide) but also from dusts collected from roasting and smelting zinc, lead, and copper sulfide concentrates. Thallium has also been recovered in hydrometallurgical processes, e.g., in purification of the zinc sulfate solutions in electrolytic zinc refining and lithopone production.[2/]

The history of thallium production in the United States is not very clear. Much of the demand has always been met by imports. Since the 1930's, ASARCO Inc.,* appears to be the only U.S. company that has isolated pure thallium or thallium sulfate on a commercial basis. Production is intermittent and has never exceeded 5 tons annually. Thallium is recovered electrolytically at the ASARCO cadmium refinery in Denver, Colorado. Thallium is not restricted to zinc ores as cadmium largely is, but thallium is most conveniently handled at the cadmium refinery operated in conjunction with zinc processing. Both thallium and cadmium appear in the flue dusts from roasting and smelting sulfide ores because of their volatility, and both are removed simultaneously from zinc sulfate solutions because of their similar electrode potentials. A large share of the thallium at the cadmium refinery handling flue dusts may actually have come from copper and/or lead concentrates rather than from zinc concentrates.

This section describes U.S. industrial processes that utilize thallium-containing raw materials and concentrate the element in their products or waste streams.

Commercial thallium production in the United States is small scale. Sulfide minerals processing for zinc and cadmium, lead, copper,

---

* Formerly, American Smelting and Refining Company, Inc. For brevity, the firm will be mentioned as ASARCO Inc., regardless of the historical period of the process.

gold, certain other metals, and sulfuric acid is the largest potential source of mobilized thallium, although not necessarily of environmental releases. Most of the wastes are held in tailings ponds, slag dumps, and other confined areas. Coal burning is probably the largest source of thallium losses to the environment. Because of possible high thallium concentrations in at least some of the raw materials used, production of iron and steel, manganese and ferroalloys, mica and feldspar, and potash are also possibly significant emitters. Information is so incomplete in these latter cases that detailed analyses of their production in the United States are not warranted at this time.

Materials balances have been attempted, but these are probably accurate only within an order of magnitude. Few U.S. data exist upon which estimates of amounts lost to the environment can be made. Because most U.S. processors are reluctant to divulge the thallium contents of their waste strems, Soviet and European data have been used for estimating the amounts of thallium in particular waste streams.

A. Commercial Thallium and Thallium Salt Production

1. Thallium metal and thallium sulfate: U.S. and world production of thallium has always been on a small scale. Actual U.S. production and sales figures have been withheld because there has long been only one U.S. producer. Table V-1 illustrates the kinds of thallium production and consumption statistics available. Even though thallium may have been recovered commercially by 1906 when commercial cadmium production from smelter flue dusts began. Thallium may also have been recovered during sulfuric acid or lithopone production. We have not identified any commercial thallium producers prior to ASARCO Inc., in the 1930's.

In the 1930's, workers at ASARCO patented methods for recovering thallium and cadmium from raw material from smelting lead and zinc sulfide ores. Figure V-1 depicts the method currently used by ASARCO Inc., as described by TRW Systems Group in 1973[3/] and/or by a 1937 ASARCO patent.[4/] Robinson[5/] estimated thallium recovery is 75%. (Various studies have reported 51-90% recoveries[6-10/] from flue dusts.)

Most of ASARCO thallium recovery was done at its Globe cadmium refinery in Denver, Colorado. ASARCO previously operated a pilot plant at Murray, Utah, for thallium recovery from the arsenical Cottrell dusts*

* The arsenical dusts, containing up to 3% Tl, were roasted to remove the arsenic. Thallium was recovered by chloridizing roasting of the residue, distilling thallium chlorides, leaching the chloride distillates, crystallizing TlCl, reducing to thallium, and remelting.[11/]

Raw Material $\xrightarrow{H_2SO_4}$ $CdSO_4$ + Tl_2SO_4 $\xrightarrow{\text{Electrolysis}}$
m Smelting Pb and Zn Sulfide Ores
(2-2,000 ppm Tl)

5-20% Tl-Cd Alloy $\xrightarrow{\text{Hot } H_2O}$ TlOH $Cd(OH)_2$ $\xrightarrow{Na_2CO_3}$

$CdCO_3$ + TlOH $\xrightarrow[\text{2) } H_2S]{\text{1) Filter, } -CdCO_3}$ $Tl_2S\downarrow$ $\xrightarrow{H_2SO_4}$

Tl_2SO_4 Solution $\xrightarrow{\text{Electrolysis}}$ Pure Tl $\xrightarrow[\text{From } H_2SO_4]{\text{Crystallization,}}$

Com. Tl_2SO_4

Figure V-1 - ASARCO Inc., Thallium Recovery Process at Its Denver, Colorado, Globe Refinery

from its lead smelter there. The Murray, Utah, plant was transferred to the Globe plant in 1949, which had installed a new thallium recovery plant (12-ton capacity) in 1948. In 1947, the Globe refinery was processing largely cadmium-containing flue dusts (5,300 ppm Tl) from the ASARCO lead smelter at Murray, Utah. The major source of the thallium in these lead smelter flue dusts was not the lead concentrates charged but converter flue dusts (1,700 ppm Tl) which were added to the lead smelter charge. These latter flue dusts were obtained from the ASARCO copper smelter at Garfield, Utah, which converted about 2,000 tons per day of concentrates (23 ppm Tl) from the Arthur and Magna mills of the Utah Copper Company. The latter was owned by Kennecott Copper Corporation.[12/] (We are belaboring this point to emphasize that Bingham Canyon copper ores were, and perhaps still are, enriched in thallium. These are still a leading source of U.S. copper.)

Shipments of thallium from the Globe refinery were probably never any greater than 2-3 tons per year. By 1968, an estimated 13 tons had been accumulated in industry stocks. (Perhaps 15-20 tons per year was in the materials processed at ASARCO's Globe refinery in the 1940's.) Shipments and imports had dwindled to about 0.5 ton per year by 1975.

Thallium sticks weighing 6.5 oz or ingots weighing about 20 lb are sealed in polyethylene plastic and shipped in wooden boxes. The sulfate is bottled in 2.5 lb amber glass and shipped in sawdust filled mailing tubes packed in wooden boxes. Orders of 25 lb can be shipped in polyethylene lined steel pails. Thallium and thallium sulfate are listed as Class B poisons by the Department of Transportation.[3/]

TRW Systems Group[3/] quotes M. Coats of ASARCO Inc., on August 10, 1972, as saying that its thallium department "operates on a very intermittent basis, having produced thallium in five short campaigns over the past 20 years. The department is not now in operation and is not scheduled for the foreseeable future." Coats also told TRW that total thallium metal sales were 10 lb, and thallium sulfate sales were 90 lb in the first seven months of 1972. On the other hand, Minerals Yearbook[13/] reports 1972 imports of 1,449 lb thallium and 936 lb thallium compounds, so that the extent of use was not so meager as ASARCO or TRW Systems Group seems to imply ASARCO also sporadically produces a small amount of thallium from the products of crude lead refining at its lead refinery in Perth Amboy, New Jersey

Unless a sizable market for thallium develops in the future, it is unlikely that environmental losses from thallium production could be more than intermittently significant. "Airborne emissions from thallium production are nil,* and the liquid effluent averages less than 0.1 ppm on delivery to the waste storage facility."[3] The thallium-containing raw material (2,000-3,000 ppm) is stockpiled indoors in anticipation of future use.[15] Refined thallium and thallium sulfate are stored in a prefabricated steel structure with asphalt paving and in a brick building with a concrete floor.

2. Production and use of other thallium compounds and alloys: Some information regarding possibilities of thallium exposure during salt manufacture is given in the subsection VII.B.1.a(2) on occupational exposures. The only complete process description found was for the still experimental (in 1947) production of lenses and windows from fused thallium halides. The processes involved purifying thallium salts by precipitation, usually as thallium sulfate with sulfuric acid, dissolving the purified thallium salt and precipitating as the bromide or iodide, drying the halide in vacuo with or without heating, weighing, mixing, melting at about 412°C (whereupon fumes begin emanating from the crucible at about 200°C and continue while the material is molten; sometimes heating was done in a sealed glass bomb rather than a furnace), and allowing it cool for several days to obtain one large crystal. The halide crystal was cut by a hacksaw, diamond wheel, or file, and ground wet.[16]

Most thallium compounds being sold today are likely from stockpiles. Probably only half of the thallium used in this country (0.5 ton currently?) is found outside academic and research circles. We did not clearly identify who these industrial users are, and methods and losses could not be identified. The environmental problem outside these plants would probably be negligible compared with the industrial hygiene aspects among workmen handling concentrated forms of thallium.**

---

* An EPA communication dated April 8?, 1971, from C. V. Spangler to R. Thompson (provided us by R. Horton) states that "almost no cadmium" was found in air emissions at the Globe refinery and "It is unlikely that you will find thallium either" because the solubilizing and electrolytic processes used there are apparently not serious emitters.

** The National Institute of Occupational Safety and Health had planned a Criteria Document on thallium to be published this spring. The plans were cancelled because it was not felt to be an industrial health threat. NIOSH identified only two or three companies using TlI for producing scintillation crystals.[17]

Table V-2 lists U.S. companies that have produced or at least sold metallic thallium, thallium compounds, or thallium alloys within the last decade.

B. Thallium Associated With Sulfide Minerals Processing

In the early 1940's, the federal government sponsored a sampling and analysis program, which was completed in the early 1950's, by the U.S. Atomic Energy Commission, looking for minor but strategic elements. Table V-3 summarizes the data published from this program regarding the occurrence of thallium in ores and mill and smelter products. Although the spectrographic analysis method was not especially sensitive, it is the most extensive set of data from which comparisons can be made. These data have been supplemented in the table by those of a 1947 study by the Salt Lake City Branch of the Bureau of Mines.[12] The thallium determinations were made by a chemical analytical method. These data form the basi for many of our estimations made below.

1. Zinc and cadmium

a. Zinc: Robinson of the U.S. Geological Survey (USGS)[5] states that U.S. zinc sulfide and lead sulfide contain 2.2 and 2 ppm thallium, respectively. Such concentrations indicate that only 2-3 tons thallium is associated with all the primary zinc mined in the United States in 1972. A somewhat better estimate can be made based on the value 10 ppm thallium in zinc concentrates. Several U.S. zinc producers state the concentrates used contain less than 10 ppm. Spectrographic data in Table V-3 indicate that most (≥ 70%) of U.S. zinc concentrates contain 10 ppm with a few concentrates containing up to 50 ppm, usually in western ores.

Figure V-2 is a flow diagram for zinc processing. Refined zinc is produced by two methods--the pyrometallurgical and the hydrometallurgical, which involves leaching and electrolysis. The U.S. plants currently producing zinc are given in Table V-4.

Roasting the zinc concentrates eliminates sulfur and most of the lead, fluorine, chlorine, and mercury. Zinc sulfide in sphalerite is converted to zinc oxide or sulfate. Common roasting methods are multiple hearth roasting, flash roasting (used prior to leaching), and fluidized-bed roasting.[19] Recovery of the roaster emissions is standard, but unrecovered thallium may contaminate the sulfuric acid produced from the roasting operation.*

---

* In 1973, only 2.6% of U.S. sulfuric acid production was from zinc smelt off-gases (only 10.9% from all smelter off-gases).[20]

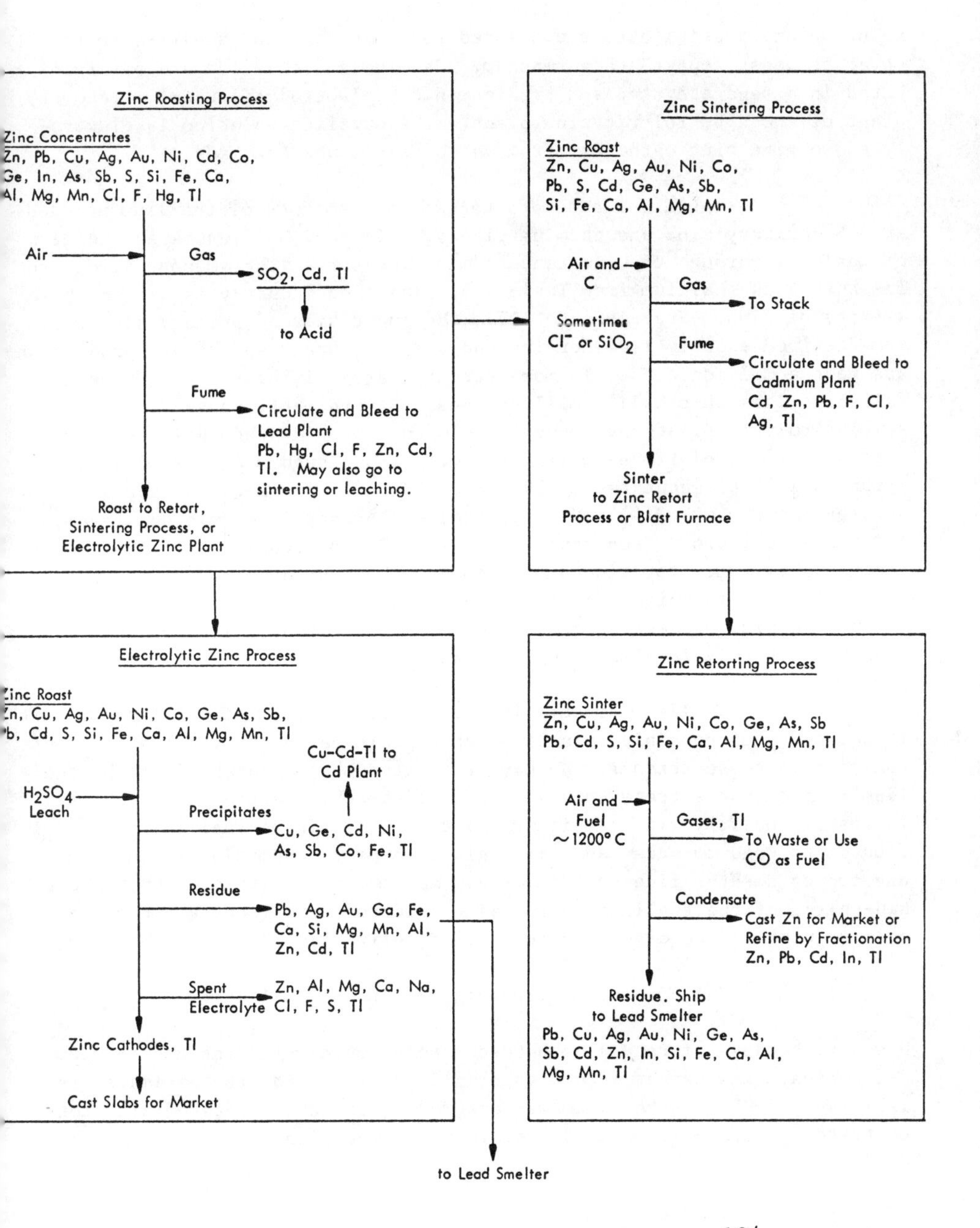

Zinc Roasting Process
Zinc Concentrates
Zn, Pb, Cu, Ag, Au, Ni, Cd, Co,
Ge, In, As, Sb, S, Si, Fe, Ca,
Al, Mg, Mn, Cl, F, Hg, Tl
Air
Gas
SO2, Cd, Tl
to Acid
Fume
Circulate and Bleed to
Lead Plant
Pb, Hg, Cl, F, Zn, Cd,
Tl. May also go to
sintering or leaching.
Roast to Retort,
Sintering Process, or
Electrolytic Zinc Plant
Zinc Sintering Process
Zinc Roast
Zn, Cu, Ag, Au, Ni, Co,
Pb, S, Cd, Ge, As, Sb,
Si, Fe, Ca, Al, Mg, Mn, Tl
Air and
C
Gas
To Stack
Sometimes
Cl⁻ or SiO2
Fume
Circulate and Bleed to
Cadmium Plant
Cd, Zn, Pb, F, Cl,
Ag, Tl
Sinter
to Zinc Retort
Process or Blast Furnace
Electrolytic Zinc Process
Zinc Roast
Zn, Cu, Ag, Au, Ni, Co, Ge, As, Sb,
Pb, Cd, S, Si, Fe, Ca, Al, Mg, Mn, Tl
Cu-Cd-Tl to
Cd Plant
H2SO4
Leach
Precipitates
Cu, Ge, Cd, Ni,
As, Sb, Co, Fe, Tl
Residue
Pb, Ag, Au, Ga, Fe,
Ca, Si, Mg, Mn, Al,
Zn, Cd, Tl
Spent
Electrolyte
Zn, Al, Mg, Ca, Na,
Cl, F, S, Tl
Zinc Cathodes, Tl
Cast Slabs for Market
Zinc Retorting Process
Zinc Sinter
Zn, Cu, Ag, Au, Ni, Co, Ge, As, Sb
Pb, Cd, S, Si, Fe, Ca, Al, Mg, Mn, Tl
Air and
Fuel
~1200° C
Gases, Tl
To Waste or Use
CO as Fuel
Condensate
Cast Zn for Market or
Refine by Fractionation
Zn, Pb, Cd, In, Tl
Residue. Ship
to Lead Smelter
Pb, Cu, Ag, Au, Ni, Ge, As,
Sb, Cd, Zn, In, Si, Fe, Ca, Al,
Mg, Mn, Tl
to Lead Smelter

Figure V-2 - Flow Diagram of Zinc Processing[18/]

solution are precipitated and shipped to copper or lead smelters to recover at least copper and germanium. Cadmium and thallium are precipitated in a separate step and treated in the electrolytic cadmium recovery plant of the electrolytic zinc plant. The purified solution is electrolyzed to give zinc cathodes, from which cast slabs are made.[18/]

Table V-5 attempts to estimate amounts of thallium handled at U.S. primary zinc and cadmium plants. Figure V-3 illustrates the flow of thallium through the zinc and cadmium industry. The amounts given are in short tons thallium from Table V-5. The zinc industry is currently in a state of flux. From 1966 to 1971, the quantities of primary slab zinc were derived equally from foreign and domestic ores, and annual production was near a million tons. In more recent years, slab zinc production has fallen to less than half a million tons. Several horizontal retorts and an electrolytic plant have recently closed, changing the ratio of distillation to electrolytic slab zinc capacity from roughly one immediately prior to 1972 to about two in 1973 and 1974.[28/] In 1972, the actual production ratio was 3:2.[29/] New electrolytic plants have come on stream, and the actual production ratio since 1974 is not available. Thus, representative averages for zinc production in the past decade were not used for Table V-5, in which the amounts of thallium involved in zinc and by-product cacmium processing are estimated. The ratio of 1:1 for distillation to electrolytic slab zinc is an assumption.

Thallium appears in almost every stream of the zinc circuit. It has not been possible to assign any certain values to the magnitude of thallium in these streams. We have estimated that about half of the thallium in zinc concentrates has been handled by the electrolytic zinc refiners and half by zinc distillers so that half of the thallium appears in flue dusts used to make cadmium. This is a poor assumption if much lead and copper smelter flue dust is used. We have also estimated that about two-thirds of the thallium in the zinc electrolyte is sent via the copper-cadmium cake to the electrolytic cadmium refinery.

(5.25 tons in concentrate x 0.67 = 3.5 tons)

However, because of recycling between the electrolytic zinc and cadmium refineries, this number is not especially useful. For the estimates in Table V-5, 5.25 tons thallium was assumed to be in the roasted zinc concentrates that are leached at electrolytic zinc plants.

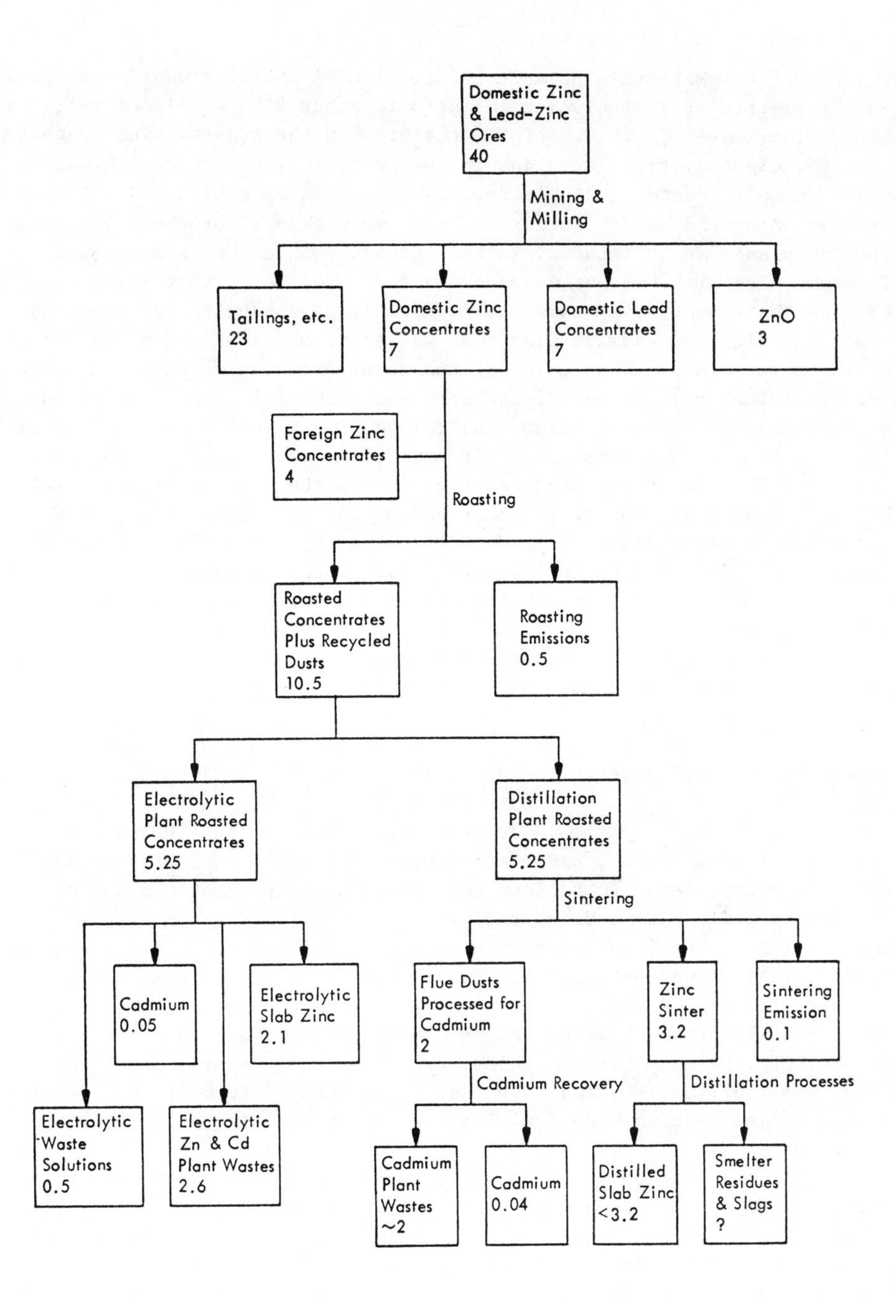

Figure V-3 - Simplified Flow of Thallium Through the Zinc and Cadmium Industry

Materials balance information on thallium was found for a Soviet electrolytic zinc plant operated together with an electrolytic cadmium refinery.[27] Up to 40% of the thallium in the roasted zinc concentrate left the electrolytic zinc plant in the product zinc. Up to 50% was removed in solid wastes (clinker from a Walez kiln operation and a lead cake from leaching the Waelz zinc oxide). No more than 1% of the thallium left the two plants as an impurity in the product cadmium or cadmium salts. The remaining thallium was unaccounted for. However, it is interesting to note that about seven times as much thallium was being recirculated between the zinc and cadmium plants as was entering in new roasted zinc concentrate (calcine). When ~ 18% of the total circulating thallium was new, the remaining thallium was distributed among the following: spent cadmium electrolyte, ≤ 46%; spent zinc electrolyte, ≤ 5%; Waelz zinc oxide leaching solution, ≤ 24%; copper cake from cadmium refining, ≤ 9%; and others, less than 2%. Solutions were recirculated to the calcine leaching step; but, presumably at some stage, they became so contaminated that they were periodically discarded and could represent much of the 9% unaccounted-for loss. Thus, we estimate in Table V-5 that about 2.1 tons thallium (~ 40% of 5.25 tons) will be found in the annual U.S. production of electrolytic slab zinc (~ 8 ppm) and about 3.1 tons will be found in solid and liquid wastes from the electrolytic zinc and cadmium plants. Only 0.05 ton thallium per year may be found in electrolytically refined cadmium (~ 20 ppm).

The estimate for 0.04 ton thallium in cadmium refined from flue dusts is developed in the next subsection.

b. Cadmium: Although thallium is recovered commercially only by the ASARCO Inc., cadmium refinery (described in Section V.A), there is every reason to believe that thallium is present but not recovered by other U.S. cadmium refineries. Thallium compounds are volatile and collect in the flue dusts from copper, lead, and zinc smelting operations. These flue dusts are the raw materials used for cadmium recovery at some plants (see Table V-4). Other thallium containing sources are copper-cadmium cake obtained at the electrolytic zinc plant. The American Society of Testing Materials (ASTM) specifies that cadmium may contain up to 30 ppm thallium.[30] Thallium must be removed if it will cause plating problems, and cadmium used for pigments must be low in thallium.[31]

In Cadmium the Dissipated Element,[32] the authors have attempted to summarize cadmium recovery methods in the U.S. based only on published literature. Their recovery flowsheet is shown as Figure V-4 (where thallium is known to be present, its symbol has been included). We have not identified any U.S. plant that operates exactly according to these flow schemes. Details of 1964 cadmium recovery practices at unidentified Western Hemisphere plants have been reviewed,[33] as well as practice at Hudson Bay Mining and Smelting Company.[34]

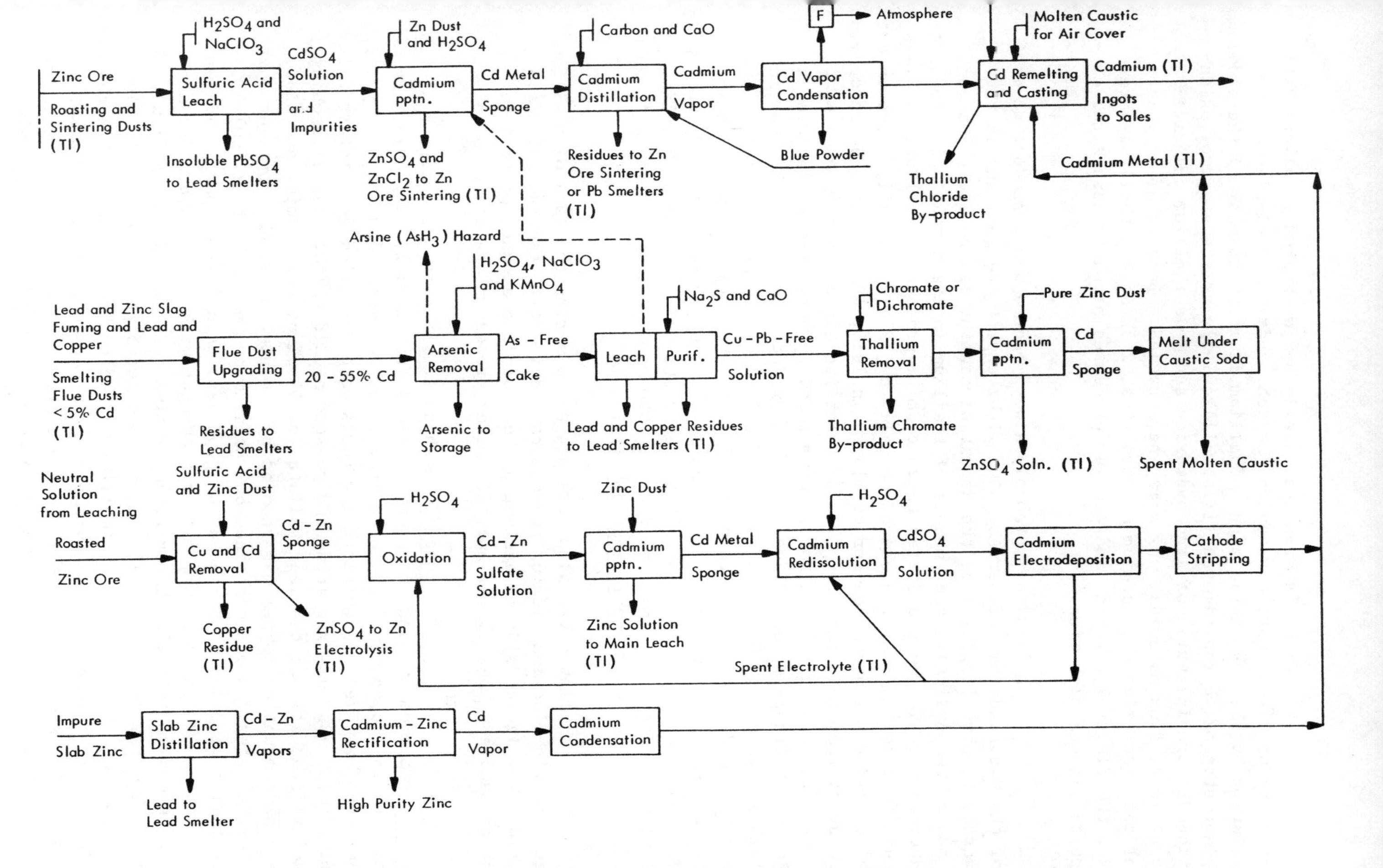

Figure V-4 - Schematic Flow Sheet for Recovery of Cadmium as By-Products of Zinc and Lead Recovery[32]

Cadmium is produced chiefly by two methods: electrolytic recovery or electromotive recovery, whereby cadmium is displaced from solution by the less noble zinc. Thallium is found in many of the cadmium plant streams, being removed (if at all) chiefly at the stage of oxidation and precipitation of iron hydroxide and in a thallium chloride-containing flux material, which may be recycled. Because of recycling of the spent zinc and cadmium electrolytes, 2-5% of the thallium entering the electrolytic zinc and cadmium processes probably leaves the circuits via the slab zinc.

Thallium has probably always been a waste product at most of the U.S. cadmium refineries. One might expect the thallium in these wastes to have increased as the amount of cadmium recovered increased (and correspondingly, the amounts of thallium in the product zinc to have decreased). The United States began cadmium production in 1906, but less than 50 tons was being produced annually by 1914. Initially, zinc retort blue powder was processed for cadmium. In 1910, ASARCO Inc., began cadmium recovery from lead smelting flue dusts at its Globe plant in Denver.[23] Hydrometallurgical plants for cadmium recovery were started in 1916 at Kennet, California, and Midvale, Utah. Production reached 250 tons in 1925 and about 1,500 tons by 1929. The increase in production was largely due to the development of electrolytic zinc recovery.[35] U.S. cadmium production probably reached its peak of about 6,500 tons in 1969, when production at electrolytic zinc plants was 2,000 tons.[36] Table V-2 lists current and recent cadmium producers. Others operating in the past can be identified in the Minerals Yearbook series.

Table V-4 also attempts to summarize the fate of thallium at specific zinc and cadmium processors and identify the raw materials used. The latter aids in estimating the absolute amounts of thallium being handled. Details of specific plant processes are reviewed below.

(1) Specific methods for recovery from zinc sintering dusts: Cadmium production recently ceased at Blackwell, Oklahoma. Practice at this plant in 1966, when production was about 600 tons per year[37], has been reviewed. Zinc sintering dusts were densified in a rotary kiln, mixed with 93% sulfuric acid, and roasted at 218-427°C. Water leaching solubilized all but the lead sulfate and 5-10% of the cadmium and zinc. A small amount of copper was removed by cementation at pH 3 on zinc dust (the cemented residue contained 80% cadmium and probably thallium). Cementation of cadmium on zinc dust at pH less than 2 gave sponge cadmium containing 30 ppm thallium. After treatment with sodium hydroxide and then ammonium chloride, the refined cadmium contained 5 ppm thallium. The cadmium oxide produced from it also contained 5 ppm thallium. At 640 tons cadmium per year, only 38 lb thallium was being "recovered" in the chloride flux and 6 lb/year remained in the refined cadmium. A former worker at the Blackwell refinery stated that the concentration of thallium in the flux was not high, and it was recycled to recover cadmium and other values.[38]

Figure V-5 gives MRI estimates of the amounts of thallium in various intermediates of cadmium recovery at St. Joe Minerals Corporation. It would appear from the flow diagram that less than half of the thallium occurring in zinc sintering fumes is removed per se in the cadmium circuit. It is assumed for the calculations that even if 16-35% of the thallium remains in the leaching residues,[22,41,42] it would eventually be recycled. The other alternative is that some fraction remains in the lead cake and is found in residues processed for precious metals.

Information regarding the degree of thallium removal attainable at unidentified electromotive cadmium plants is given in Table V-6. The average thallium removal from leach solution to product cadmium at these six Western Hemisphere plants is 97%. About 3%, or an average 17 ppm thallium, remained in the cadmium. If these numbers are typical for thallium removal at all U.S. cadmium plants processing flue dusts except ASARCO's Denver plant, which also once processed imported flue dusts for cadmium, less than 1.5 tons thallium is in the raw materials used.* However, dissolution of the thallium in the zinc smelter dust into the leaching solution may not be 100%. Japanese workers found that only 65% of the thallium (but 93.6% of the cadmium) was solubilized in a hot water solution that was used to leach flue dusts from an electrothermic zinc smelter sintering plant after they had been roasted with concentrated sulfuric acid at 450°C.[42] On this basis, 2 tons thallium may be a better estimate.

(2) <u>Cadmium production and thallium removal in the electrolytic zinc plant</u>: We have unavoidably mentioned cadmium electrolytic recovery within the zinc electrolytic recovery discussion because of extensive recycling between the two systems. In electrolytic zinc plants, the raw material for cadmium production is precipitated during the purification of the zinc sulfate solution. Zinc dust precipitates elements more electropositive than zinc by cementation from solution as a metallic powder or "cake." Thallium along with cadmium, zinc, copper, cobalt, lead, nickel, arsenic, antimony, and iron are normally present in the cake in relatively large amounts. Certain other elements are found

---

$$* \quad \frac{0.5\ (5{,}000 \text{ tons U.S. cadmium production})(17 \text{ ppm})}{3\%} = 1.4 \text{ tons}$$

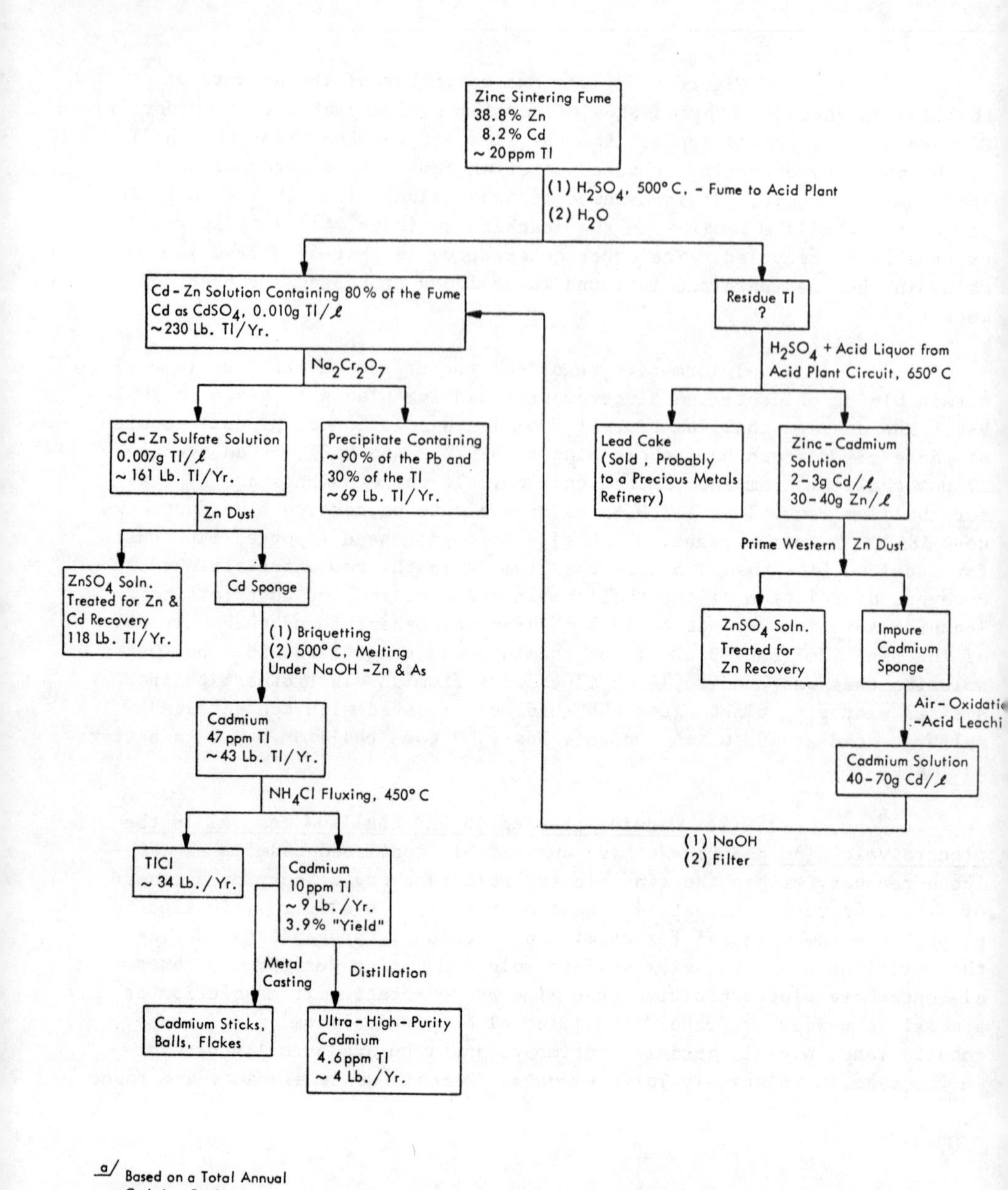

a/ Based on a Total Annual Cadmium Production of 437 Tons, 94% Cd Recovery

Figure V-5 - Flow Sheet for Cadmium Recovery at St. Joe Minerals Corporation, Monaca, Pennsylvania [39,40] [a]

in minor concentrations. Ten to 25% (20-30%[27]) of the thallium may remain initially in the concentrate leaching residues,[22] but these can be recycled by treatment in the Waelz kiln and leaching.[27]

Varying amounts of the named elements in the cadmium cake or sponge are redissolved in sulfuric acid solution (such as spent zinc electrolyte). Arsenic, antimony, and iron are often removed from cadmium sulfate solutions by hydrolytic precipitation. The free acid is neutralized with freshly precipitated $Zn(OH)_2$ or, rarely, CaO. This method only partly removes thallium. Strong oxidizing agents will convert Tl(I) to Tl(III) so that $Tl(OH)_3$ can be precipitated quantitatively. For example, at about pH 5, the solubility product of $Tl(OH)_3$ is about 10^{-18}; and it is precipitated along with cobalt, lead, iron, arsenic, and antimony.[43] Thallium can be removed separately if Fe(II) and As(III) are first air oxidized and then hydrolyzed at pH 5.0-5.3 followed by increasing the pH to 5.5-5.6 with $Ca(OH)_2$ and oxidizing Tl(I) with $KMnO_4$.[22] Lund and Sheppard of St. Joseph Lead Company[33] reported that a $KMnO_4$ treatment reduced the thallium content of a leaching solution from 200 mg/liter to less than 1 mg/liter.

Sometimes chromate or dichromate is added to precipitate thallium(I) as Tl_2CrO_4 (solubility product 10^{-12}), but some thallium is oxidized so that $Tl(OH)_3$ is precipitated too. Fugleberg (1973)[43] calculates that at equilibrium, pH 5, and dichromate concentration 10^{-3}, $[Tl^+]$ = about 1 g/liter. Similarly at a chromate concentration 10^{-3} and pH 5, $[Tl^+]$ = 0.037 g/liter. Thus, these agents, especially dichromate, do not give good thallium removal.

Cadmium is reprecipitated from the purified sulfate solution by cementation on zinc dust again. The standard electrode potential of thallium is 0.336 v; that of cadmium is 0.403 v. This is not a sufficiently large potential difference for good separation. (If a deficiency of zinc dust is used for the cementation step, good separation can be achieved.) Since the zinc sulfate solution with higher thallium content is further treated for zinc and cadmium recovery, cadmium cementation is not an outlet for thallium from the system.[43]

The cadmium sponge from the cementation is dissolved, and cadmium is electrodeposited onto aluminum cathodes (normally, the insoluble anodes are lead). Thallium can be separated well in this step. For example, a solution containing 300 mg Tl/liter upon electrolysis (at current density 100-200 ampere per square meter, ~ 100 g H_2SO_4/liter, and 50-100 g Cd per liter) will give a cadmium with only 26 ppm thallium. (However, thallium does reduce the current efficiency.) To obtain cadmium containing less than 10 ppm thallium, the cadmium electrolyte can contain no more than about 200 mg Tl/liter.[43] At Hudson Bay Mining and Smelting

Company, an electrolyte containing about 800 mg Tl/liter gave cadmium containing 39 ppm thallium.[34]

Melt refining of the electrodeposited cadmium using NH_4Cl or an NH_4Cl-$ZnCl_2$ mixture can also bring the thallium concentration to less than 10 ppm.[43] Lund and Sheppard report 66-95% thallium elimination using $ZnCl_2$ or a mixture of $ZnCl_2$ with NH_4Cl.[33]

A consultant[44] at National Zinc Company's electrolytic zinc and cadmium plants at Bartlesville, Oklahoma, indicated that the processes used there are among those described above by Fugleberg.[43] He stated that thallium must be removed because it causes plating problems. Refining residues go to copper or lead smelters. He felt that some thallium appearing in the roaster gases was precipiated with lime [it would have to be $Tl(OH)_3$] in the acid plant.

On the other hand, the AMAX Zinc Company plant at East St. Louis, does not at present remove thallium in its cadmium circuit. If only sponge cadmium is marketed,[45] it might contain as much as 50 ppm thallium. (However, this cadmium would not meet the ASTM specifications of $\leq$ 30 ppm Tl.)[30]

An electrolytic process for cadmium refining at ASARCO Inc.'s Denver globe refinery has been described in Figure V-1. (An electromotive method is also indicated in Table V-4.) Product cadmium contains only 1 to 2 ppm thallium. Environmental losses of thallium are not expected at the Globe plant. The most probable losses of thallium are as gases and fumes escaping control systems in roasting and especially sintering operations that produce the flue dusts used as raw materials. These operations are not performed at Denver.

Bunker Hill has operated electrolytic plants to recover cadmium not only from zinc plant copper-cadmium residues but also from lead sintering and smelting dusts. No intentional outlet was provided for thallium in the electrolytic cadmium method used before 1962.[22] However, the thallium content of the electrolytic cadmium was probably less than 30 ppm, the amount in the cadmium produced from flue dusts. (Here thallium was removed in the $NaClO_3$-$KMnO_4$ steps and as a TlCl flux.) Table V-3 gives 50 ppm thallium for four Shoshone County zinc concentrates (~ 1943), but these did not include Bunker Hill mines and mills.

c. Lithopone: Several review articles encountered in this study mentioned lithopone production as a source of thallium. The pigment lithopone, a coprecipitated zinc sulfate-barium sulfate, has been widely used in paints, floor coverings, and coated fabrics. Currently, production is negligible.[46]

In decades past, lithopone production may have been an environmental source of thallium. The "concentrated" "desulfurized" zinc ore was leached with sulfuric acid and the zinc sulfate solution was purified in two steps that would have removed practically all of the thallium in solids waste (which has been used as a commercial source of rare metals in the past in the United States[47] and Europe[48,49]). These steps were permanganate or chlorate oxidation-precipitation and zinc dust or flue dust cementation.[50]

The Calspan report[46] states that the largest U.S. lithopone plant (American Zinc Company, Inc.) is in Columbus, Ohio. (Magorian,[51] an author of the draft report,[52] mentioned that a lithopone plant defunct for 10 to 15 years in the northeast side of Columbus has been built over by a housing development.) The runoff from the plant referred to in the final report "apparently runs into a nearby cemetery and percolates into the soil." It was not sampled.

2. Lead: The processing of lead concentrates, containing copper, silver, gold, and other valuable elements, generally follows a sequence of sintering, smelting, drossing, and refining as shown in Figure V-6.

In sintering, the temperature is raised to the point where incipient fusion of the lead ore occurs to give a hard porous clinker for blast furnace treatment. Thallium oxides and chloride volatilize at ~ 750°C. About half of the thallium is transferred into the sintering dusts.[22] Savichev (1964)[53] states that thallium in lead dusts is in the form of the sulfide, selenide, telluride, and metal. Grigorovich et al.[54] state that the thallium in lead ores passes almost completely into the smelting dusts, chiefly as oxides and sulfides, the biggest fraction of thallium being in the sintering dust.

Naimark (1960)[24] studied the volatilization of thallium from galena concentrates. Up to about 900°C, thallium and mercury volatilization were about the same; but less cadmium volatilized than thallium until about 900°C, when thallium and cadmium behaved alike. Heating lead concentrates containing 50 ppm thallium at 950°C left a residue containing only 10 ppm, while the dusts contained 3,000 ppm thallium.

The lead sinter is smelted in a blast furnace at ~ 1400°C, where the major separation of gangue material is made. The lead smelter feed, in addition to the lead concentrate, includes coke (whose presence increases the volatility of thallium[24]), silica or limestone fluxing materials and may also include secondary products such as lead-bearing dusts, fumes, fines, slags, drosses, and residues shipped from copper and zinc smelters. The liquid lead as it flows from the blast furnace may contain copper,

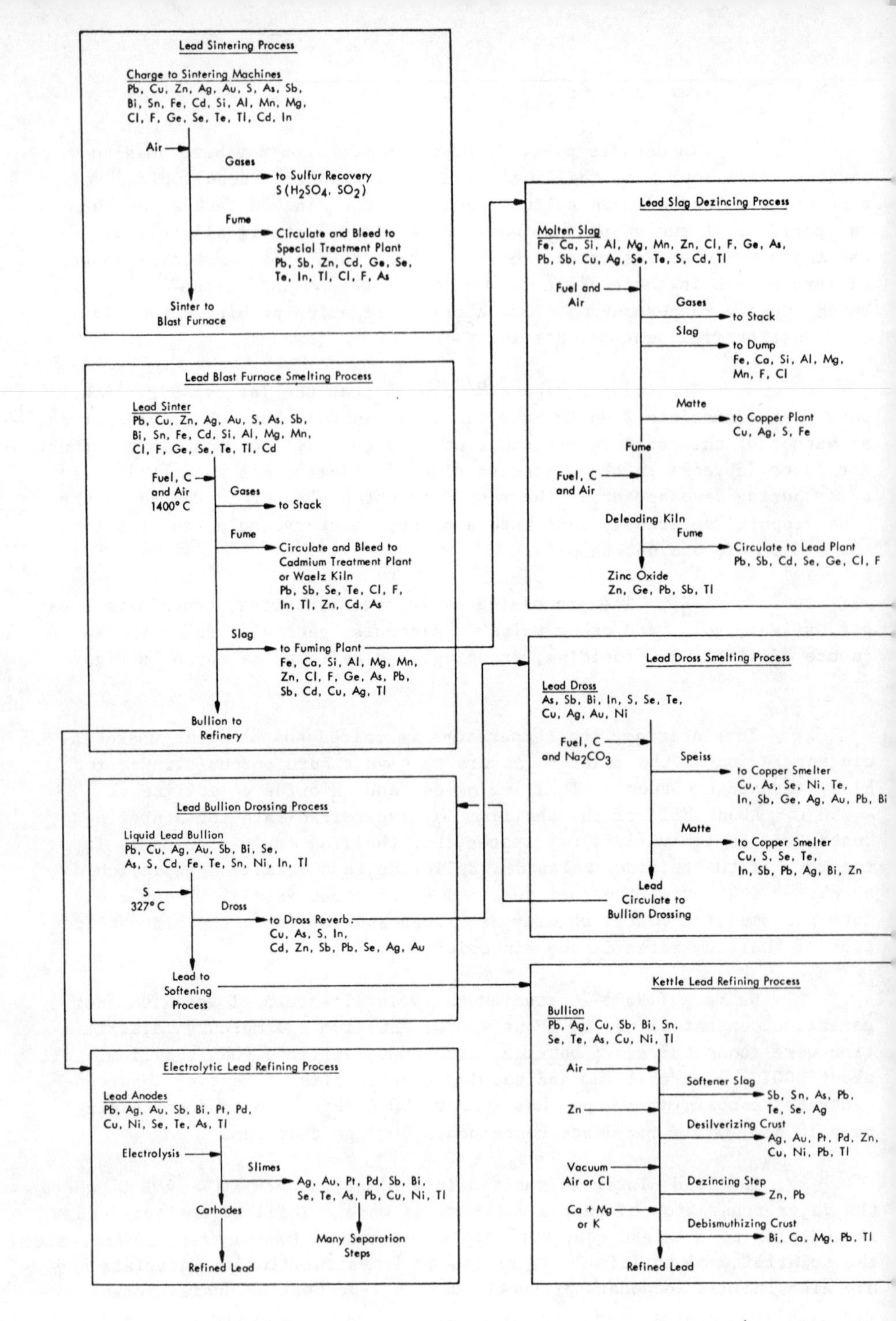

Figure V-6 - Flow Diagram of Lead Processing[18]

silver, and other elements; upon cooling, many elements separate from the lead as a dross.[18,26,55] The blast furnace dust may contain 30-50% of the thallium in the feed. Fifteen percent of the thallium transfers to the slag. The remaining 35-55% of the thallium is distributed between the matte, if any, and bullion.[22,56] J. G. Parker of the Bureau of Mines[57] has calculated amounts of thallium in the United States appearing in blast furnace fume, slag, and lead bullion annually to be 0.3, 0.4, and 1 ton. The distribution would be ~ 20%, ~ 20%, and 60%, respectively. Parker further deduces that the total thallium content of lead concentrates is 6 to 7 tons annually, which compares rather well with our estimate of 11 tons for thallium in zinc processing. (About the same amounts of sulfide raw materials are used in each case.) Parker's estimate implies that three-fourths of the thallium in the lead concentrates has been transferred into the sintering dusts.

The slag from the lead smelting may undergo a dezincing process and be charged while molten to the zinc fuming furnace; it may be recycled to the sinter process; or it may be discarded. The blast furnace dust may be recycled to build up its cadmium concentration to the point "where the total cadmium losses with the slags and in the gases and other products become equal to the amount of cadmium entering the process in the raw material" (practice at the Bunker Hill Company).[22] Such a situation has been described for thallium in subsection V.B.1.a, at an electrolytic zinc-cadmium recovery plant.[27]

Lead bullion, from the blast furnace smelting process, next undergoes a drossing procedure. The lead bullion drossing process comprises holding molten bullion at a temperature just above the melting point of lead (327.4°C). Copper dross rises to the top and is skimmed off. The last traces of copper are removed by sulfur. Melting at 303.5°C, the thallium present probably remains in the molten lead.

In lead refining, electrolytic and kettle processes are the two most commonly used procedures. Some bullion, especially if high in bismuth, is processed by the Betts electrolytic processes, in which impurities in the bullion concentrate in the anode slimes. These impurities are separated and recovered by methods similar to those in the kettle lead refining process. Thallium probably accumulates in the slimes and may be removed from the circuit in product bismuth.

The first step in kettle lead refining is softening of the drossed lead bullion. Arsenic, antimony, and tin are oxidized into a litharge slag in a reverberatory furnace, or, as in the Harris process, are oxidized in a caustic soda slag in heated steel kettles. Sometimes, stepwise softening concentrates arsenic, antimony, and tin into separate slags, which are processed into alloys, pure metals, or metal oxides.[18]

In the desilverizing step, using the Parke's process, 1 to 2% metallic zinc is added to the softened lead at a temperature above the melting point of zinc (419.5°C) to give, on cooling, a dross or crust of silver-zinc alloy containing noble metals and any copper or nickel present. The residual lead is reheated and scavenged with more zinc to give a zinc crust containing a little silver, which is used as the zinc addition to a new kettle of lead bullion.[58] Conducting desilverizing in two stages allows isolation of high-gold and high-silver products, which facilitates recovery of these metals.[19] Any thallium should be removed by the zinc during desilverizing.

The passage of materials between lead and zinc smelters greatly complicates balance calculations. Lead smelter dusts were reported to contain up to 5,300 ppm thallium;[59] however, these dusts are continually recycled and do not allow calculation of the magnitude of thallium in the original lead concentrate. Table V-3 lists values for thallium in several lead bearing materials, but none appears representative of the whole industry. Most of the lead smelter dusts with high thallium content were ultimately processed by ASARCO Inc., for cadmium recovery at Denver. In 1947, copper ores of the Utah Copper Company were really the major source of the thallium in these dusts.[12] With no better data, we are inclined to agree with Parker's estimate of 6-7 tons thallium in lead processing. However, since sintering dusts are recycled until their cadmium content builds up, uncontrolled atmospheric emissions of thallium at higher concentrations are expected at lead smelters than at zinc plants.

3. <u>Copper</u>: The processing of copper ore follows the sequence of roasting, smelting, converting, and refining as shown in Figure V-7.

In the roasting process, the ore is treated below the fusion point to drive off uncombined moisture (drying); chemically combined water and carbon dioxide (calcining); or sulfur, arsenic, antimony, tellurium, etc. (roasting). Any flue dust that can be segregated from the fume is normally returned to the charge.

Smelting of the roasted ore in a blast furnace was once a common practice, but this process of smelting is not suitable for the finely divided sulfide concentrates now available. Therefore, reverberatory furnaces are now generally used to produce matte and sulfur dioxide, which results from excess sulfur. The copper smelter feeds contain 25-35% copper, 27-33% sulfur, 20-30% iron, and 10-20% gangue constituents.[26] During reverberatory smelting, high-sulfur minerals decompose into sulfur, which burns to SO_2, and matte-forming cuprous and ferrous sulfides. The end result of the smelting process is a molten matte with a slag floating atop the matte. Soviet data indicate that the relative amounts of thallium transferred during reverberatory smelting from the roasted concentrate

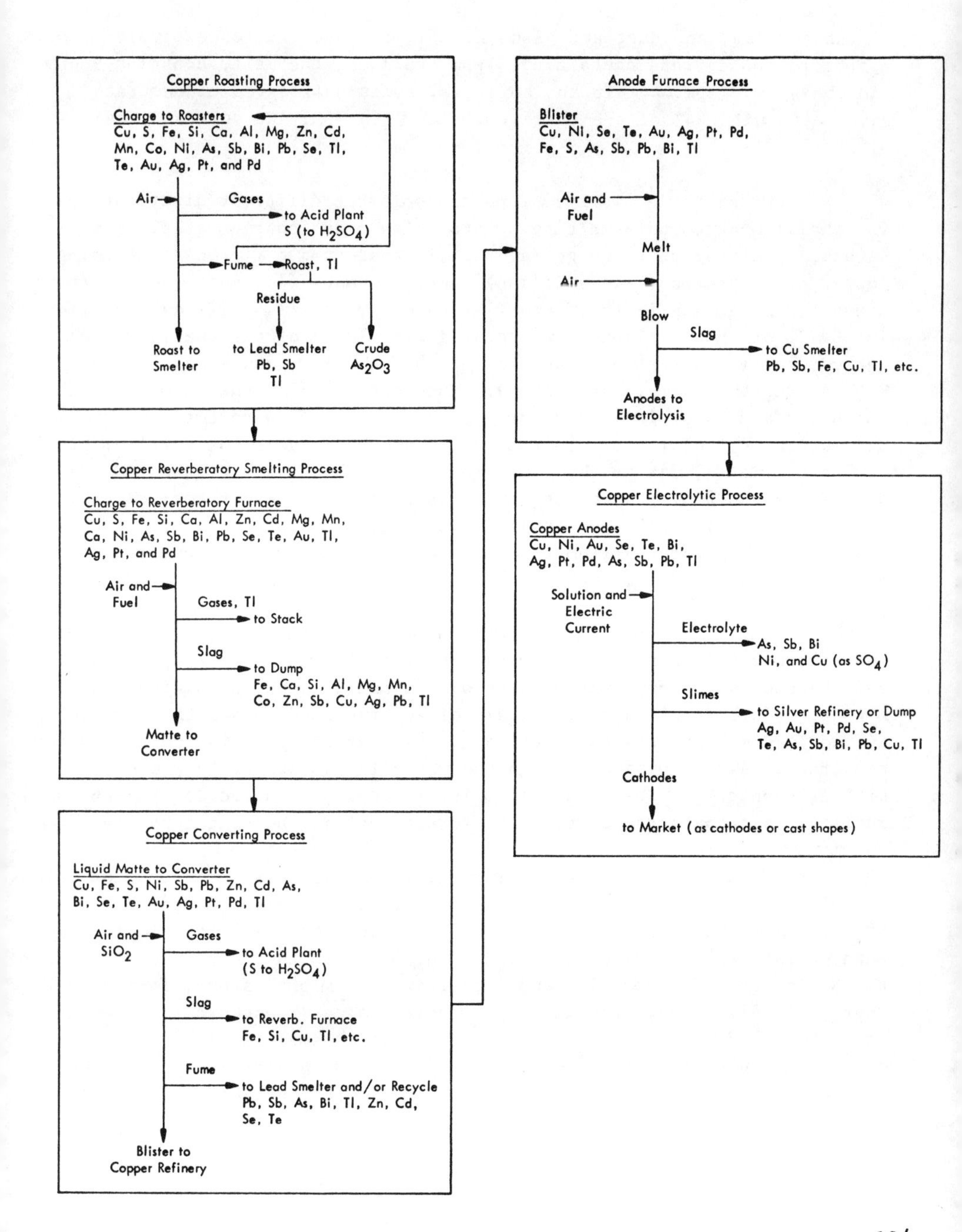

Figure V-7 - Flow Diagram of Copper Smelting and Refining Processes[18]

to matte, slag, and dust and gases are 31, 62, and 7%, respectively. For a roasted concentrate containing 4 ppm thallium, the matte contains 9 ppm and the slag 3 ppm.[22] In another case, smelting roasted concentrates gave matte with 24% of the thallium; slag with 45%; and dust and gases, 31%.[60]

Other Soviet workers reported somewhat different distributions for thallium at copper-smelting plants using crude concentrates. Reverberatory smelting of a charge (average Tl content 8 g Tl/ton) containing copper concentrates (7-13 g Tl/ton), shaft furnace flue dust (90 g Tl/ton), converter slag (3 g Tl/ton), and flux gave a matte (14 g Tl/ton) containing 69.9% of the thallium, a waste slag with 18.1% of the thallium, collected dust (30 g Tl/ton) with 2.6% of the thallium, and a gas loss of 9.4% of the thallium (34 g Tl/ton). Much more (~ 60%) thallium goes into the dusts and gases in oxygen-enrichment or shaft-furnace smelting.[60-62] Reverberatory smelting of other crude concentrates gave 46% thallium transfer to the matte, 34% to the slag, and 20% to the dust and gases.[60] As of 1974, seven U.S. copper smelters roasted their concentrates; eight did not.[63] Thus, for U.S. copper smelters, extrapolation would give an average 20% transfer of thallium from the reverberatory furnace charge to the dusts and gases and 40% each transfer to matte and slag.

During converting, the molten matte is changed to blister copper. The molten matte is periodically removed from the reverberatory furnace and charged to the copper converter with a small amount of silica flux.[58] A copper converter is a cylindrical refractory-lined vessel that can be tilted to discharge its contents and is similar in operation to a Bessemer converter. Air is passed through the matte in two stages to produce the blister copper.[25] The resultant blister copper, so named due to its rough surface, is transferred to refining furnaces or may be cast into pigs.[64,65]

Soviet data show an average (for two smelters) of about 28% transfer of thallium into the black copper (only 90-95% copper compared with 98-99% copper in blister copper), about 38% into the recycled slag, less than 4% into the recycled dust, and about 30% into the lost gases.[61] From the Soviet data for smelting and converting, we might assume, for every 10 tons of thallium appearing in copper concentrates (roasted or nonroasted), that about 3.2 tons thallium is lost as gases in smelting (~ 2 tons) and converting operations (1.2 tons), about 1.1 tons appears in the blister copper, and 5.7 tons is in slags and dusts that are recycled or sent to dumps.

The final step in copper refining involves electro- or fire-refining of the blister copper. Fire-refining capacity in the United States in 1968 was 365,000 tons (at two-thirds of the U.S. copper smelters), whereas 2.3 million tons of copper was electrolytically refined. Approximately 60% of the U.S. electrolytic-refining capacity is in six Atlantic Coast refineries while fire-refining is done in Michigan, New York, New Jersey, New Mexico, and Texas.

Fire-refining may be done in reverberatory furnaces, which are smaller than those used in matte smelting, or in cylindrical furnaces that can be tilted on trunnions. With the reverberatory furnace, the charge comprises solidified blister copper plus high-grade scrap while the cylindrical furnaces accept molten blister copper directly from the converters.[19] In this process, the molten metal is agitated by compressed air to oxidize sulfur and metal impurities. The metal impurities form oxides which accumulate as slag and are removed by skimming.[19,65] The copper, containing cuprous oxides, is deoxidized by adding coke and by "poling," in which green wood poles decompose into reducing gases. Natural gas is replacing wood as a purifying agent. The fire-refined copper may be cast directly into forms for industrial use; or if recovery of silver and other precious metals is warranted, refining proceeds only far enough to insure homogeneous anodes for electrolytic refining.[19,25,64,65]

Anode furnace refining purifies and degases the blister copper to produce copper anodes. In the electrolytic copper refining process, the anodes are consumed to 10-20% of their original weight and then returned to the anode furnace for reprocessing into new anodes. Anode slimes comprise 1-15% of the original anode weight and contain all of the silver, gold, lead, selenium, and tellurium from the anode.

The cathode copper may be sold without further treatment but is usually melted, given a form of fire-refining, and cast into refinery shapes.[19]

We have no data on the fate of all the thallium in blister copper. Some thallium is likely lost in the anode-furnace refining. Some does occur in the anode slimes. (A flotation product from Japanese copper electrorefining anode slimes contained 27.86% silver [99.6% recovery] and 1.55% thallium [80.9% recovery].)[66] About 760 short tons of primary silver was recovered in 1972 as a by-product of copper ores and copper-mixed ores. If the Ag-Tl ratio was the same as in the Japanese case cited above, 54 tons thallium would appear at the primary silver refineries annually. This would imply $\sim$ 500 tons thallium is in the original copper concentrates (or has been added to the copper circuit from other nonferrous metal operations). A number of this magnitude would be likely if the average concentration in the ore was 2 ppm and all the thallium appeared in the concentrates.

Presumably, it is not unusual for thallium to appear in silver (and other precious metals recovered at the refinery). For example, the ALFA Division of Ventron Corporation furnished a specification sheet with its latest catalog for 99+% ACS-grade silver nitrate, stating that the maximum level for thallium in a typical analysis is 10 ppm. (However, if all the copper-derived primary silver--760 tons--were converted to silver nitrate, the maximum would allow only 0.012 ton thallium.) Probably most of the thallium in the anode slimes is lost in gases and slags during their upgrading. Anode slimes are also sometimes discarded.[51/] Calspan Corporation[46/] found 20-26 ppm thallium in a sludge dump considered to be from the Anaconda copper refinery at Perth Amboy, New Jersey.

The meager data on thallium in U.S. copper-bearing materials, first given in Table V-3 are summarized in Table V-7. These data do not appear to lend themselves readily to generalizations. Both Eastern copper ores and Utah Copper Company concentrates containing 0-10 and 23 ppm thallium, respectively, are reported to give reverberatory furnace flue dusts of approximately the same thallium content (perhaps one concentrate had been roasted and one had not). Table V-8 shows that a material balance for thallium in Utah Copper Company ore processing (1940's) is thwarted by the fact that the converter dust must have been recycled many times prior to the analyses of 1,500-1,700 ppm thallium and its being sent to the Murray lead smelter. We have estimated in Table V-9 that 530 tons thallium occurs in U.S. copper ores at current production rates. Approximately 110 tons thallium may then occur in the copper concentrates (small amount imported) with the remainder in tailings and other concentrates.

Calspan Corporation[46/] has found that thallium is higher in streams receiving wastewater and runoff from certain mines, mills, or smelters processing copper-, lead-, and zinc-bearing materials than upstream or downstream concentrations. Since the Calspan report is in press at the time of this writing, we have asked permission only to summarize their data as it appears in Table VII-9. Again, no clear-cut pattern emerges as to which types of ores are most likely to be enriched in thallium.

Zitko et al. (1975)[67/] have also detected thallium in the environment. Tailings pond effluents from base metal (chiefly copper and zinc) operations in northeastern New Brunswick give thallium concentrations of 0.0007-0.0883 ppm in the rivers receiving the discharges. Only 28% of the thallium present in one of the rivers could be precipitated by liming at pH 9.6 and centrifugation, possibly indicating the majority of the thallium is Tl^{+}.

4. Gold: Table V-3 shows that thallium is found in very high concentration in some gold tailings and concentrates and not at all in others. Thallium has definitely been associated with the gold-bearing materials at Lead, South Dakota, and Carlin, Nevada, two of the top three gold-ore mines. All three practice cyanidation, although the third does not to a great extent. (About 80% of all gold in the United States is treated by cyanidation.)

The most famous cases are the dumps in Mercur and Manning, Utah. Mercur was estimated to have about 2,000 tons of thallium. The Salt Lake City Bureau of Mines report[12/] by Zimmerley states "there is available an enormous quantity of mines and mill dump material, plus underground ore, with a thallium content ranging from 0.8 to more than 1 lb [363 to greater than 454 g] per ton." Snyder Mines, Inc., the owners of the dumps, felt that thallium is associated with gold; Zimmerley was "not entirely convinced of the gold-thallium correlation." It may be that, even today, some of the reluctance of major companies to divulge thallium concentrations in their ores is due to a believed association between thallium and some other metal.[68/] The extensive data provided us by the USGS always had analyses for indium and frequently for mercury, gold, selenium, and tellurium. It is interesting to note that among the 1972 Minerals Yearbook list of 25 leading copper-producing ores, Utah Copper Mine is first; and among the leading gold producers, Utah Copper Mine is second, the source of gold being both copper and gold-silver ores. Among the other 12 copper mines that are leading gold producers is Phelps Dodge's New Cornelia Mine in Pima County, Arizona, the only other mine with gold-silver ores, too. Pima County copper concentrates were the only Arizona copper concentrates in which thallium was reported in Reference 59. Magma Copper Company, San Manuel Division, Pinal County, recovers gold by cyanidation of molybdenite concentrates from copper ore flotation. (Kennecott Copper Corporation is an even larger molybdenum producer in Pinal County. Magma Copper Company[69/] claims there is "no measurable amount of thallium in the [copper] concentrate that is processed at the San Manuel smelter.") Reference 59 found thallium in Pinal County Mo-V concentrates but not in molybdenite ore.

The Salt Lake City report[12/] details unsuccessful attempts to concentrate thallium from the Manning and/or Mercur materials. The gold and thallium could not be removed by lengthy direct cyanidation. At less than 300 g H_2SO_4/liter, "virtually no thallium was taken in solution." At 300 g/liter, only 44% was dissolved. The high acid consumption was attributed to the calcareous nature of the ore. It was possible to volatilize more than 90% of the thallium from the Manning dump ores at 950°C in the presence of coke, pyrite, or sodium chloride. However, the latter volatilized 44% of the gold, too.

The United States produces about 3 million tons of gold ore per year with most of the production at Lead, South Dakota, and Carlin and Cortez, Nevada. The thallium concentration in the ores is probably no more than 10 ppm so that 30 tons thallium per year is the maximum expected in the tailings ponds, but the concentrations there may be more than 100 ppm. Some tailings and process water have been discharged to a creek at Homestake.[7]

5. Other metals from sulfide minerals: European and Soviet literature indicates that thallium occurs in several metal sulfide concentrates. For example, References 71-78 describe recovery of thallium from arsenic-, mercury-, and/or antimony-bearing materials. In the United States, low tonnages but high concentrations of thallium might be expected in such materials handled at copper, lead, and zinc smelters.

Table V-3 shows 10-40 ppm thallium in two molybdenum-vanadium concentrates and 0 in several other molybdenum containing materials. None was found in mercury-bearing materials from California and Arkansas. Of five samples of tungsten concentrates and tailings, 20 ppm thallium was found in only a ferberite concentrate.

Roasted iron sulfide materials from Wisconsin contained 0-20 ppm thallium.

6. Pyrites and smelter sulfuric acid: The United States produces about 0.8 million tons pyrites (including cupreous pyrites) annually. At 2 ppm thallium, this would be 1.6 tons thallium per year. The sulfuric acid produced from pyrites is primarily used in cellulose fiber manufacturing, ore leaching, and nonferrous metal processing.

Even less sulfuric acid is produced from smelter fumes. The chief end uses of smelter acid are nonferrous metal processing and ore leaching with some use for fertilizers, explosives, pulp and paper, petroleum refining, etc.[79] Together, these sources account for less than 10% of the U.S. sulfur. No data for thallium contents of smelter acid were found, but it may often be expected to be present if cadmium is. Versar, Inc., in 1975[20] estimated that only 1-3% of the total cadmium in the zinc concentrate is lost during roasting to the by-product 76% sulfuric acid if, as reported by Fulkerson and Goeller,[32] the acid contains 20-60 ppm cadmium. However, two contacts Versar made stated the cadmium content is 0.005-0.006 ppm. Reference 79 lists values of 0.001-1.06 ppm cadmium in smelter-produced sulfuric acid from 12 smelters. On the basis of our estimate of 11 tons thallium in zinc concentrates, a 3% loss into the 76% smelter acid (about 1.8 million tons 100% sulfuric acid produced in 1972) would give a thallium content of 0.3 tons or 0.2 ppm. Another approach is to calculate emissions from operations at copper smelters with acid plants. About 11% of copper concentrates are roasted to give roaster gases that are treated at an acid plant; about 13% of converter emissions are treated at acid plants. Thus, up to about 4 tons thallium might be trapped in smelter acid. The only analysis found for thallium in sulfuric acid is 460 x 10^{-12} g/ml (0.00025 ppm) in reagent grade acid (analytically pure, Merck).[80]

C. Thallium Releases in Burning of Coal and Other Fossil Fuels

1. Coal: In the years 1968 to 1972, the United States consumed an average of about 500 million tons of coal annually, most of which is bituminous coal. Most (95%) U.S. bituminous coal production is east of the Mississippi River.[81] Table V-11 shows that 0.7 ppm thallium is a fairly good estimate for the average thallium content of most coal burned or coked in the United States. On this basis, the coal used annually contains about 350 tons thallium. Levels of environmental thallium near defunct anthracite coal-mining or coal-burning operations have been reported[46] to be near normal; therefore, anthracite coal is neglected here as a thallium source.

The thallium appears to be present in bituminous coal in the form of sulfide inclusions, chiefly pyrites, rather than in the organic coal mass.[82] In the United States, the coal can be washed and mechanically cleaned to remove some of the pyrite and marcasite.[83]

The highest content reported for thallium in atmospheric emissions from coal burning is 76 ppm in the finest particulates.[84] The higher concentrations for thallium, like those for other recondensed volatile metals, were on the surface of the fine particulates.[85,86,88] Thallium, as well as cadmium, lead, arsenic, selenium, potassium, and other volatile metals, are preferentially associated with α-quartz and mullite particles in the fly ash rather than with the glassy aluminosilicates.[87] Based on estimates[89] for uncontrolled particulate emissions from coal burning [electrical utilities (3.1 million tons), industrial boilers (2.6 million tons), and coke manufacture (0.2 million tons)], an average of 60 ppm thallium might be found in the particulates if the entire thallium content volatilized.

A rather complete set of data for making a material balance for thallium in coal burning was obtained from an MRI project examining trace element contents (spark source mass spectroscopic analysis) in materials at a coal-burning electrical utility. Our analysis of these data is shown in Figure V-7.

Davis, Natusch et al.,[84] have published data, from which a material balance for thallium in coal burning can be made only if certain assumptions are made. Figure V-8 is our analysis of the data of Reference 84* with the assumptions that the thallium content of the coal is 1.5 ppm and that only 75% of the fly ash is collected (e.g., if 90% of all coal burners had controls with 85% efficiency). [A cyclone was used in the actual case; these usually have 45-90% efficiency. If 95% retention is assumed, the coal used would have had to contain about 1 ppm thallium. Only 30% of the thallium would be lost in airborne emission in this hypothetical situation.]

Radian Corporation has published information regarding thallium flow in western coal-burning plants.[90,91] This is reproduced in Table VII-12. No mass balance was attempted for thallium in the references. At Station I, the ratio of thallium in bottom ash: collected ash: airborne ash is ~ 1: less than 6:2. At 0.006 lb thallium per hour in the flue gas and 165 lb fly ash per hour, the thallium content of the airborne fly ash would be about 36 ppm (if 100% in the particulates) from burning coal containing less than 0.1 ppm thallium. The total flow from the stack at standard conditions is 8.35×10^5 scfm or 71,000 lb/hr gas. We calculate that the thallium content of the flue gas is 0.08 ppm (the reference gives 0.012 and less than 0.0016 ppm). The error of the analyses (spark source mass spectrometry) was not estimated in the reference since the accuracy was so strongly dependent on the matrix. At Station III, it appears that the ratio of collected to airborne thallium is 1:11.

At the Allen coal-fired steam plant in Tennessee, Bolton et al. (1973)[92] reported that the mass flow of thallium in grams per minute was coal, less than 2.5 (less than 2 ppm Tl); slag tank solids, 0.22 (2 ppm); precipitator inlet, 1.9 (40 ppm); and precipitator outlet 0.059 (30 ppm). These and other data indicated the precipitator efficiency for thallium was 97-99%.

An average loss of about half of the thallium content to the air during coal burning appears to be a reasonable estimate. Thus, about 180 tons thallium emitted to the atmosphere is the biggest anthropogenic source we have identified.

The areas that are most concentrated in coal-burning and coking plants are EPA Regions III, IV, and V where about 22%, 24%, and 40%, respectively, of U.S. coal is burned. (See Table VII-12.)

---

* More of these data are given in the Appendix.

** The TLV is 0.1 mg/m^3; 0.08 ppm = 0.7 mg/m^3. A good estimate for ground-level concentration, however, would be 0.0007 mg/m^3 or 700 ng/m^3.

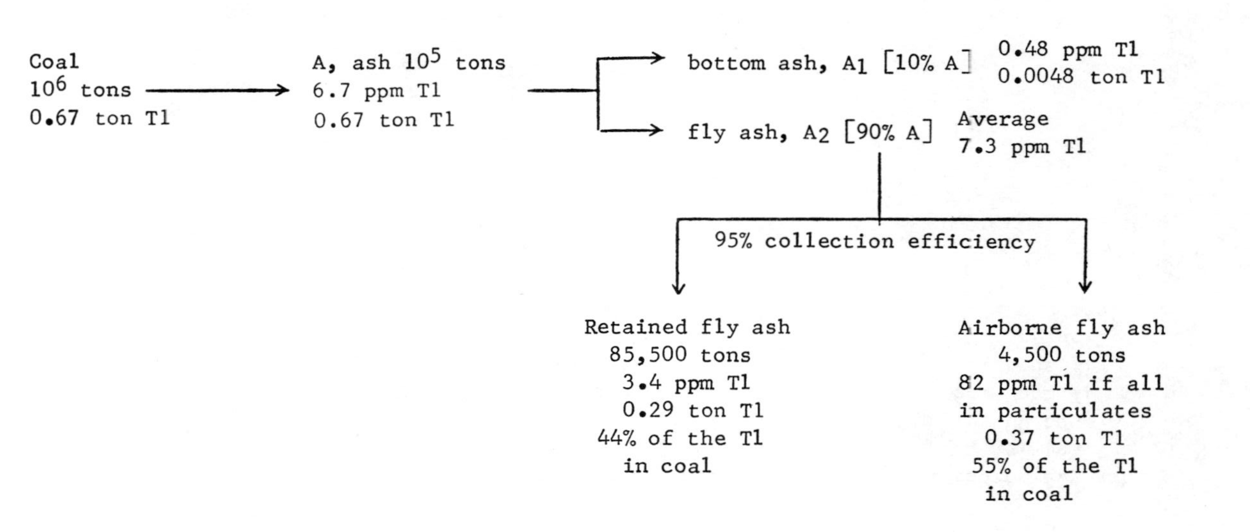

Figure V-8 - Thallium Material Balance at a Coal-Burning Electrical Utility

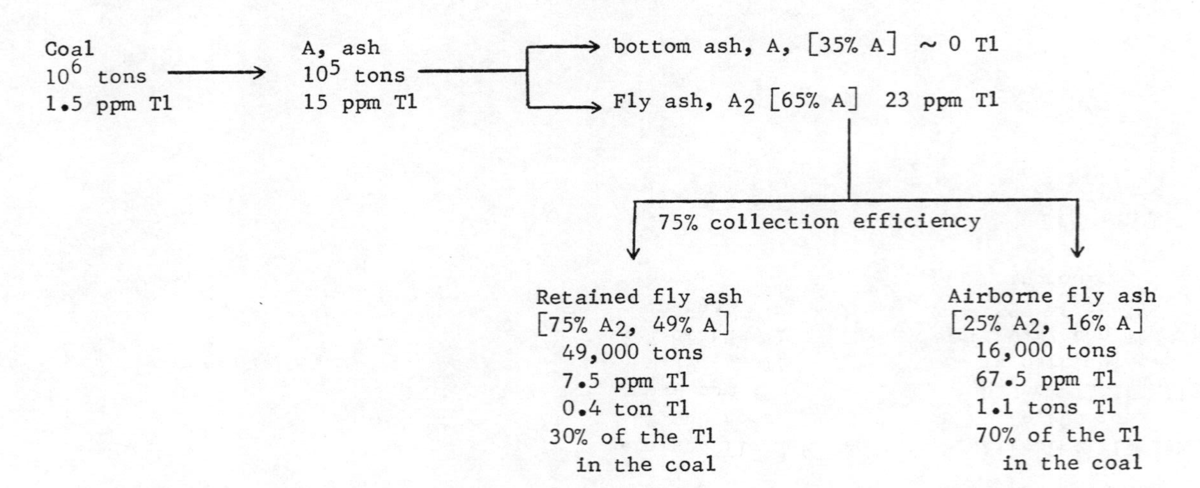

Figure V-9 - Thallium Material Balance in Coal Burning, Hypothetical Case Based on Data of Reference 84

The fate of the collected fly ash in these areas should be followed, too. Only 7% of the 20.5 million tons of collected fly ash in 1970 was used in processes such as cement making,[93] but research aimed at more complete utilization of fly ash continues. In uses such as soil stabilization, the potential for leaching thallium is higher than for most of the other contained heavy metals.

2. Petroleum and oil shale: We are not aware of any determinations of thallium in U.S. petroleum. Yen[94] cited a report by Ball et al. (1956) listing Tl as one of 24 metals found by spectrographic analysis in U.S. crude oils. A 1960 paper[95] by the same authors on this topic lists Ti, not Tl. Some USSR petroleum contained 0.4-0.5 ppb thallium.[96] Since ~ 90% of the heavy metal content of petroleums concentrates in asphalts and residual fuel oils,[97] the amounts of thallium present should be ascertained. The fuel oils especially would be a potential atmospheric source of thallium on burning. We calculate about 16 tons thallium in the annual domestic demand for petroleum oils based on the assumptions of 5 billion 42-gal. barrels domestic demand,[98] 7 lb/gal,[99] and a thallium concentration of 0.5 ppb.

The Appendix lists values up to 25.8 ppm for thallium in U.S. shales with an average cited by one author of 0.9 ppm. Since the Unita Basin of Utah is a part of the Green River Formation, where shale oils as a possible fuel source have received attention, and thallium levels in Utah minerals are known to be high, the potential for thallium release from shale oils should be studied.

D. Thallium in Ferrous Metal Production

1. Iron and steel: Most iron ores processed in the United States contain iron oxides and are not expected to be as enriched in thallium as sulfides might be. In 1968, 269 million short tons iron ore was smelted in the United States including only about 0.8 million tons as a by-product from nonferrous sulfide ores and 0.27 million tons manganiferous ore.[100] Even if it is not an enriched source, iron ore may have contained about the crustal abundance of thallium (1 ppm) or close to 300 tons of thallium per year since 1968. Comparatively minor sources of thallium in the iron and steel industry would be the content in cadmium- and zinc-plated scrap.

Pig iron production probably does not concentrate thallium in its emissions. For example, a Soviet source[101] indicates that sintering iron ore in a low-reducing atmosphere prevents volatilization of thallium compounds so that the recirculated furnace dust contains less than 1% of the thallium. Blast furnace smelting of the sintered iron ore, containing most of the thallium, gives pig iron containing 72% of the thallium. The vent gases and dusts are washed to give slimes containing traces of thallium e.g., 10 ppm. (Reference 102 gives a procedure for recovering thallium, indium, and gallium from such dusts.) Extrapolation of this information to the United States indicates that 80-90 tons thallium per year are emitted in iron blast furnace gases. The blast furnace particulate emissions are well-controlled[89] so that only about 1% or 0.8-0.9 ton thallium would be emitted to the atmosphere, calculated at less than 20 ppm if it appears in the particulates. At an estimated 20 ppm in steelmaking furnace emissions,* ~ 5 tons thallium would be emitted by current steelmaking practices with 37 tons in the controlled emissions. The ratio of scrap to pig iron was about 1:1 in 1972 and the scrap steel may well contain half of the thallium charged to the furnaces. Perhaps ~ 40% of the thallium charged is lost to the slags as in copper smelting.

2. Manganese and ferroalloys: As Table VII-6 shows, certain U.S. manganese ores contain thousands of parts per million thallium. Soviet references report that thallium is found in gravity-magnetic concentrates,[104] sinter dust,[105] dust from reductive electrosmelting of manganese ores (e.g., 3,000 ppm Tl,** 32% Mn, ~ 30% Pb),[106] and emissions from a silicomanganese electric furnace (major thallium distillation at 1150°C and during the first hour).[107]

About 1 million tons ferromanganese and silicomanganese are produced annually in the United States from about 2.3 million tons manganese ores (~ 98% from foreign sources).

Dusts produced by the ferroalloy furnaces would be about 0.12 million ton;[89] with 40% net control, there would be 48,000 tons of dusts emitted. Thallium concentrations of 3,000 ppm in these dusts would indicate a total thallium content of about 360 tons. With only 40% control, about 140 tons could be emitted to the atmosphere. The ferroalloys chapter of the 1972 Minerals Yearbook[108] indicates that manganese alloys are produced in up to 24 cities. It is possible that manganese ore processing sites could be the most concentrated sources of atmospheric thallium emissions.

On the other hand, we have found no data for thallium in imported manganese ores. Almost 40% of 1972 world manganese production was in the USSR, but the 7% of world production imported into the United States came largely from Africa and Brazil.

---

* Emissions would be 232,000 tons based on emission factors in Sax (1974)[8] and on 1972 steel production and steelmaking methods.[103]

** The thallium in laboratory-produced dusts could be only slightly dissolved (~ 5%) by 10% sulfuric acid at 90°C. It was concluded that thallium(III) was probably adsorbed on the hydrated manganese oxides because when a reductant was added, manganese and thallium were easily extracted under these conditions.

Crittenden[109/] of the USGS, an author of the manganese chapter in United States Mineral Resources, feels that neither manganese nodules or U.S. manganese ores are representative of the types of imported ores. Thallium characteristically originates in hypogene settings in near-surface, quite young deposits. The imports are syngenetic. He felt that whatever thallium was in the ores might be largely transferred into slags in preslagging and further operations.

Fleischer[47/] of the USGS (a long time editor of and contributor to the mineralogical and geological chemistry section of Chemical Abstracts) was not aware of any thallium determinations in manganese ores except those found in the United States.

Parfet,[110/] Vice President of Caemi International, Inc., dealers in manganese ores and alloys, never heard of it in manganese ores.

The only thallium containing manganese ore known to Krecker,[111/] of an independent analytical firm, came from Sierra Leone. The ores are preliminarily examined before X-ray fluorescence of atomic absorption analysis by emission arc spectrography to look for interfering elements. He had never seen a thallium line.

Ward,[112/] a manganese ore specialist at Union Carbide, probably the largest producer of manganese ferroalloys, was not aware that thallium was ever analyzed for in manganese ores nor were the laboratory people with whom he conferred.

The National Academy of Sciences (NAS)[113/] has estimated that compared with world thallium resources worth \$11 million (U.S. \$1.99 million), manganese nodules represent a source of thallium worth \$62 billion. Present world thallium demand is only valued at \$100,000 so that intentional thallium recovery from the nodules is very unlikely. The seabed will be disturbed during deep-sea mining, but the extent has not yet been determined because the particular mining method has not been selected. Based on the NAS assumption of ~ 5,000 metric tons mined nodules (30% water) per day for a 300-day year and an assumption of 100 ppm thallium in the nodules, about 50 short tons thallium will be mobilized annually. If ore processing is done at sea, tailings disposal will be a problem At present, however, economical processing can only be done on shore.

The NAS foresees that near-bottom water may become enriched in heavy metals by leaching of resuspended sediments, but the academy concluded that "important effects seem unlikely both in view of the relative low density of the near-bottom prowlers and the fact that the sedimentary material was previously settled on the sea floor as a result of natural sedimentation processes." Natural ocean phenomena scatter sediments on a much larger scale than ocean mining ever would.

E. Thallium in Nonmetallic Mineral Processing

1. Pegmatitic minerals: Minerals common in U.S. granitic pegmatites contain up to 385 ppm thallium, the highest concentration found in a biotite in North Carolina, the major U.S. mica producer. New Mexico (105-210 ppm) and Colorado (6-250 ppm thallium) each have about one-sixth the production volume of North Carolina. U.S. potassium feldspars contain up to 50 ppm thallium.

a. Mica: Micas are complex hydrous aluminum silicate minerals. Of the three principal mica minerals, U.S. muscovites (white mica) and biotites (black mica) usually are enriched in thallium. Biotite and phlogopite, however, have little commercial value. Wedepohl[114] summarized that muscovites from granite pegmatites generally contain 2-60 ppm thallium.

At 50 ppm, the ~ 300,000 short tons mica used in the United States in 1972[115] would contain about 15 tons thallium. Mica is not changed chemically in its uses. The thallium leachability might be slightly enhanced in the ground-up state (about 6 tons; about 1 ton wet-ground). Environmental problems due to thallium are not expected from mica production or use.

b. Feldspar: About 740,000 tons feldspar is produced annually in the United States.[116] This would contain 37 tons thallium at 50 ppm. No environmental problems are expected from feldspar production, but about 80% is used for glass and pottery making, a furnacing operation. However, feldspar is probably diluted considerably by other glass-making ingredients so that any thermally induced thallium emissions would also be diluted. The nonleachable content in discarded beverage and food glass in urban refuse may total up to 30 tons thallium. (MRI spark source spectrographic analysis of urban refuse-derived fuel shows 0.16 ppm thallium; at 125 million tons urban refuse in the United States per year, this would be 20 tons.)

2. Potash: Frequent references in the literature to high thallium in potash minerals are to the potassium minerals of mica and potassium feldspars mentioned above. We have not found any indication that thallium is even as high in U.S. fertilizer potash minerals as the crustal abundance.

About 87% of U.S. production of potassium compounds in 1970 came from crystalline deposits in southeastern New Mexico and eastern Utah. Most of the rest comes from brine operations in northwestern Utah, California, and Michigan. The potassium minerals of the crystalline deposits (sylvite, langbeinite, carnallite, et al.) are much richer in potassium (about 43% KCl) than are the brines.[117]

If the nearly 5 million short tons potassium salts used annually in the United States (95% of which are used for fertilizer) contained 1 ppm thallium, only 5 tons thallium would be spread over agricultural land in 49 states.

Some of the thallium present would be water soluble since potassium salts other than the chloride (e.g., the phosphate, carbonate, nitrate, and sulfate) are used in fertilizer formulations. Processing losses of thallium to water during initial potassium chloride isolation by flotation or crystallization are not expected to be significant because of the low water solubility of TlCl (1 g/260 g cold water, 1 g/70 g boiling water[118]) compared with that of KCl unless considerable oxidation to Tl(III) would occur.

TABLE V-1

COMMERCIAL THALLIUM "STATISTICS"[a]

Year	Domestic Production and Shipments	Imports	Domestic Demand or Consumption	Remarks
1922			~ 25-50 lb/yr[119]	
1924	["total world production" ≤ 500 lb[120]]			Rodenticide use in U.S. has begun.
1925	"annual production is very small"[121]			U.S. and Germany were apparently the only commercial producers.
1926			"brisk demand"[122]	
1930		Major part of consumed thallium is imported.[123]		
1934	"recovered in considerable quantities"[124]			
1937		from Germany, Poland, and Belgium		"...Quite small amounts" are used for special alloys. "...A few hundred pounds annually have long been used in optical glass, photoelectric cells, and medicine..."

Year	Domestic Production and Shipments	Imports	Domestic Demand or Consumption	Remarks
1939	-[b/]			
1940	-[b/]			
1941	-[b/] < 0.5 ton[14/]		3,500 lb	
1942	Output "stepped up" < 0.5 ton[14/]	0	7,500 lb	Requirement estimated by Fish and Wildlife Service
1943	Output "increased somewhat" < 0.5 ton[14/]	0	10,000 lb	"Supply was only one-third of demand."
1944	Production "increased slightly." < 0.5 ton[14/]			
1945	"Several thousand lb" "Declined somewhat compared with 1944." ~ 2 tons[14/]	~ 3,000 lb (from Begium and France)	6,000 lb	Fish and Wildlife Service estimates. Supply-demand deficit so large, the War Production Board put thallium chemicals under allocation. Rodenticide used in war theaters.
1946	One-fifth 1945 output [0.4 ton?][c/] 2.7 tons[14/]		7,500 lb	Fish and Wildlife Service estimate.
1947	~ 3 times 1946 output [1.2 ton?][c/]		< 7,500 lb	Fish and Wildlife Service estimate. Use of new organic pesticides expected to reduce the thallium requirement.
1949	Shipments "a few percent above 1948" [1.6 ton?][c/]			
1950	"Few thousand lb" Shipments increased		Marked increase in rodenticide use	
1951	"Few thousand lb"		Same as 1950?	Rodenticide and insect control use declined. Use in scintillator counter crystals and as a phosphor activator increased.
1952			"Few thousand lb"	
1953	"Few thousand lb"			Cerro de Pasco Copper Corporation, Brooklyn, N.Y., produced small amounts of salts and metal.

TABLE V-1 (continued)

Year	Domestic Production and Shipments	Imports	Domestic Demand or Consumption	Remarks
1954	"Greater" than in 1953			
1955	"About the same...as in 1954"			
1956	Production: Less metal, more compound Shipments: Both greater than in 1955			
1957	Production: Less metal and compound than in 1957			
1958	About same as 1957			
1959	"Shipments...increased slightly"			
1960	Shipments of metal and compounds decreased			
1961	Production about the same as in 1960			
1962	Production: More metal, less sulfate Shipments: Much more sulfate than in 1961			
1963	Production: More metal, less sulfate Shipments: Both less than in 1962			
1964	Production: More metal, more sulfate Shipments: Less metal, more sulfate than in 1963		12,308 lb[125]	Uses: Babitzke (1976)[125] gives elec. 4,332 lb, agric. 6,876 lb, pharmaceuticals 300 lb, and others 800 lb.
1965	No production		9,393 lb[125]	Private use of thallium products to control insects and rodents prohibited in 1965. Uses: elec. 2,550 lb, agric. 2,935 lb, pharmaceuticals 200 lb.[125]
1966	-[b]. No production		6,285 lb[125]	
1967	No production	117 lb	5,360 lb[125]	Use for electronics and components of solder and fusible

Year	Domestic Production and Shipments	Imports	Domestic Demand or Consumption	Remarks
1968	First production since 1964. 5,300 lb (~ 2.7 tons)[126/] Shipments: More metal and compounds than in 1967 Industry Stocks 26,300 lb (~ 13.2 tons)	141 lb metal 100 lb[126/]	8,277 lb[125/]	Greenspoon[126/] estimated thallium production from smelter residues by country of origin: U.S. 2,600 lb, Canada 1,600 lb, Mexico 700 lb, Peru 200 lb, Australia 10 lb, and others 190 lb. "All data are estimated..." "There are virtually no recorded data on domestic mine production of thallium or quantities contained in imported ore." Uses: Elec. 5,162 lb, agric. 2,165 lb, pharm. 300 lb[125/]
1969	Production: More metal and compound than in 1968 Shipments: Both less than in 1968	~ 600 lb	9,058 lb[125/]	ASARCO said to have produced and consumed more metal.[125/] Uses: Elec. 3,148 lb, agric. 4,910 lb, pharm. 300 lb[125/]
1970	Production and shipments of metal and compounds less than in 1969	metal 1,250 lb compounds 2,050 lb	6,187 lb[125/]	Uses: Electronics ~ 40%, low-melting alloys ~ 40%, agricultural pesticides 10%, organometallic chemistry ~ 5%, development research ~ 5%. Babitzke[125/] gives elec. 53%, agric. 32%
1971	Production more than in 1970, shipments less [~ 0.9 ton?][c/]	2,131 lb[127/] (metal plus compounds)	3,949 lb[127/]	Uses: Electronics and low-melting alloys ~ 25%, academic purposes and development research ~ 50%, miscellaneous 25% (e.g., less than 50 lb as metallic iodide additions to Hg arc lamps). Babitzke[125/] gives elec. 36%, agric. 50%.
1972	First 7 months: metal shipments 10 lb, sulfate 90 lb[3/]	2,200 lb[127/]	2,463 lb[127/]	Reference 3 cites a Globe refinery spokesman as saying that thallium had been produced in 5 short campaigns within the last 20 years. Uses: Elec. 1,133 lb, agric. 830 lb, pharm. 100 lb[125/]
1973	[Probably no ASARCO shipments]	1,894[127/]	1,400 lb[127/]	Uses: Elec. 1,000 lb, agric. 150 lb, pharm. 150 lb[125/]
1974	[Probably no ASARCO shipments]	1,345 lb[127/]	1,800 lb[127/]	Uses: Elec. 1,300 lb, agric. 165 lb, pharm. 165 lb[125/]
1975	[Probably no ASARCO shipments]	1,200 lb[127/]	1,000 lb[127/]	

a/ The source of the information is from the Minerals Yearbook of the appropriate year except where cited.
b/ Thallium is not covered in these Minerals Yearbooks.
c/ MRI estimate.

TABLE V-2

U.S. COMPANIES HANDLING THALLIUM[14,128]

Company	Products
American Hoechst Corporation	TlCl, TlI
ASARCO Inc.	Metallic Tl of various degrees of purity, TlI, Tl_2SO_4
Apache Chemicals, Inc.	Tl acetate, TCl, TlI, Tl metal, Tl sulfate
Atomergic Chemicals Company, Division of Gallard-Schlesinger Chemical Manufacturing Company	Tl compounds, Tl metal
Belmont Smelting and Refining Works, Inc. (Brooklyn, New York)	Tl
F. W. Berk and Company, Inc.[a] (Woodridge, New Jersey)	Tl salts, Tl amalgam
Bram Metallurgical-Chemical Company	Tl
City Chemical Corporation[a]	Organic and inorganic Tl compounds including the acetate, arsenate, iodide, fluosilicate, lactate, molybdate, and tartrate
Cominco Products, Inc. Electron Materials Division[a] (Jersey City, New Jersey)	High purity Tl (99.999% and 99.9999%)
Eastman Organic Chemicals, Eastman Kodak	Tl acetate, formate
Electronic Space Products, Inc. (Los Angeles, California)	Tl oxide (99.999% pure), TlI, Tl sulfate, Tl metal
Elmat Corporation (Roseville, California)	Liquid metallic alloys including Hg-In-Tl eutectic composition

TABLE V-2 (continued)

Company	Products
Fairmount Chemical Company, Inc. (Newark, New Jersey)	Tl, Tl compounds
Fielding Chemical Company	TlCl, TlI, Tl acetate, Tl formate
Great Western Inorganics	TlI
Harchen Division of Wallace and Tiernan, Inc.[a] (Belleville, New Jersey)	Tl
Harshaw Chemical Company, Inc.[a]	Scintillation and optical crystals, esp. Tl bromide-iodide; TlI-activated CsI; and TI-activated KI
ICN-K&K Laboratories, Inc.	Metal (99.98%) and various compounds: acetate, arsenide, borate, bromide (99.9%) carbonate, chloride (99.0-99.9%), ethoxide, fluoride (99.9%), formate, hydroxide (15% solution), iodide (99.6%), malonate, nitrate (99.9%), oxide (99.2-99.5%), perchlorate, Tl_2Cl_3, succinate, sulfate (99.0%), sulfide (99.9%), telluride, thiocyanate, thallium triselenide, and complex salts: KCl-TlCl, $TlNO_3$-$AgNO_3$[b]
Keru Chemical Corporation[a]	Metallic thallium of 99.999% purity in the form of wire, rods, etc.; selenide, sulfide, and oxide of thallium of purity 99.9 and 99.995%; TlCl, TlBr; powder of purity 99.999%; optical monocrystals (TlCl, TlBr, TlBr-TlI)
A. D. MacKay, Inc. (New York, New York)[a]	Thallium compounds
National Biochemicals Company	Tl formate

TABLE V-2 (continued)

Company	Products
Research Organic/Inorganic Chemical Corporation	Tl acetates, chloride, formate, iodide, sulfate; cyclopentadienylthallium(I); Tl acetylacetonate, benzoate, diethylmalona[te] etc.
Semi-Elements, Inc.[a,c]	Tl metal of purity 99.9995%
United Mineral and Chemical Corporation (New York, New York)	Tl metal (99.999+% pure), ingots, rods, wire, bars, powders, flakes, and granules
Var-Lac-Oid Chemical Company	Tl acetate, TlCl, Tl formate, TlI, Tl metal, Tl sulfate
Ventron Corporation--Alfa Products	Tl acetate, TlCl, Tl formate, TlI, Tl metal, Tl sulfate

a/ Not in 1975.

b/ All products listed--for laboratory purposes according to a 1964 bullet[in] of the company.

c/ Company specialized in production of mono- and polycrystalline articles for the electronic industry.

THALLIUM IN ORES AND MILL AND SMELTER PRODUCTS[59]

Source	Location	Spectrographic Determination Tl_2O_3 Concentration, ppm	Detection Limit, ppm	Chemical Determination of Thallium Concentration, ppm[12]	Remarks
Copper-bearing Materials					
Ore, mill heads	Arizona				
Cu mill heads	Greenlee County	0	10		
Schist ore	Pinal County	0	1%		
Cu mill heads	Utah copper mine ore, Arthur plant	0	10		But see below.
Utah Copper Company ore				2	About half the Tl in the ore remained in the tailings. If pyrite and other sulfides were removed, the siliceous gangue was Tl-free. The molybdenite conc. was low in Tl. Zimmerley(1947)[12] concluded that the bornite, covellite, and chalcocite minerals bore Tl but not the pyrite, chalcopyrite, and molybdenite.
Copper concentrates	Arizona Gila County, Greenlee County, Santa Cruz County, Yavapai County	0	10		
Copper concentrates	Pima County	10	10		
Copper concentrates	Brittania Beach, British Columbia	20-30	10		Different ore from the source of the zinc concentrate. The pyrite conc. contained 90 ppm Tl and the Cu-pyrite tailings, 10 ppm.

TABLE V-3 (continued)

Source	Location	Spectrographic Determination Tl_2O_3 Concentration, ppm	Detection Limit, ppm	Chemical Determination of Thallium Concentration, ppm[12/]	Remarks
Copper-bearing Materials (continued)					
Cu conc. from -8+20 tables	Pyrrhotite deposit, Eastern U.S.	10	10		No Tl detected in the conc. from the -3+8 tables, mill heads, or tails.
Cu conc. from -20+48 tables	Pyrrhotite deposit, Eastern U.S.	10	10		
Cu conc. from -48 tables	Pyrrhotite deposit, Eastern U.S.	10	10		
Pyrrhotite concentrates	Galax, VA, and Maine			10	
Copper concentrates	Idaho Lemhi County	0	10		
Copper concentrates	Blackbird			10	The cobalt concentrates also contained 10.
Copper concentrates	Michigan	0	10		
Copper concentrates	Montana (Butte mines)	0	10		
Cu-Au-Ag concentrates	Sierra County New Mexico	0	10		None detected in tailings either.
Copper concentrate	Ore from Holden mine, Chelan County, WA	30	10		Less than 1 ton Tl in 1945 in Cu concentrates. Zn conc. contained 60 ppm Calc. ~ 3 ppm in ore. 10 ppm in the Cu-Zn tailings. Leading Au & Cu producer in WA in 1945. Cu smelted in Tacoma; cyanidation ppts. sent to Selby; Zn concs. to Kellogg, ID.

TABLE V-3 (continued)

Source	Location	Spectrographic Determination Tl_2O_3 Concentration, ppm	Detection Limit, ppm	Chemical Determination of Thallium Concentration, ppm[12/]	Remarks
Copper-bearing Materials (continued)					
Copper concentrate	Bingham, Utah	0		23	Pyrite concentrates made from the tailings contained 18-21 ppm. Pyrite concs. from Park City, Midvale, & Bauer, Utah contained 10 ppm Tl. Principal feed to Garfield smelter.
Copper tailings	Arizona Gila County, Morenci, Greenlee County; Maricopa County; Pima County; Yavapai County	0	10		
Smelter and Refinery Products					
Reverberatory smelter slags	Arizona Douglas smelter (Copper Queen and other Cochise county ores), Morenci smelter (Greenlee County ores)	0	10		
Slag	United Verde smelter (Verde [Jerome] district ores)	0	10		
Blast furnace slag	Eastern U.S. Cu-Zn ores	0	10		Zn and iron sulfide concs. from these ores contain 10 ppm (Tl_2O_3)
Reverberatory slag	Eastern U.S. Cu-Zn ores	0	10		
Converter slag	Eastern U.S. Cu-Zn ores	0	10		

TABLE V-3 (continued)

Source	Location	Spectrographic Determination Tl_2O_3 Concentration, ppm	Detection Limit, ppm	Chemical Determination of Thallium Concentration, ppm[12/]	Remarks
Copper-bearing Materials (continued)					
Smelter and Refinery Products (continued)					
Wet and dry flue dusts from iron roaster	Eastern U.S. Cu-Zn ores	0	10		
Cu flue dust from reverberatory furnace	Eastern U.S. Cu-Zn ores	40	10		
Cu flue dust from blast furnace	Eastern U.S. Cu-Zn ores	0	10		
Cu flue dust from converter	Eastern U.S. Cu-Zn ores	0	10		
Residue from Cu sulfate tank	Eastern U.S. Cu-Zn ores	0	10		
Acid plant sludge from chamber plant	Eastern U.S. Cu-Zn ores	50	10		
Cu slags, flue dust from dump	Idaho Mackay smelter (Custer County ore)	0	10		No Tl in Cu tailings from same ore.
Copper flue dust (heads for arsenic plant)	Montana Washoe Reduction Works, Anaconda, Montana. Ore from Butte mines, Silver Bow, Montana	0	10		No Tl detected in concs. or tailings from the complex ore.

TABLE V-3 (continued)

Source	Location	Spectrographic Determination Tl_2O_3 Concentration, ppm	Detection Limit, ppm	Chemical Determination of Thallium Concentration, ppm[12]	Remarks
Copper-bearing Materials (continued)					
Smelter and Refinery Products (continued)					
Arsenic roaster residue	Montana Washoe Reduction Works, Anaconda, Montana. Ore from Butte mines, Silver Bow, Montana	200	10		See Murray, Utah, lead smelter
Converter dust	Garfield, Utah			1700	Zimmerley (1947)[12] concluded the bulk of the thallium received by the Murray Pb smelter (whose Cd flue dust was the Globe plant's chief source of thallium) came from dust or fume from the Garfield Cu smelter. Garfield operated "almost exclusively" on Cu concs. from the Arthur & Magna mills of the Utah Copper Co. The concs. contained 23 ppm Tl. At ~ 2000 tons conc./day: ~ 0.05 ton Tl or ~ 17 tons/yr.
Reverberatory furnace stack dust	McGill, Nevada			1500	At the time of sampling, the smelter was operating on a large amount of Utah Copper Co. conc. Thus, the porphyry Cu of the Ely district was not necessarily enriched in Tl.
Converter stack flue dust	McGill, Nevada			50	

TABLE V-3 (continued)

Source	Location	Spectrographic Determination Tl_2O_3 Concentration, ppm	Detection Limit, ppm	Chemical Determination of Thallium Concentration, ppm[12]	Remarks
Lead and/or Zinc-bearing Materials (continued)					
Ore, mill heads Zn ores	Arkansas Boone County, Searcy County, Newton County	0	10		
Pb-Zn-Cu ore (siliceous)	Colorado San Juan area	0.0X%	10		Not detected in the tailings. Not detected in a few other Colorado Pb-Zn ores.
Pb-Zn and Zn-Pb ores	Grafton and Coos Counties, New Hampshire	0	10		
Zn and Zn-Pb ores	Grant County, New Mexico	0	10		
Pb/Zn Concentrates					
Zinc concentrates	Arizona Santa Cruz County ore	10	10		
Zinc concentrates	Yavapai County ore	20	10		
Zinc concentrates	Yavapai County ore (Iron King mine)	0	10		
Zinc concentrates	Mohave County ore	40	10		
Zinc concentrates	Pinal County ore	50	10		
Zinc concentrates	Boone County, Arkansas	0	10		

TABLE V-3 (continued)

Source	Location	Spectrographic Determination Tl_2O_3 Concentration, ppm	Detection Limit, ppm	Chemical Determination of Thallium Concentration, ppm[12]	Remarks
Lead and/or Zinc-bearing Materials (continued)					
Pb/Zn Concentrates (continued)					
Zinc concentrates	Brittania Beach, British Columbia	60-70	10		
Zinc concentrates	Cu-Zn deposits, Eastern U.S.	10	10		Iron sulfide concs. from same ores contain 0-10 ppm (Tl_2O_3).
Lead concentrates	Arizona Santa Cruz County, Yavapai County, Mohave County, Pinal County	0	10		
Pb-Zn materials from numerous mines and mills	Idaho	0	10		
Except:					
Zinc concentrate	Morningmine and mill, Shoshone County	50	10		Associated Pb-Zn-Ag tailings: 10 ppm
Zinc concentrate	Frisco mine, Shoshone County	50	10		Associated Pb-Zn-Ag tailings: 10 ppm
Zinc concentrate	Page and Blackhawk mines, Page Mill, Page, Idaho	40	10		Associated Pb-Zn-Ag tailings: 10 ppm
Pb concentrate	Page and Blackhawk mines, Page Mill, Page, Idaho	0	10		
Zinc concentrate	Montana Ore from Broadwater mine, Cascade County	30	10		None in Pb-Zn tailings.

TABLE V-3 (continued)

Source	Location	Spectrographic Determination Tl_2O_3 Concentration, ppm	Detection Limit, ppm	Chemical Determination of Thallium Concentration, ppm[12/]	Remarks
Pb/Zn Concentrates					
Pb-Zn bulk concentrate	Ores from 5 other mines, Cascade County	10	10		None in Pb-Zn tailings.
Lead smelter ore sample, largely concentrate	Murray, Utah, smelter Utah ores			20	
Zinc concentrate	Grant County, New Mexico	0	10		None detected in tailings.
Lead concentrate	Grant County, New Mexico	0	10		
Zinc concentrate	Hyatt mine, Edwards District, St. Lawrence County, New York	10	10		None detected in tailings.
Zinc jig concentrate	Tri-State District, Ottawa County, Cardin, OK	10	10		20 ppm in tailings.
Lead jig or flotation concentrates	Tri-State District, Ottawa County, Cardin, OK	0	10		
Zinc flotation concentrates	Tri-State District, Ottawa County, Cardin, OK	0	10		
Zinc table concentrates	Hockerville, Oklahoma	10	10		
Zinc concentrate coarse mill sample	Kansas Bilharz Mine, Cherokee County, KS	10	10		None detected in the Pb concentrate.
Flotation conc.	Kansas Bilharz Mine, Cherokee County, Trecce, KS	10	10		At Treece, feed was from old tailings.

TABLE V-3 (continued)

Source	Location	Spectrographic Determination Tl_2O_3 Concentration, ppm	Detection Limit, ppm	Chemical Determination of Thallium Concentration, ppm[12/]	Remarks
Pb/Zn Concentrates					
Zinc flotation concentrate	Kentucky Gratz County	10	10		None found in Pb concentrate or tailings.
Zinc flotation concentrate	Missouri Tri-State District, Jasper County	10	10		Coarse concentrate contained 20 ppm. None in Pb concentrate or tailings
Coarse zinc jig concentrate	Tri-State District, Moniteau County	10	10		None in Pb concentrate or tailings.
Zinc concentrate	Tri-State District, Newton County	0	10		
Zinc flotation and jig concentrates	Mascot, Knox County, Tennessee	0	10		None detected in tailings.
Zn and Pb flotation concentrates	Claiborne County, Tennessee	0	10		
Pb concentrates from carbonate ore	Embree mine, Unicoi and Washington Counties, Tennessee	0	10		See the Mn concentrate also (miscellaneous heading)
Zn concentrate from carbonate ore	Embree mine, Unicoi and Washington Counties, Tennessee	60	10		None in Zn or Pb slimes in old tailings pond, but 10 ppm in a mill sample of Pb-Zn tailings slimes.
Zn concentrate	Spotsylvania County, Virginia	20	10		
Zn concentrate	Holden mine, Chelan County, Washington	60	10		30 ppm in the Cu concentrate, 10 ppm in the Cu-Zn tailings.

TABLE V-3 (continued)

Source	Location	Spectrographic Determination Tl_2O_3 Concentration, ppm	Detection Limit, ppm	Chemical Determination of Thallium Concentration, ppm[12/]	Remarks
Pb-Zn Concentrates					
Zn concentrate	Stevens County, Okanagan County, Washington	0	10		
Zn concentrate	Grandview and Bella May mines, Pend Oreille County, Washington	30	10		None in Zn-Pb tailings.
Zn flotation concentrate	Pend Oreille Mine, Pend Oreille County, Washington	40	10		None in Zn-Pb tailings.
Zn-Pb feed to tailings mill	Benton, Wisconsin Lafayette County	10	10		
Tailings					
Cu-Pb-Zn-Ag tailings	Arizona, Yavapai County	0	10		
Pb-Zn-Au-Ag tailings	Arizona, Yavapai County	0	10		
Zn-Pb tailings	Mohave County	0	10		
Pb-Zn-Cu tailings	Maricopa County	0	10		
Zn tailings	Boone County, Arkansas	0	10		
Cu-Zn tailings	Eastern U.S. Cu-Zn deposits	0	10		
Zn-Pb tailings from tailings pile	Ottawa County, Oklahoma	20	10		10 ppm in jig Zn concentrate, but none in flotation Zn concentrate. None detected in coarse Pb-Zn mill feed.

TABLE V-3 (continued)

Source	Location	Spectrographic Determination Tl_2O_3 Concentration, ppm	Detection Limit, ppm	Chemical Determination of Thallium Concentration, ppm[12/]	Remarks
Lead and/or Zinc-bearing Materials (continued)					
Tailings (continued)					
Pb-Zn tailings slimes, mill sample	Embree mine, Unicoi and Washington Counties, Tennessee	10	10		None found in slimes in old tailings pond.
Smelter and refinery products					
Zn smelter slag	Concentrates from Tri-State District and northern Arkansas	10	10		
Zn smelter residue	Concentrates from Tri-State District and northern Arkansas	40	10		
Zn smelter flue dusts	LaSalle, Illinois (ores and concentrates from Tri-State, TN, WI, WA, and B.C.)	10	10		
Zn smelter flue dusts	LaSalle, Illinois (coarse Missouri ores)	10	10		
Zinc oxide plant ZnO	LaSalle, Illinois ores and concentrates from Tri-State, etc.	20	10		
Residue		40	10		
Zn Cottrell dust from sintering	E. St. Louis	30	10		
Waelz plant Zn oxide	E. St. Louis	30	10		
Waelz plant clinker	E. St. Louis	0	10		
Zn furnace residue	E. St. Louis	10	10		
Zn Cottrell dust from roasting	E. St. Louis	40	10		

TABLE V-3 (continued)

Source	Location	Spectrographic Determination Tl_2O_3 Concentration, ppm	Detection Limit, ppm	Chemical Determination of Thallium Concentration, ppm12/	Remarks
Lead and/or Zinc-bearing Materials (continued)					
Smelter and refinery products (continued)					
Zinc oxide plant calcines and concentrates	E. St. Louis [Columbus, OH & Mascot, TN]	10	10		
Cottrell dust from above	E. St. Louis [Columbus, OH & Mascot, TN]	40	10		
Zinc oxide plant ZnO Residue discharge	Danville [Concentrates from Tri-State District & Wisconsin]	10	10		
Zinc calcines (smelter heads for Monsanto, IL)	Concentrate from Tri-State District	20	10		
Zinc calcines (smelter heads for Monsanto, IL)	Concentrates from Newfoundland	10	10		
Zinc calcines (smelter heads for Monsanto, IL)	Concentrates from Fresnillo, Mexico	20	10		
Purification cake (after addition of Zn dust to Zn sulfate solution)	Monsanto, Illinois	10	10		
Oxide plant ZnO clinker	Columbus, Ohio (ore from Mascot, Tennessee)	0	10		

TABLE V-3 (continued)

Source	Location	Spectrographic Determination Tl_2O_3 Concentration, ppm	Detection Limit, ppm	Chemical Determination of Thallium Concentration, ppm[12]	Remarks
Lead and/or Zinc-bearing Materials (continued)					
Smelter and refinery products (continued)					
Zinc calcines	Concentrates from St. Eulalia, Mexico (calcined in Dumas, Texas)	30	10		Not detected in the concentrate.
Cadmium furnace residue	E. St. Louis	10	10		Not detected in any of the foreign zinc concentrates analyzed.
Lead sludge from cadmium plant	E. St. Louis	10	10		
Lead sludge from cadmium plant	Dumas, Texas	10	10		
Zinc Cottrell dust from sintering	Blackwell, Oklahoma, and Dumas, Texas	30 40	10		
Lead smelter flue dust	Danville, Illinois (concentrates from Tri-State District and Wisconsin)	10	10		
Lead refinery products	E. Chicago, Indiana	0	10		
Lead smelter Cottrell dust	East Helena, Montana (ASARCO)	400	10		None in smelter slag or lead bullion.
Lead smelter baghouse dust	East Helena, Montana (ASARCO)	300	10		
Lead smelter flue dust	East Helena, Montana			650	Shipped to Murray, Utah Pb smelter. Feed: Pb-Ag concentrates from Coeur d'Alene region, residues from electrolytic Zn plants at Anaconda and Great Falls, and Montana ores and concentrates.

TABLE V-3 (continued)

Source	Location	Spectrographic Determination Tl_2O_3 Concentration, ppm	Detection Limit, ppm	Chemical Determination of Thallium Concentration, ppm[12]	Remarks
Lead and/or Zinc-bearing Materials (continued)					
Smelter and refinery products (continued)					
Lead roaster baghouse dust	Midvale, Utah			3800	Houses have been built over the waste pile at the remains at the Midvale smelter[51]
Lead blast furnace baghouse dust	Midvale, Utah			240	
Lead blast furnace baghouse dust	Tooele, Utah			70	
Lead blast furnace baghouse dust	Murray, Utah			700	Smelter concentrates plus ore: 20 ppm Tl. Among the feeds were the Garfield copper converter dust, 1700 ppm Tl, and E. Helena flue dust, 650 ppm. At 20 ppm Tl, the Pb concentrates could not be the source of most of the Tl at the Murray smelter. Only about 60,000 tons/yr of lead concentrates were produced in Utah in the late 1940's.
Lead smelter cadmium dust sent to Globe smelter	Murray, Utah			5300	
Lead smelter, bag dust	Selby, California			1050	One of the feeds to the Murray, Utah, Pb smelter.
Flue dust pile from "old operations" [< 1947]	Kennett (near Shasta), California			210-220	Possibly thallium had been leached by years of exposure to rain.
Lead plant Cottrell dust	Kellogg, Idaho			70	
Lead blast furnace baghouse	Kellogg, Idaho			20	

TABLE V-3 (continued)

Source	Location	Spectrographic Determination Tl_2O_3 Concentration, ppm	Detection Limit, ppm	Chemical Determination of Thallium Concentration, ppm[12/]	Remarks
Lead and/or Zinc-bearing Materials (continued)					
Smelter and refinery products (continued)					
Lead smelter Cottrell-precipitator dust	Murray, Utah			> 3%[11/]	Until this dust was stockpiled instead of recycled, much of the Tl appeared in the arsenic products shipped from the plant; e.g., before: 2100 ppm in "white arsenic," after: 300 ppm.
Lead blast furnace dust after removal of Cd	Alton, Illinois			100	
Lead blast furnace baghouse dust	Alton, Illinois			70	
Lead hearth baghouse dust	Alton, Illinois			< 10	
Lead smelter Cottrell dust	Galena, Kansas			20	Sent to Blackwell, Oklahoma
Lead smelter baghouse dust	Herculaneum, Missouri			90	
Lead smelter sinter plant flue dust	Herculaneum, Missouri			20	
Lead smelter blast furnace flue dust	Herculaneum, Missouri			10	
Lead smelter stack dust	Frederickstown, Missouri			10	
Lead smelter flue dust	Frederickstown, Missouri			< 10	

TABLE V-3 (continued)

Source	Location	Spectrographic Determination Tl_2O_3 Concentration, ppm	Detection Limit, ppm	Chemical Determination of Thallium Concentration, ppm[12/]	Remarks
Tungsten-bearing Materials					
W (ferberite) concentrates (3 samples from jigs & tables or flotation)	Colorado Pride mine et al., Boulder County	20	10		Not detected in the flotation tails and slimes
W concentrates	Yellow Pine District, Valley County, Idaho	0	10		No Tl in the Sb concentrate either.
W (scheelite and sulfide) concentrates and tailings	Lincoln County, Nevada	0	10		
W tailings	Vance County, North Carolina	0	10		
W concentrates, slimes, tailings	Stevens County, Washington	0	10		
Gold-bearing Materials					
Gold-processing materials	Alabama	0	10		
Au tailings	Arizona Maricopa County, Mohave County (several different mills), Yavapai County	0	10		
Au-Ag tailings	Yavapai County	0	10		
Au tailings	California	0	10		
Au cyaniding residues (2)	Argonaut mill, Jackson, California Ore from Amador County	10-20	10		

TABLE V-3 (continued)

Source	Location	Spectrographic Determination Tl_2O_3 Concentration, ppm	Detection Limit, ppm	Chemical Determination of Thallium Concentration, ppm[12]	Remarks
Gold-bearing Materials					
Au slimes, tailings, concentrates, mill heads, roaster slag	Colorado	0	10		
Au tailings	Montana Fergus County	200	10		Barnes-King Mine. None in ore from Spotted Horse mine.
Au tailings	Phillips County, Montana	0	10		
Flue dust from Au-As ore roaster	Ore from Getchell Mine, Humboldt County, Nevada	50		10	None in tailings at mill.
Cyanide charge and Au tailings	Lead, South Dakota (Lawrence County)	0-10	10		Only detected in a stope filling. (residue from sand leach vats)
Au tailings	Lawrence County, South Dakota Maitland and Trojan	0	10		
Au-arsenopyrite concentrates	Pennington County, South Dakota	10	10		
Au-table and jig concentrates	Pennington County, South Dakota	100	10		Different mine from above.
Gold tailings	Mercur, Utah Tooele County Ore from Ophir and adjacent mines	100-200			
Sand dump gold ore by rock type	Manning, Utah			50-550	No effective method was found to beneficiate these ores. Cyanidation dissolves both Tl and Au.
Pyrite, arsenopyrite, stibnite, gold	Yellow Pine, Idaho			10	
Gold, arsenopyrite concentrate	Brittania Beach, British Columbia			10	

TABLE V-3 (continued)

Source	Location	Spectrographic Determination Tl_2O_3 Concentration, ppm	Detection Limit, ppm	Chemical Determination of Thallium Concentration, ppm[12/]	Remarks
Molybdenum-bearing Materials					
Mo-V flotation concentrates	Arizona Pinal County	40	10		
Mo-V table concentrate	Pinal County	10	10		
Cu-Mo tailings	Pinal County	0	10		
Mo concentrates	Yavapai County	0	10		
Molybdenite ores	Pinal County	0	10		
Mo concentrates	Hurley, New Mexico (ore from Grant County)	0	10		None in Cu ore or tailings.
Molybdenite concentrates, tailings, roaster flue dust	Utah Copper mine ores, Arthur mill, Utah	0	10		
Miscellaneous					
Bauxite-processing materials	Alabama	0	10		
	Arkansas	0	10		
Graphite-processing materials	Alabama Arkansas	0	10		
Manganese ores	Arkansas	0-10	10		
Burned Hg rock from tailings piles	Arkansas	0	10		
Burned Hg ores, flue dust from Hg furnace	California	0	10		

TABLE V-3 (continued)

Source	Location	Spectrographic Determination Tl_2O_3 Concentration, ppm	Detection Limit, ppm	Determination of Thallium Concentration, ppm[12]	Remarks
Miscellaneous (continued)					
Manganese tailings	Georgia	0-10	10		
Ti-Fe-V tailings	Essex County, New York	20	10		
Mn concentrate metallurgical grade	Mn ore from Washington and Unicoi Counties, Tennessee	40	10		10 ppm in low-grade concentrate and jig tailings in dump.
Iron sulfide and oxide magnetic roast (roasted once)	Zn-Fe jig concentrates from Wisconsin and northern Illinois, Vinegar Hill custom mill, Cuba City, Grant County, Wisconsin	10	10		None detected in the twice-roasted roast.
Iron oxide calcines (4 samples)		0-20	10		
Iron oxide calcines from flash roasted flotation concentrate of iron sulfide (2 samples)	Zn-Fe jig concentrates from Wisconsin and northern Illinois, Vinegar Hill custom mill, Cuba City, Grant County, Wisconsin	20	10		

TABLE V-4 U.S. ZINC AND CADMIUM PRODUCERS

Company and Location[a] (Dates of Operation)	Zinc Production Process	Zinc Capacity[63] conc. tons/day	Zinc Capacity thousands of short tons/yr	Raw Materials of Zinc Production	Materials for Cadmium Production	Cadmium Production Process	Annual Cadmium Production, Short tons[36]	Thallium Content of Cadmium Shipped, ppm	Fate of Thallium
AMAX Zinc Co., Inc. Blackwell, OK (1907-1974?)	Horizontal Retort	275		Galena, KS, Pb smelter Cottrell dust 20 ppm Tl[12]	Zn sintering dusts[37]	Cd refinery now closed. See text.	~ 600 (1966) Avg. 1962-1970 ~ 640	5	TlCl recycled (about 38 lb Tl/yr) 6 lb Tl in the Cd
Sauget (E. St. Louis), IL (1941-present)	Electrolytic See Reference T702		84	Until ~ 1972, American Zinc of Illinois operated a custom smelter; concs. varied, largely from S. America & Canada. Since 1973, smelter uses concs. mostly from MO, some from Newfoundland and TN. Concs. < 10 ppm Tl[130]	Cd cemented from $ZnSO_4$ solution	Electrolytic? "Tl would follow Cd." Facilities for Tl removal not needed presently.[130]	Avg. ~ 300 1962-1971	Sponge Cd? ~ 50	None isolated. In 1970, only sponge Cd was marketed.[45] ~ 30 lb/yr in the Cd?
ASARCO Inc. Amarillo, TX (1923-May 1975)	Horizontal Retort	135							
Corpus Christi, TX (1941-present)	Electrolytic	300							
Denver, CO (< 1929-present)	-	-		-	Pb dust with high As content. Flue dusts from own operations. Used to use flue dusts imported from ASARCO's Mexican partner. Also used flue dusts from Anaconda et al. Corpus Christi sent	Electrolytic for Tl recovery. See Figure VII-1. A method used < 1962 involved distillation of briquetted sponge, not electrolysis. No outlet for Tl. (The Midvale, UT, Pb dusts circulated until 15-20% Cd	Figures not published.	1-2	Tl-contg. raw materials and thallium products are stockpiled and/or sold.

TABLE V-4 (continued)

Company and Location[a] (Dates of Operation)	Zinc Production Process	Zinc Capacity: conc. tons/day[63]	Zinc Capacity: thousands of short tons/yr	Raw Materials of Zinc Production	Materials for Cadmium Production	Cadmium Production Process	Annual Cadmium Production, Short tons[36]	Thallium Content of Cadmium Shipped, ppm	Fate of Thallium
ASARCO Inc. Denver, CO (< 1929-present) (continued)					its Cd-contg. dusts [roasting?] to Denver.[131] [~ 20 tons Tl/yr in 1947?]	Cd recovery by electrolysis. No Tl recovery.)[22]			Stockpiled or sold.
Stephensport, KY (building[132])	Electrolytic		180						
National Zinc Co. Bartlesville, OK (1907-August 1976)	Horizontal Retort	140	56				Figures not published.		
(Dec. 1976-present)	Electrolytic	About same[44]	62[133]	Zn concs. contg. ≤ 10 ppm Tl[44]		Electrolytic. Tl must be removed since it causes plating problems.[44]		Probably < 10	Refining residues go to Cu or Pb smelter.
The Bunker Hill Co. Kellogg, ID (1928-present)	Electrolytic	300	104	Underground mines.	Electrolytic Zn plant Cu-Cd residue, 20-28% Cd.[22]	Electrolytic since 1945. No route for Tl removal in < 1962 description.[22] Cd production since 1945.	Avg. 1962-1971 ~ 540	Probably < 30 ppm	? Calspan[46] found some of the highest environmental levels of thallium at Kellogg.

TABLE V-4 (continued)

Company and Location[a] (Dates of Operation)	Zinc Production Process	Zinc Capacity: conc. tons/day[63]	Zinc Capacity: thousands of short tons/yr	Raw Materials of Zinc Production	Materials for Cadmium Production	Cadmium Production Process	Annual Cadmium Production, Short tons[36]	Thallium Content of Cadmium Shipped, ppm	Fate of Thallium
The Bunker Hill Co. Kellogg, ID (1928-present) (continued)					Pb sintering and smelting dusts recycled from 0.05% Cd in the charge until 3-4% Cd in the dusts.[22]	Electromotive. Dusts & siliceous flux furnaced twice. Second sublimate 44.5% Cd. Treated with $NaClO_3$ and $KMnO_4$. Removes As, Fe, & Tl. Sponge distilled at 1150°C. Refined Cd melted under caustic and then NH_4Cl.[22]		30	Tl removed with Fe & As and as TlCl flux. Tl in distn. residue.
The New Jersey Zinc Co. Palmerton, PA (1899-present)	Vertical Retort	315		Mostly from company mines (N.J., CO, PA, VA, TN) 460 lb Tl/yr at ~ 2 ppm	Pb smelter flue dusts ?	Cd solution 1) - $PbSO_4$ → 2) $KMnO_4$, aeration. As, Fe oxides. probably $Tl(OH)_3$↓[43,134]	Figures not published		Iron press residue taken out of system. (Recycled to Pb smelter?[32]) Maybe as dilute in residues as in the concentrates. Skimmings recycled.[134]
Clarksville, TN (planned[132])	Electrolytic		90			Electrolytic ?			(see above)

Company and Location[a] (Dates of Operation)	Zinc Production Process	Zinc Capacity: conc. tons/day[63]	Zinc Capacity: thousands of short tons/yr	Raw Materials of Zinc Production	Materials for Cadmium Production	Cadmium Production Process	Annual Cadmium Production, Short tons[36]	Thallium Content of Cadmium Shipped, ppm	Fate of Thallium
St. Joe Minerals Corp. Monaca, PA (ZnO 1930-present) (Zn 1938-present)	Electrothermic $Zn:ZnO$ 6:1 See Reference T703	640	245	1930-1940 Smelter used Zn concs. from own NY mines. By 1970, < 50% concs. from own mines. Other domestic & foreign concs. ~ 25% of the feed is other than concs. 230 lb Tl would be only ~ 0.5 ppm in the concs. If ~ 2 ppm assumed in the concs, ~ 75% is lost during Zn roasting and smelting.	Sintering fume. > 230 lb Tl content in 1971. ~ 20 ppm in the heat-treated fume	Electromotive See Figure VII-2	516 in 1970 437 in 1971 Avg. 1962-1971 ~ 400	3-10 ppm	See Figure VII-2
Anaconda Great Falls, MT (closed August 1972)	Electrolytic				Cu-Cd residues from Electrolytic Zn production.[22]	Electrolytic. At least in early (1938) methods, Tl removed from dissolved Cd sponge by $KMnO_4$.[22]	~ 580 1962-1971	?	Cu residues to Anaconda Copper Works

a/ The locations of several other zinc plants still producing zinc in 1968 are given in Reference 129, p. 949.

TABLE V-5

ESTIMATED AMOUNTS OF THALLIUM IN U.S. PRIMARY ZINC AND CADMIUM PROCESSING, 1974

Material	Amount, 10^6 Short Tons	Estimated Thallium Concentration (ppm)	Thallium Content, Short Tons
Domestic Zinc and Lead-Zinc Ores	20[135/]	2	40
Waste Rock, Rock Products, Concentrator Tailings	~ 18	1.3	~ 23
Lead Concentrates	1.3[135/]	~ 5 ?	6-7
Domestic Zinc Concentrates	1.0[135/]		
For Slab Zinc Production	0.7	10	7
For ZnO and Other Chemicals	0.3	10-30[59/] in the ZnO	~ 3
Foreign Zinc Concentrates	0.4	10	4
Total Zinc Concentrates For Slab Zinc (minus roasting loss)	1.1	10	11
Processed by Distillation After Roasting	0.55		5.25
Processed by Electrolysis After Roasting	0.55		5.25
Roasting Dust Emission	0.0125[20/]	40[a/]	0.5

TABLE V-5 (continued)

Material	Amount 10^6 Short Tons	Estimated Thallium Concentration (ppm)	Thallium Content, Short Tons
Roasted Zinc Concentrates Plus Recycled Roasting Dust	1.1	10[b]/	10.5
Sintering Dust Emission	0.0045[20]/	20-40[c]/	0.1
"Zinc Smelter Flue Dust" Processed for Cadmium	0.013[d]/	10[e]/-? (avg. 150?)	2
[Imported Flue Dust Processed for Cadmium by ASARCO Inc.[131]/(not current)]	[0.004]	[1,000]	[4]
Distilled Slab Zinc	~ 0.25	< 13	3.2
Zinc Smelter and Zinc Oxide Residue	?	10-40[20]/	?
Zinc Smelter Slag	?	10[f]/	?
Cadmium From Flue Dusts			0.04
Copper-Cadmium Cake to Cadmium Electrolytic Refinery (most recycled to zinc electrolytic plant)			3-4

TABLE V-5 (continued)

Material	Amount 10^6 Short Tons	Estimated Thallium Concentration (ppm)	Thallium Content, Short Tons
Electrolytic Slab Zinc	0.25	8	2.1
Electrolytic Zinc and Cadmium Plant Solid Wastes			2.6
Cadmium From Copper-Cadmium Cake			0.05
Electrolytic Waste Solutions			0.5

a/ East St. Louis, Illinois. 59/

b/ Spectrographic analyses of Tri-State, Mexican, Newfoundland calcines 10-30 ppm. 59/

c/ 30-44 ppm, Blackwell, Oklahoma; Dumas, Texas, East St. Louis. 59/

d/ Approximately 24 tons flue dust/1,000 tons Zn conc. 136/

e/ Concs. from Tri-State, TN, WI, WA, B.C., MO

f/ Tri-State District and Arkansas. 59/

TABLE V-6

THALLIUM REMOVAL FROM STREAMS AT ELECTROMOTIVE CADMIUM PLANTS[33/]

	Plant G	Plant H	Plant L	Plant M	Plant N	St. Joe Minerals
Leach solution from dust						
Tl content, g/ℓ	0.01	0.008	0.04	0.1-0.2	0.01 (water leach)	0.01
Tl/Cd x 10^6	125-250	267-533	1,000	~ 6,000	250	250
Purified leach solution						
Tl content g/ℓ	0.009	0.0008	0.01	< 0.01	0.008	0.007
Tl/Cd x 10^6	113-225	27-53	250	≲ 90	130	175
Percent Tl removal at this stage	~ 10	~ 90	75	~ 98.5	52	30
Electromotive cadmium						
Tl content, ppm	5	< 5-10	< 30	40	< 10	10
Percent Tl removal from the cadmium at this stage (based on original amount in leach solution)	~ 97	~ 98	> 97	~ 99.3	> 96	96

TABLE V-7

THALLIUM IN COPPER BEARING MATERIALS[59/]

Material	Thallium Content, ppm
Ely, NV (~ Kennecott ore ?) rocks	~ 10
Utah Copper Co. ore	2
Utah Copper Co. conc.	23
Arizona concs.	0-10
Arizona tailings	0
Eastern Cu concs.	0-10
Cu concs., ID, MI, MT,* NM	0
Cu concs., Chelan, WA	30
Arizona slags	0
Eastern copper slags	0
Idaho slags	0
Reverberatory furnace stack dust, McGill, NV	50
Reverberatory furnace flue dust, Eastern copper	40
Converter flue dust, Eastern copper	0
Converter flue dust, Garfield, UT	1700 } Utah Copper concs.
Converter flue dust, McGill, NV	1500 } Utah Copper concs.
Acid plant sludge, Eastern copper (from roasting & converting gases)	50

* The arsenic roaster residue worked up from the Anaconda copper flue dust(0 ppm Tl) had

TABLE V-8

THALLIUM IN UTAH COPPER COMPANY ORE-PROCESSING MATERIALS[a/]

Material	10^6 Short Tons	Thallium Concentration, ppm	Thallium Content, tons
Ore	40	2	80
Concentrate	0.8	23	18
Reverberatory dust	0.18	50 (11)	9 (calc. 2 at 12% loss)
Converter dust	0.54	1,500 (at some stage)	810 ?

a/ The thallium concentrations are based on 1940's Minerals Yearbook data. The amount of ore is based on current Utah Copper Division copper production data.

TABLE V-9

ESTIMATION OF THALLIUM IN COPPER PROCESSING IN 1972[a]

	Amount, 10^6 Short Tons	Thallium Concentration, ppm	Thallium Content Short Tons
U.S. copper ores	267	2	530
U.S. copper concentrate	5.6[b]	20	110
Tailings and other concentrates	245	1.7	420
Foreign and domestic concentrates smelted in the United States	6.2[b]	20	120
Slags and dusts recycled or dumped			(5.7)(12)=68
Atmospheric losses	0.067[89]	570[c]	(3.2)(12)=38
Blister copper			(1.1)(12)=13
Copper and silver refining wastes		?	< 13
Refined copper		< 7	< 13
Refined silver		< 10	< 0.012

a/ Based on copper statistics from Minerals Yearbook 1972.[137]

b/ Assumed 30% copper in concentrates; based on refinery, not mine production.

c/ This assumed the loss was in the form of particulates not gases. The discrepancy between this value and the values on order of magnitude less in most copper smelter ducts (Table V-3) may be explained by loss of thallium as gas.

TABLE V-10

WATER POLLUTION BY THALLIUM NEAR SULFIDE MINERALS PROCESSORS[46]

Company	Type of Plant	Location	Thallium Concentration, ppm[a]		
			Upstream	Wastewater, Runoff	Downstream
ASARCO Inc.	Copper Smelter	Tacoma, Wash.	<0.01 <0.001	<0.01-2.4 0.00015-0.002	-
The Bunker Hill Co.	Lead, Zinc, Copper Processing	Kellogg, Idaho	<0.0001	0.0014-0.11 (< 15 in mud)	0.00018-0.0006
ASARCO Inc.	Lead and Zinc Smelters	East Helena, Mont.		(8.6 in mud)	
Anaconda Co.	Copper Smelter and Mill	Butte-Anaconda, Mont.	N.D.[c] <0.00025	0.2 (10 in mud) <0.0001-0.00055	<0.01 <0.01
Kennecott Copper Corp.	Copper Mine and Smelter	Ray-Hayden, Ariz.	<0.003 (below mine)	<0.003-0.005	<0.003
Magma Copper Co.	Copper Mine and Mill	Superior, Ariz.		0.003	
Magma Copper Co.	Copper Mine, Mill, and Smelter	San Manuel, Ariz.		<0.003	
Phelps Dodge Corp.	Copper Smelter	Morenci, Ariz.		-	
Inspiration Consolidated Copper Co.	Copper Smelter	Globe-Miami, Ariz.		0.005 (recycled)	

TABLE V-10 (continued)

Company	Type of Plant	Location	Thallium Concentration, ppm[a/]		
			Upstream	Wastewater, Runoff	Downstream
United States Smelting and Refining Co.	Copper Pit and Mill	Grant County, N.M.		<0.003 (recycled)	<0.003
Kennecott Copper Corp.	Santa Rita Pit (copper)	Grant County, N.M.		0.005	0.0037
Kennecott Copper Corp.	Hurley Copper Mill and Smelter	Hurley, N.M.		<0.003	
Kennecott Copper Corp.	Planet Mine (closed)	Near Parker, Ariz.	0.0021[d/]	(5.1 in sediment)	
American Zinc Co., Inc. (now an AMAX Zinc Co., Inc., plant)	Zinc Smelter and Electrolytic Refinery	East St. Louis, (Sauget), Ill.		<0.09-0.28	
Anaconda Co.	Copper Refinery	Perth Amboy, N.J.		(20-26 in sludge dumps)	
Copper Pigments and Chemical Works, Inc.	Copper Pigments Manufacturer	Sewaren, N.J.		(17 in sludge)	
ASARCO Inc.	Copper Smelter	El Paso, Texas		(0.3 in sediment)	
St. Joe Minerals Corp.	Lead Smelter	Herculaneum, Mo.		N.D.?	

Company	Type of Plant	Location	Thallium Concentration, ppm[a]		
			Upstream	Wastewater, Runoff	Downstream
Homestake Mining Co. AMAX Lead Co.	Lead Smelter	Near Bixby, Mo.		<1 ppm	
ASARCO Inc.	Lead Smelter	Glover, Mo.		0.003	
Copper Range Co.	Copper Mine and Smelter	White Pine, Mich.		(N.D. even in recycled flue dusts)	
Cities Service	Copperhill Copper Smelter and Acid Plant	Copper Basin Region, Tenn.		(N.D. in the ore and slags but $<$ 1.4 in mill tailings, pond sediments, etc.)	

a/ Analysis by extraction and atomic absorption spectroscopy by Calspan Corporation.

b/ The smelter handles concentrates from the Phillippines (mainly enargite Cu AsS), which is high in gold, silver, and pyrite. Arsenical materials from other ASARCO Inc., plants are also smelted here.

c/ ND = not detected.

d/ Highest natural background level fround in the study.

TABLE V-11

THALLIUM IN U.S. COALS 138/

	Coal Source		
	Eastern Province	Interior Province	Western States
Frequency of detection, %	43	49	9
Number of samples	600	123	104
Average ash content of dry coal, %	9.3	10.5	9.8
Content in ash when detected, ppm	19	8	5
Average thallium content of all ashes, where zero used for contents below the detection limit (5 ppm), ppm	8	4	0.5
Average thallium content of dry coal, ppm	0.73	0.38	0.05

TABLE V-12

MATERIAL FLOW ANALYSIS WESTERN COAL-BURNING PLANTS[90,91]

	Station I (sub-bituminous)		Station II (sub-bituminous)		Station III (lignite)	
	Tl, lb/hr	Tl, ppm	Tl, lb/hr	Tl, ppm	Tl, lb/hr	Tl, ppm
Coal	< 0.028	< 0.1	< 0.028	< 0.1	< 0.028	
Bottom ash sluice water	< 0.001				< 0.0003	
Cooling tower blowdown	< 0.0004					
Make-up water	< 0.004		< 0.0005			
Lime	< 0.0004	< 0.1				
Total in	< 0.034		< 0.028		< 0.028	
Bottom ash	0.002	0.19			< 0.0017	
Precipitator ash			0.006	0.57 0.13		
Bottom ash sluice water	< 0.001				0.0005	
Sluice ash			< 0.0004	< 0.10		
Scrubber solids	0.013	0.25				
Sluice ash filtrate			< 0.0005			
Scrubber liquds	< 0.004					
Wet electrostatic precipitators			< 0.0004	< 0.0007		
Economizer ash	0.000008	avg. 0.34			< 0.00001	
Economizer ash sluice water					< 0.00008	
Cyclone ash					0.0050	
Flue gas	0.006	0.012				
Flue gas south duct	(53 lb/yr)	<0.0016			0.075 (657 lb/yr)	
Flue gas north duct					< 0.0021	
Total out	0.026		0.007		0.085	

TABLE V-13

COAL-BURNING PLANTS IN THE U.S.[139]

EPA Region	Number of Plants	10^6 Tons Coal and Lights
I	23	1.5
II	33	14.5
III	111	114.7
IV δ	39	125.0
V	146	206.5
VI	20	7.8
VII	5	25.1
VIII + X	2 + 7	15.1
IX	29	6.3
Total	415	516.5

BIBLIOGRAPHY. SECTION V.

1. Mellor, J. W., A Comprehensive Treatise on Inorganic and Theoretical Chemistry, Vol. 5, Longmans, Green and Company, New York, N.Y., 1946, pp. 406ff.

2. Zelikman, A. N., O. E. Krein, and G. V. Samsonov, Metallurgy of Rare Metals, Israel Program for Scientific Translations, Jerusalem, 1966 (original 2nd ed. publ. 1964).

3. Ottinger, R. S., J. L. Blumenthal, D. F. Dal Porto, G. I. Gruber, M. J. Santy, and C. C. Shih, Recommended Methods of Reduction, Neutralization, Recovery, or Disposal of Hazardous Waste. Vol. XIII. Industrial and Municipal Disposal Candidate Waste Stream Constituent Profile Reports. Inorganic Compounds (Continued), TRW Systems Group for the U.S. Environmental Protection Agency, PB-224 592, National Technical Information Service, U.S. Department of Commerce, Springfield, Va., 1973.

4. Waggaman, W. H., G. G. Heffner, and E. A. Gee, Thallium. Properties, Sources, Recovery, and Uses of the Element and Its Compounds, Bureau of Mines Information Circular 7553, U.S. Department of the Interior, Washington, D.C., 1950.

5. Robinson, K., "Thallium" in United States Mineral Resources, D. A. Brobst and W. P. Pratt, Eds., U.S. Geological Survey Professional Paper 820,631, U.S. Government Printing Office, Washington, D.C., 1973.

6. Getskin, L. S., A. G. Batyuk, P. P. Tsyb, V. P. Savraev, R. J. Gorokhvodatskaya, and V. P. Zinov'ev, "Treatment of Lead Smelter Dust by Sulfation," Sbornik Nauch. Trudov, Vsesoyuz. Nauch.-Issledovatel Gorno-Met. Inst. Tsvetnykh Metal., 3, 44-68 (1958); Chem. Abstr., 54, 18236h (1960).

7. Ponomareva, E. I., P. P. Tsyb, E. L. Shalavina, A. G. Batyuk, and Yu. N. Menzhulin, "Extraction of Nonferrous and Rare Metals from Melting Furnace Dusts in the Chimkent Lead Works," Tr. Inst. Met. i Obogashchen Akad. Nauk Kazakh. S.S.R., 1, 76-87 (1959); Chem. Abstr., 54, 15152e (1960).

8. Okonishnikov, A. N., P. P. Tsyb, and G. I. Kuzebnaya, "Recovery of Thallium from the Agglomeration Dusts of Lead Plants," Rudnyi Altai, Sovet. Narod. Khoz. Vostoch. Kazakhstan. Ekon. Admin. Raiona, 1, 29-33 (1958); Chem. Abstr., 54, 24220i (1960).

9. Kozlovskii, M. T., M. V. Nosek, S. P. Bukhman, P. I. Zabotin, and V. M. Ilyushchenko, "Separation of Thallium from Agglomerated Dust of the Chimkent Lead Plant by Water Treatment," Tr. Inst. Khim. Akad. Nauk Kazakh S.S.R., 3, 5-15 (1958); Chem. Abstr., 53, 18806d (1959).

10. Ponomarev, V. D., E. I. Stolyarova, Yu. A. Koz'min, L. V. Favorskaya, and E. I. Shalavina, "Alkaline Extraction of Flue Dusts from Lead Plants," Izvest. Akad. Nauk Kazakh. S.S.R., Ser. Gornogo Dela, Met., Stroitel. i Stroimaterial, No. 4, 3-17 (1957); Chem. Abstr., 52, 1978 (1958).

11. Prater, J. D., D. Schlain, and S. F. Ravitz, Recovery of Thallium from Smelter Products, Bureau of Mines Report of Investigations 4900, U.S. Department of the Interior, Washington, D.C., 1952, 9 pp.

12. Zimmerley, S. R., Thallium, Salt Lake City Division, Metallurgical Bran U.S. Bureau of Mines, Salt Lake City, Utah, 1947.

13. Staff, Division of Nonferrous Metals, Bureau of Mines, "Minor Metals" i Minerals Yearbook 1972, Vol. I, Metals, Minerals, and Fuels, Bureau of Mines, U.S. Department of the Interior, Washington, D.C., 1974, pp. 1347-1358.

14. Kogan, B. I., "O Tallii" ["Thallium"], Redk. Elem., No. 5, 27-48 (1970) Chem. Abstr., 76, 153 (1972).

15. Keston, S. N., Assistant Vice-President, Environmental Affairs, ASARCO Inc., New York, N.Y., personal communication (telephone), January 19

16. Sessions, H. K., and S. Goren, "Report of Investigation of Health Hazards in Connection with the Industrial Handling of Thallium," U.S. Naval Med. Bull., 47, 545-550 (1947).

17. Chandler, J., National Institute of Occupational Safety and Health, personal communication (telephone), March 1977.

18. Phillips, A. J., "The World's Most Complex Metallurgy (Copper, Lead and Zinc)," Trans. Met. Soc. AIME, 224, 657-668 (1962).

19. Hallowell, J. B., J. F. Shea, G. R. Smithson, Jr., A. B. Tripler, and B. W. Gonser, Water-Pollution Control in the Primary Nonferrous-Metal Industry. Vol. I. Copper, Zinc, and Lead Industries, Battelle Memorial Institute, Prepared for Office of Research and Monitoring, U.S. Environmental Protection Agency, U.S. Government Printing Office, Washington, D.C., 1973.

20. Sargent, D. H., and J. R. Metz (Versar, Inc.), Technical and Microeconomic Analysis of Cadmium and Its Compounds, Final Report--Task 1, EPA 560/3-75-005, Office of Toxic Substances, U.S. Environmental Protection Agency, Washington, D.C., 1975.

21. Lee, A. G., The Chemistry of Thallium, Elsevier Publishing Company, New York, N.Y., 1971.

22. Chizhikov, D. M., Cadmium, [translation by D. E. Hayler of Kadmii, Izv. Akad. Nauk S.S.S.R., Moscow, 1962], Pergamon Press, New York, N.Y., 1966.

23. Howe, H. E., "Cadmium and Cadmium Alloys" in Kirk-Othmer Encyclopedia of Chemical Technology, Vol. 3, 2nd ed., A. Standen, executive editor, Interscience Publishers, a division of John Wiley and Sons, New York, 1964, pp. 884-899.

24. Naimark, E. L., "Issledovanie Protsessa Vozgonki Primesei iz Cyr'ya i Poluproduktov Svintsogo Proizvodstva pri Pomoshchi Spektral'nogo Analiza" ["Study of the Volatilization Process of Ingredients from Raw Material and Intermediates of a Lead Smelting Plant with the Aid of Spectral Analysis"], Tr. Inst. Met. i Obogashch. Akad. Nauk Kaz. S.S.R., **2**, 47-57 (1960); Chem. Abstr., **54**, 20737b (1960).

25. Salsbury, M. H. W. H. Kerns, F. B. Fulkerson, and G. C. Branner, Marketing Ores and Concentrates of Gold, Silver, Copper, Lead, and Zinc in the United States, Bureau of Mines Information Circular 8206, Bureau of Mines, U.S. Government Printing Office, Washington, D.C., 1964.

26. Schack, C. H., and B. H. Clemmons, Review and Evaluation of Silver-Production Techniques, U.S. Bureau of Mines Information Circular 8266, U.S. Government Printing Office, Washington, D.C., 1965.

27. Leitsin, V. A., "Povedenie Talliya v Protsessakh Gidrometallurgicheskogo Proizvodstva Tsinka" ["Thallium in Hydrometallurgical Zinc Production"], Sb. Nauch.-Tekh. Trudov Nauch.-Issled. Inst. Met. Chelyabinsk. Sovnarkhoza, No. 1, 120-129 (1960).

28. Roditi, E., Ed., Metal Statistics 1976, American Metal Market, Fairchild Publications, Inc., New York, N.Y., 1976.

29. McMahon, A. D., J. M. Hague, and H. R. Babitzke, "Zinc" in Minerals Yearbook 1972, Vol. I, Metals, Minerals, and Fuels, Bureau of Mines, U.S. Department of the Interior, Washington, D.C., 1974, pp. 1299-1333.

30. ASTM, 1974 Annual Book of ASTM Standards, Part 12. Chemical Analysis of Metals; Sampling and Analysis of Metal Bearing Ores, American Society for Testing and Materials, Philadelphia, Pa., 1974.

31. Heindl, R. A., "Cadmium" in Minerals Facts and·Problems, Bureau of Mines Bulletin 650, U.S. Department of the Interior, U.S. Government Printing Office, Washington, D.C., 1970, pp. 515-526.

32. Fulkerson, W., and H. E. Goeller, Eds., Cadmium. The Dissipated Element ORNL NSF-EP-21, Oak Ridge National Laboratory, Oak Ridge, Tennessee, National Science Foundation Environmental Program (RANN), 1973.

33. Lund, R. E., and R. E. Sheppard, "Cadmium Purification Practice in Zinc Smelting," J. Metals, 9(10?), 724-730 (1964).

34. Stickney, W. J., "Cadmium Extraction from the Ores of the Hudson Bay Mining and Smelting Co.," Canadian Institute of Mining and Metallurgy, Transactions, 69, 334-338 (1966).

35. Anonymous, "Minor Metals. Thallium," Mineral Industry 1929, 700 (1930).

36. Year Book of the American Bureau of Metal Statistics, 51st Annual Issue for the Year 1971, American Bureau of Metal Statistics, New York, N.Y. 1972.

37. Nauert, R. L., "Cadmium Preparation at Blackwell," J. Metals, 1, 15-17 (1966).

38. Gage, W. L., Smelter Superintendent, Phelps Dodge Corp., Douglas, Arizona personal communication (telephone), January 1977.

39. Sheppard, R. E., and A. O. Martel, "Cadmium Extraction from Zinc Sinter Plant Fume. St. Joe Process" in International Symposium on Hydrometallurgy, Chicago, Ill., February 25 - March 1, 1973, D. J. I. Evans and R. S. Shoemaker, Eds., The American Institute of Mining, Metallurgical, and Petroleum Engineers, New York, N.Y., 1973, pp. 859-867.

40. Lund, R. E., J. F. Winters, B. E. Hoffacker, T. M. Fusco, and D. E. Warnes, "Josephtown Electrothermic Zinc Smelter of St. Joe Minerals Corporation" in A.I.M.E. World Symposium on Mining and Metallurgy of Lead and Zinc, Vol. 2, C. H. Cotterill and J. M. Cigan, Eds., Am. Inst. of Mining, Metallurgical, and Petroleum Engineers, New York, N.Y., 1970, pp. 549-580.

41. Gaprindashvili, V. N., N. D. Kalandadze, and R. P. Gogorishvili, "Sulfuric Acid Treatment of Base-Metal Sulfide Ores," Issled. po Khim. Pererabotke Rud, Akad. Nauk Gruz. S.S.R., Inst. Neorgan. Khim. i Elektrokhim., 5-12 (1966); Chem. Abstr., 65, 11847f (1966).

42. Mimura, H., "Fundamental Researches for Construction of New Cd Plant. I," Nippon Kogyo Kaishi, 78, 603-608 (1962); Chem. Abstr., 58, 3162c (1963).

43. Fugleberg, S., "Principles of Cadmium Production and Their Application at the Outokumpu Zinc Plant in Kokkola" in International Symposium on Hydrometallurgy, Chicago, Ill., February 25 - March 1, 1973, D. J. I. Evans and R. S. Shoemaker, Eds., The American Institute of Mining, Metallurgical, and Petroleum Engineers, New York, N.Y., 1973, pp. 1145-1167.

44. Moore, T. I., Consultant, National Zinc Company, Bartlesville, Oklahoma, personal communication (telephone), January 1977.

45. Banes, O. H., R. K. Carpenter, and C. E. Paden, "Electrolytic Zinc Plant of American Zinc Company" in A.I.M.E. World Symposium on Mining and Metallurgy of Lead and Zinc, Vol. 2, C. H. Cotterill and J. M. Cigan, Eds., The American Institute of Mining, Metallurgical, and Petroleum Engineers, New York, N.Y., 1970, pp. 308-328.

46. Calspan Corporation, Heavy Metal Pollution from Spillage at Ore Smelters and Mills, preliminary copy of the final report version provided in March 1977, Calspan Corporation, Buffalo, N.Y., for the U.S. Environmental Protection Agency, Cincinnati, Ohio (in press).

47. Fleischer, M., U.S. Geological Survey, Reston, Virginia, personal communication (telephone), January 1977.

48. Hirschel, W. N., "Cadmium, Thallium, Indium, and Gallium as By-Products of the Lithopone Industry," Chem. and Ind., 52, 797-798 (1933).

49. Downie, C. C., "Metallurgical Fume and Flue Dust," Mining Mag., 86, 80-83 (1952).

50. Shreve, R. N., Chemical Process Industries, 3rd ed., McGraw-Hill Book Company, New York, N.Y., 1967, pp. 437-438.

51. Magorian, T. R., Ecology and Environment, Inc., Buffalo, New York, personal communication (telephone), January 1977.

52. Margorian, T. R., K. G. Wood, J. G. Michalovic, S. L. Pek, and M. W. Van Lier, Water Pollution by Thallium and Related Metals, Calspan Corporation for the U.S. Environmental Protection Agency, Cincinnati, Ohio, 1974. [This is the draft version of the final report issued under another title in 1977 and was obtained from the National Technical Information Service as PB-253 333.]

53. Savichev, E. I., "Fazovyi Analiz Produktov Gidrometallurgicheskogo Proizvodstva na Soedineniya Talliya" ["Phase Analysis of the Products in the Hydrometallurgical Production of Thallium Compounds"], Spektral'n. i Khim. Materialov, Sb. Metodik, 172-175 (1964).

54. Grigorovich, A. N., F. L. Shalavina, N. A. Milyutina, E. G. Svirchevskaya and T. D. Gorina, "Kompleksnoe Izvlechenie Kadmiya, Indiya, Talliya, i Tsinka iz Pylei Svintsovoi Plavki" ["Complex Extraction of Cadmium, Indium, Thallium, and Zinc from the Dust of Lead Smelting"], Tr. Inst. Met. i Obogashch., Akad. Nauk Kazakh. S.S.R., 1, 65-75 (1959); Chem. Abstr., 54, 11899 (1960).

55. Paone, J., "Lead" in Minerals Facts and Problems, Bureau of Mines Bulletin 650, U.S. Department of the Interior, U.S. Government Printing Office, Washington, D.C., 1970, pp. 603-620.

56. Polyvyannyi, I. R., V. P. Ovcharenko, and I. I. Elyakov, "Distribution of Metals Between Products of Shaft-Furnace Lead Smelting with Preheating and Oxygen Enriching of the Blast," Tr. Inst. Met. Obogashch., Akad. Nauk Kaz. S.S.R., 34, 71-76 (1968); Chem. Abstr., 70, 22030g (1969).

57. Parker, J. G., U.S. Bureau of Mines, Denver, Colorado, personal communication (telephone), November 1976.

58. Schack, C. H., and B. H. Clemmons, "Chapter 4. Extractive Processes," in Silver. Economics, Metallurgy, and Use, A. Butts and C. D. Coxe, Eds., D. Van Nostrand Co., Inc., Princeton, N.J., 1967, pp. 57-77.

59. Kaiser, E. P., B. F. Herring, and J. C. Rabbitt, Minor Elements in Some Rocks, Ores, and Mill and Smelter Products, Trace Elements Investigations Report TEI-415, U.S. Atomic Energy Commission, Oak Ridge, Tenn., 1954.

60. Deev, V. I., M. I. Kochnev, A. I. Okunev, and B. I. Sergin, "Distribution of Cadmium, Rare, and Uncommon Elements Among Products of Oxygen-Flash Smelting Copper-Zinc Concentrates," Tsvetnye Metally, 33(4), 19-24 (1960); Soviet J. Non-Ferrous Metals, 33(4), 23-29 (1960); Chem. Abstr., 55, 1337c (1961).

61. Ryabov, Y. F., A. I. Okunev, L. D. Kirr, and V. A. Oshman, "Raspredeleni Nekotorykh Redkikh i Rasseyannykh Elementov v Produktakh Pererabotki Mednykh Rud i Kontsentratov" ["Distribution of Certain Rare and Disseminated Elements in the Products Obtained from Processing of Copper Ore and Concentrates"], Byull. Tsvetnoi Met., No. 22, 24-27 (1957).

62. Deev, V. I., A. I. Okunev, M. I. Kochnev, S. A. Vermenichev, and B. I. Sergin, "Povedenie Redkikh i Rasseyannykh Elementov pri Plavke Sul'fidnykh Kontsentratov na Kislorode" ["Behavior of Rare and Trace Elements During the Smelting of Sulfide Concentrates with Oxygen"], Tr. Inst. Met., Akad. Nauk S.S.R., Ural'sk. Filial, No. 8, 43-50 (1963); Chem. Abstr., 60, 10272h (1964).

63. Environmental Protection Agency Office of Air and Waste Management, Office of Air Quality Planning and Standards, Background Information for New Source Performance Standards: Primary Copper, Zinc, and Lead Smelters. Vol. 1: Proposed Standards, EPA-450/2-74-002a, National Technical Information Service, U.S. Department of Commerce, Springfield, Va., 1974.

64. Ageton, R. W., and G. N. Greenspoon, "Copper" in Minerals Facts and Problems, Bureau of Mines Bulletin 650, U.S. Department of the Interior, U.S. Government Printing Office, Washington, D.C., 1970, pp. 535-554.

65. McKee, Arthur G., and Company, "Va. Copper, Zinc and Lead Smelting Practice," Systems Study for Control of Emissions. Primary Nonferrous Smelting Industry, Vol. I, National Air Pollution Control Administration, Public Health Service, U.S. Department of Health, Education, and Welfare, 1969, pp. Va-1 to Vb-5.

66. Saito, A., and F. Kaneko, "Flotation of the Anode Slime from the Electrorefining of Copper," Japan. Kokai 74 123, 427, November 26, 1974, 4 pp.; Chem. Abstr., 83, 208639m (1975).

67. Zitko, V., W. V. Carson, and W. G. Carson, "Thallium Occurrence in the Environment and Toxicity to Fish," Bull. Environ. Contam. Toxicol., 13, 23-30 (1975).

68. Angino, E. E., Consultant in Geochemistry, Department of Geology, University of Kansas, personal communications, 1976, 1977.

69. Ridinger, D. C., Director of Environmental Affairs, Magma Copper Company, personal communication (letter), January 1977.

70. Wilkinson, R. R., and G. R. Cooper, Study on Chemical Substances from Information Concerning the Manufacture, Distribution, Use, Disposal, Alternatives, and Magnitude of Exposure to the Environment and Man, Task III - The Manufacture and Use of Selected Inorganic Cyanides, Midwest Research Institute for the U.S. Environmental Protection Agency Office of Toxic Substances, Washington, D.C., 1976.

71. Rozlovskii, A. A., A. A. Til'ga, and A. I. Ustinov, "Oxidizing Roasting in Fluidized Bed of Antimony and Mercury-Antimony Flotation-Concentrate," Tsvetn. Metal., 36(12), 34-37 (1963); Chem. Abstr., 61, 334g (1964).

72. Glotko, E. D., E. I. Savichev, V. V. Malakhov, and E. I. Golovin, "Treating Mercury Residue," U.S.S.R. Patent 123,700, November 9, 1959; Chem. Abstr., 54, 8570a (1960).

73. Glotko, E. D., E. I. Golovin, V. V. Malakhov, and E. I. Savichev, "Processing of Roasted Stupp [Intermediate Hg Product]," Tr. Altaisk. Gorno-Met. Nauchn.-Issled. Inst., Akad. Nauk Kas. S.S.R., 11, 164-167 (1961); Chem. Abstr., 57, 4395i (1962).

74. Solozhenkin, P. M., L. P. Orlova, and R. M. Davydovskaya, "Behavior of Thallium During Concentration of Arsenic and Mercury-Antimony Ores," Uch. Zap. Tsentr. Nauchn.-Issled. Inst. Olovyan. Prom., 3, 37-43 (1964); Chem. Abstr., 64, 7735d (1966).

75. Sosnovskii, G. N., Z. A. Auezov, and A. A. Burba, "Search for a Process for Treating an Arsenic-Thallium Filter Cake," Tr. Irkutsk. Politekh. Inst., 47, 12-19 (1969); Chem. Abstr., 74, 129238r and 75, 131989a (1971).

76. Spasic, M., B. Durkovic, and D. Sinadinovic, "Pyrometallurgical Separation of Thallium from Antimony Concentrates from Alsar Mine," Tehnika (Belgrade), 22(1), 40-42 (1967); Chem. Abstr., 68, 116561s (1968).

77. Spasic, M., B. Durkovic, and D. Sinadinovic, "Hydrometallurgical Separation of Thallium from Antimony Ore Concentrate from Alsar Mine Deposits," Tehnika (Belgrade), 22(2), 238-241 (1967); Chem. Abstr., 68, 61803q (1968).

78. Spasic, M., B. Djurkovic, and D. Sinadinovic, "Metallurgical Concentration and Separation of Thallium from Arsenic Concentrates of the Alsar Mine," Tehnika (Belgrade), 22(6), 973-975 (1967); Chem. Abstr., 69, 88933q (1968).

79. Hedley, W. H., S. M. Mehta, and P. L. Sherman, Determination of Hazardous Elements in Smelter-Produced Sulfuric Acid, Environmental Protection Agency, Research Triangle Park, N.C., PB-240 343, National Technical Information Service, U.S. Department of Commerce, Springfield, Va., 1974.

80. Weinig, E., and P. Zink, "Über die quantitative massenspektrometrische Bestimmung des normalen Thallium-Gehalts im menschlichen Organismus" ["Quantitative Mass Spectrometric Determination of the Normal Thallium Content in the Human Body"], Arch Toxikol., 22(4), 255-274 (1967); Chem. Abstr., 67(1), 723w (1967).

81. Westerstrom, L. W., "Coal--Bituminous and Lignite" in Minerals Yearbook 1972, Vol. I, Metals, Minerals, and Fuels, Bureau of Mines, U.S. Department of the Interior, Washington, D.C., 1974, pp. 329-393.

82. Voskresenskaya, N. T., "Thallium in Coal," Geochemistry (USSR), No. 2, 158-167 (1968); Geokhimiya, No. 2, 207-216 (1968).

83. Lowe, W., "Origin and Characteristics of Toxic Wastes with Particular Reference to the Metal Industries," J. Water Pollut. Control Fed., 69(3), 270-280 (1970).

84. Davison, R. L., D. F. S. Natusch, J. R. Wallace, and C. A. Evans, Jr., "Trace Elements in Fly Ash. Dependence of Concentration on Particle Size," Environ. Sci. Technol., 8(13), 1107-1113 (1974).

85. Linton, R. W., A. Loh, D. F. S. Natusch, C. A. Evans, and P. Williams, "Surface Predominance of Trace Elements in Airborne Particles," Science, 191(4229), 852-854 (1976).

86. Natusch, D. F. S., "Physico-Chemical Associations of Trace Contaminants in Coal Fly Ash" in Preprints of Papers Presented at the 169th National Meeting before the Division of Environmental Chemistry, American Chemical Society, Philadelphia, Pa., April 6-10, 1975, pp. 139-141.

87. Natusch, D. F. S., C. A. Evans, P. K. Hopke, A. Loh, and R. Linton, "Characterization of Trace Elements in Coal Fly Ash" in International Conference on Heavy Metals in the Environment, Toronto, Ontario, Canada, October 27-31, 1975, pp. C-73 to C-75.

88. Schwitzgebel, K., F. B. Meserole, R. G. Oldham, R. A. Magee, F. G. Mesich, and T. L. Thoem, "Trace Element Discharge from Coal-Fired Power Plants" in International Conference on Heavy Metals in the Environment, Toronto, Ontario, Canada, October 27-31, 1975, pp. C-69 to C-70.

89. Sax, N. I., Ed., Industrial Pollution, Van Nostrand Reinhold Company, New York, N.Y., 1974.

90. Radian Corporation, Coal Fired Power Plant Trace Element Study, Vol. I, A Three Station Comparison, Prepared for the Environmental Protection Agency, Denver, Colo., 1975.

91. Radian Corporation, Coal Fired Power Plant Trace Element Study, Vol. II, Station I, Prepared for the Environmental Protection Agency, Denver, Colo., 1975.

92. Bolton, N. E., R. I. Van Hook, W. Fulkerson, W. S. Lyon, A. W. Andren, J. A. Carter, and J. F. Emery, Trace Element Measurements at the Coal-Fired Allen Steam Plant, ORNL-NSF-EP-43, Progress Report June 1971 - January 1973, Oak Ridge National Laboratory, Oak Ridge, Tenn., 1973.

93. Jackson, J., "Total Utilization of Fly Ash" in Proceedings of the Third Annual Waste Utilization Symposium, M. A. Schwartz, Ed., Cosponsored by the U.S. Bureau of Mines and IIT Research Institute, Chicago, Ill., March 14-16, 1972, pp. 85-93.

94. Yen, T. F., "Chemical Aspects of Metals in Native Petroleum" in The Role of Trace Metals in Petroleum, T. F. Yen, Ed., Ann Arbor Science Publishers, Inc., Ann Arbor, Mich., 1975, pp. 1-30.

95. Ball, J. S., W. J. Wenger, H. J. Hyden, C. A. Horr, and A. T. Myers, "Metal Content of Twenty-Four Petroleums," J. Chem. Eng. Data, **5**(4), 553-557 (1960).

96. Nuriev, A. I., and G. K. Efendiev, "Thallium and Gallium Contents in Stratal Waters, Crude Oils, and Surrounding Rocks," Azerb. Khim. Zh., No. 2, 112-116 (1966); Chem. Abstr., **65**, 15099 (1966).

97. Smith, I. C., T. L. Ferguson, and B. L. Carson, "Metals in New and Used Petroleum Products and By-Products--Quantities and Consequences" in The Role of Trace Metals in Petroleum, T. F. Yen, Ed., Ann Arbor Science Publishers, Inc., Ann Arbor, Michigan, 1975, pp. 123-148.

98. Kirby, J. G., D. A. Carleton, and B. M. Moore, "Crude Petroleum and Petroleum Products" in Minerals Yearbook 1972, Vol. I, Metals, Minerals, and Fuels, Bureau of Mines, U.S. Department of the Interior, Washington, D.C., 1974, pp. 909-1026.

99. Anonymous, Compilation of Air Pollutant Emission Factors (Revised), Office of Air Programs Publication No. AP-42, U.S. Environmental Protection Agency, Research Triangle Park, N.C., 1972.

100. Reno, H. T., and F. E. Brantley, "Iron" in Minerals Facts and Problems, Bureau of Mines Bulletin 650, U.S. Department of the Interior, U.S. Government Printing Office, Washington, D.C., 1970, pp. 291-314.

101. Makarova, T. I., and K. I. Marinina, "Behavior of Trace Elements in the Blast-Furnace Process," Sb. Nauchn. Tr. Vses. Nauchn.-Issled. Gorno-Met. Inst. Tsvetn. Metal, 7, 205-209 (1961); Chem. Abstr., 59, 13630d (1963).

102. Mishin, V. D., V. I. Smirnov, and V. V. Fokin, "Extraction of Zinc from the Dust of Blast Furnaces," Byull. Tsvetnoi Met., 10, 16-20 (1958); Chem. Abstr., 55, 13226f (1961).

103. Brantley, F. E., "Iron and Steel" in Minerals Yearbook 1972, Vol. I, Metals, Minerals, and Fuels, Bureau of Mines, U.S. Department of the Interior, Washington, D. C., 1974, pp. 641-666.

104. Litvinyuk, A. F., "Combined Utilization of Dzhezdinsk Manganese Ores," Gornyi Zhur., 3, 9-11 (1960); Chem. Abstr., 54, 14019g (1960).

105. Dadabaev, A. Yu., M. A. Milusheva, and S. N. Sushchenko, "Extraction of Thallium from Industrial Solutions by Means of Ion Exchangers," Tr. Inst. Met. i Obogashch., Akad. Nauk Kaz. S.S.R., 11, 129-136 (1964); Chem. Abstr., 62, 8723e (1965).

106. Bakeev, M. I., E. A. Buketov, and R. S. Andamasov, "Kompleksnaya Pererabotka Pyli ot Elektroplavki Margantsevykh Rud" ["Complex Treatment of Dust from the Electric Smelting of Manganese Ores"], Tr. Khim.-Met. Inst., Akad. Nauk Kaz. S.S.R., 3, 60-74 (1967); Chem. Abstr., 69(8), 29432c (1968).

107. Medvedev, G. V., E. R. Svadkovskaya, and T. Gabdulin, "Complex Utilization of Dzhezda Manganese Ore During Smelting of Silicomanganese from It," Bol'shoi Dzhezkazgan. Dobycha i Pererabota Rud (Alma-Ata, Akad. Nauk Daz. S.S.R.) Sb., 452-458 (1963): Chem. Abstr., 61, 2769 (1964).

108. Matthews, N. A., "Ferroalloys" in Minerals Yearbook 1972, Vol. I, Metals, Minerals, and Fuels, Bureau of Mines, U.S. Department of the Interior, Washington, D.C., 1974, pp. 525-533.

109. Crittenden, M. D., Jr., U.S. Geological Survey, Menlo Park, Calif., personal communication (telephone), January 1977.

110. Parfet, H. B., Vice-President, Caemi International, Inc. (dealer in manganese ores), New York, N.Y., personal communication (telephone), December 1976.

111. Krecker, B., Andrew S. McCreath and Son, Harrisburg, Pa., personal communication (telephone), December 1976.

112. Ward, J. M., Ore Specialist, Union Carbide, New York, N.Y., personal communication (telephone), January 1977.

113. Panel on Operational Safety in Marine Mining, Mining in the Outer Continental Shelf and in the Deep Ocean, National Academy of Sciences, Washington, D.C., 1975.

114. Wedepohl, K. H., Ed., Handbook of Geochemistry, Element 81, Vol. 2, No. 3, Springer-Verlag, Berlin-Heidelberg, 1972, pp. 81-B-1 to 81-O-1.

115. Petkof, B., "Mica" in Minerals Yearbook 1972, Vol. I, Metals, Minerals, and Fuels, Bureau of Mines, U.S. Department of the Interior, Washington, D.C., 1974, pp. 783-791.

116. Wells, J. R., "Feldspar, Nepheline Syenite, and Aplite" in Minerals Yearbook 1972, Vol. I, Metals, Minerals, and Fuels, Bureau of Mines, U.S. Department of the Interior, Washington, D.C., 1974, pp. 515-523.

117. Smith, G. I., C. L. Jones, W. C. Culbertson, G. E. Ericksen, and J. R. Dyni, "Evaporites and Brines" in United States Mineral Resources, Geological Survey Professional Paper 820, D. A. Brobst and W. P. Pratt, Eds., U.S. Department of the Interior, U.S. Government Printing Office, Washington, D.C., pp. 197-216.

118. Windholz, M., S. Budavari, L. Y. Stroumtsos, and M. N. Fertig, Eds., The Merck Index. An Encyclopedia of Chemicals and Drugs, 9th ed., Merck and Co., Inc., Rahway, N.J., 1976.

119. Meyer, H. C., "Uncommon Ores and Metals," Eng. Mining J.-Press, **115**(3), 105-106 (1923).

120. Meyer, H. C., "Uncommon Ores and Metals," Eng. Mining J.-Press, **119**(3), 93-94 (1925).

121. Meyer, H. C., "Uncommon Ores and Metals," Eng. Mining J.-Press, **121**(3), 94-95 (1926).

122. Meyer, H. C., "Uncommon Ores and Metals," Eng. Mining J., **123**(4), 136-137 (1927).

123. Anonymous, "Minor Metals. Thallium," Mineral Industry 1930, **39**, 660-661 (1931).

124. Roush, G. A., Ed., The Mineral Industry. Its Statistics, Technology and Trade During 1934, Vol. 43, McGraw-Hill Book Co., Inc., New York, N.Y., 1935.

125. Babitzke, H. R., "Thallium" in Mineral Facts and Problems, 1975 ed., Bureau of Mines Bulletin 667, U.S. Department of the Interior, Washington, D.C., 1976, pp. 1109-1114.

126. Greenspoon, G. N., "Thallium" in Mineral Facts and Problems, 1970 ed., Bureau of Mines Bulletin 650, U.S. Department of the Interior, Washington, D.C., 1970, pp. 749-758.

127. DeFilippo, R. J., "Thallium" in Commodity Data Summaries 1976, Bureau of Mines, U.S. Department of the Interior, 1976, p. 172.

128. Anonymous, Chemical Week 1976 Buyers' Guide Issue, McGraw-Hill, Inc., New York, N.Y., 1975, pp. 590-591.

129. Rausch, D. O., and B. C. Mariacher, Eds., AIME World Symposium on Mining and Metallurgy of Lead and Zinc. Vol. I. Mining and Concentrating of Lead and Zinc, The American Institute of Mining, Metallurgical, and Petroleum Engineers, Inc., New York, N.Y., 1970.

130. Carpenter, R. K., AMAX Zinc Co., Inc., E. St. Louis, Ill., personal communication (telephone), January 1977.

131. Dunn, C., Ore Purchasing Department, ASARCO Inc., New York, N.Y., personal communication (telephone), January 1977.

132. Anonymous, "Asarco Axes Texas Plant; Pollution Rule Casualty," Am. Metal Market, **82**(10), 6 (1975).

133. Smith, A., "National Zinc Plans July Start-Up for New Electrolytic Refinery," Am. Metal Market, **83**(121), 21 (1976).

134. Smith, Jerome, New Jersey Zinc, Allentown, Pa., personal communication (telephone), January 1977.

135. Midwest Research Institute, A Study of Waste Generation, Treatment and Disposal in the Metals Mining Industry, Final Report, Hazardous Waste Management Division, Office of Solid Waste Management Programs, U.S. Environmental Protection Agency, Washington, D.C., 1976.

136. Petrick, A., Jr., H. J. Bennett, K. E. Starch, and R. C. Weisner, The Economics of Byproduct Metals (in Two Parts). 2. Lead, Zinc, Uranium, Rare-Earth, Iron, Aluminum, Titanium, and Lithium Systems, Information Circular 8570, U.S. Department of the Interior, Bureau of Mines, U.S. Government Printing Office Stock No. 2404-01342, Washington, D.C., 1973.

137. Schroeder, H. J., "Copper" in Minerals Yearbook 1972, Vol. I, Metals, Minerals, and Fuels, Bureau of Mines, U.S. Department of the Interior Washington, D.C., pp. 473-509.

138. Magee, E. M., H. J. Hall, and G. M. Varga, Jr., Potential Pollutants in Fossil Fuels, EPA-R2-73-249, Office of Research and Monitoring, U.S. Environmental Protection Agency, Washington, D.C., 1973.

139. GCA Corporation, GCA Technology Division, National Emissions Inventory of Sources and Emissions of Silver, Contract No. 68-02-0601, Prepared for the Monitoring and Data Analysis Division, Office of Air Quality Planning and Standards, U.S. Environmental Protection Agency, Washington, D.C., 1973.

VI. OCCURRENCE OF THALLIUM IN LIVING ORGANISMS

In this section are tabulated concentrations and, sometimes, concentration factors* of thallium in aquatic animals, terrestrial plants and animals, and humans. The literature is replete with thallium tissue concentrations of poisoned animals and humans, but only a few of these data are included for comparison with "normal background" values. A human body burden is calculated, and the daily dietary intake is discussed. The sparse data do not indicate biomagnification in the food chain.

A. Aquatic Plants

No data for background thallium concentrations in aquatic plants were found. The marine alga *Ulva rigida* shows an accumulation coefficient for radiothallous ions at 14°C of 28.4 (at 24°C, 43.5) in the light.[1,2] Calspan Corporation[3] found 0.043 to less than 0.85 ppm in algae growing in contaminated waters. From the data, we calculated concentration factors of 9.3 to greater than 430.

A Lawrence Livermore Laboratory compilation[4] gives no measured or derived concentration factors for marine and freshwater plants, although values of 15,000 for invertebrates and 10,000 for fish were derived for both marine and freshwater animals. The seawater concentration used was 10^{-5} ppm[5], and the same value was assumed for freshwater. With 1.00 ppm as the assumed thallium concentration for both marine and freshwater plants, a calculated concentration factor for aquatic plants would be 100,000. Such a value is not borne out by the data of Calspan Corporation,[3,6] or Polikarpov.[1,2]

For rooted plants growing in contaminated water (grass, cattails), Calspan Corporation[3] reported 0.59 to 1.0 ppm thallium in the roots.

B. Terrestrial Plants

Values for thallium when detected in terrestrial plants range from 0.01 to 3,800 ppm, based on the ash. The values reported are cited in Table VI-1. A value close to ~ 0.5 ppm based on the ash is probably

---

* The terms "concentration factor" and "accumulation coefficient" are most often used by authors to denote the quotient between the concentration of the element in the organism divided by the concentration of that element in the surrounding soil or water.

typical for many of the species. Highest values are apparently obtained in areas of unusual thallium enrichment (e.g., near Rozden in the Alsar ore region of Macedonia, Yugoslavia). U.S. data are sparse. Probably the most indicative data of thallium in terrestrial U.S. plants are those of King,[7/] who detected thallium (by 6-step semiquantitative spectrography) in only 1% of several hundred Colorado plants and somewhat more often in Idaho plants, both groups being within or adjacent to mineral belts. King considers values of 2 to 7 ppm (ash) as anomalous and greater than 10 ppm as highly anomalous. Other U.S.G.S. analysts found 0.4 to 0.5 ppm in plant tissues, by atomic absorption (ash weight ?).[8/]

The concentration-factor data for thallium are largely Soviet. The range for all the plants (0.08 to 20 based on the plant ash) is closely duplicated by the ranges for two species: Artemisia campestris (wormwood) and Achillea setacea. U.S.G.S. analyses for two U.S. plant species indicate concentration factors of 1.8 to 10 (presumably based on ash weight).[8/] In the Alsar region, plants containing up to 3,800 ppm thallium (ash) have proved toxic to cattle.[9,10/] (For example, a 1,000-lb animal might ingest thallium on the order of 5 mg/kg if its day's forage contained 3,800 ppm).

The highest value seen for U.S. soils is 5 ppm. If plants grown on such soils served as an entire human diet and had a concentration factor of 10, then this "worst case imaginable" would provide a daily dietary intake of 80 mg or $\sim$ 1 mg/kg.

At certain U.S. smelter sites and at a bituminous coal fly ash pond, grass, horsetails, helianthus, and poplar sampled by Calspan Corporation contained 0.22 to 2.8 ppm thallium.[3/] Since these are wet weight concentrations, all would be in the range considered "highly anomalous" by King.

C. Aquatic Animals

Table VI-2 gives the values reported in the literature for thallium concentrations in shellfish, fish, and other aquatic animals. Noddack (1939)[1] reported 0.001 to 0.03 ppm thallium in aquatic species when detected. At Minamata Bay, in Japan the average for shellfish and other fish was 0.14 ppm (bottom mud and sand 0.55 ppm); but the meat contained only about 0.08 ppm. Zitko and Carson (1975)[13/] give concentration factors of 12.4 and 18.6 (wet weight) for marine mussels and clams. They conclude that thallium accumulation in the food chain is not a danger.

At a control site in Butte/Anaconda, Montana, Calspan Corporation noted that the larvae of large mayflies contained 0.003 ppm thallium (concentration factor 12). In the fall, however, rarer small mayfly larvae showed a concentration factor of almost 1,000. At 0.004 ppm, a fish (sculpin) showed a concentration factor of 16. The highest concentration of thallium in fauna reported by Calspan is 2.8 ppm in a cranefly in water at Kellogg, Idaho.[3]

D. Terrestrial Animals

The data in Table VI-3 do not include any values for the flesh of normal food animals. Hair and hoof would be expected to be much higher in thallium than would soft tissues, based on available human data.

Table VI-4 gives an organ concentration factor for fallout radio-thallium in deer and elk, based on the stomach contents; for the deer, the stomach contained mostly apples and the concentration factor for the shoulder muscle was 0.04. Only bone showed a concentration factor greater than 1.

E. Human Tissues

There are discrepancies in Table VI-5 for normal values for thallium in human tissues (e.g., Reference 14 gives an average value for thallium in the liver corresponding to about 0.14 ppm wet weight compared with 0.00056 to 0.00285 ppm in Reference 15.

The data of Weinig and Zink (1967)[15] for thallium in normal human (German) tissues are by far the most extensive. Mass spectrometric determinations were made by the isotope dilution method. Their data for two individuals are given in Table VI-6. An estimate for a human body burden of 0.14 mg thallium was calculated with the data for the 16-year-old male; organ weights of Standard Man;[16] and estimates of 0.2 ppb for fat, 2.5 ppb for bone plus marrow, 2 ppb for the gastrointestinal tract, 1 ppb for brain, and 3.9 ppb for kidneys. In Subsection VIII.A.2b, we calculate that the average daily human excretion of thallium in urine and feces is about 1 μg. Hamilton and Minski[17] estimated that thallium daily dietary intake is less than 2 μg. We have made a crude calculation of 2 μg/day based on the sparse cereal, flesh, and vegetable concentration data in the tables of this section.

Thus, if normal humans are in thallium balance and intake equals elimination, then 2 μg/day appears to be of the correct order of magnitude. This value is 1.5% of the calculated body burden. By comparison, potassium daily intake is about 2 to 6 g[18], elimination in urine and feces is 2.9 g (84% in the urine)[19], and the body burden is about 60 g[20] or 140 g[16] (70 kg man). Thus, about 2 to 5% of the potassium body burden is ingested and excreted daily.

At about 2 ppb thallium in the human body and about 1 ppb thallium in the diet (2 μg/1600 g food/day), the human tissue "concentration factor" is about 2.

TABLE VI-1

CONTENT OF THALLIUM IN TERRESTRIAL PLANT TISSUE AND DERIVATIVES

Plant Species, Organ	Average Thallium Concentration	Soil Concentration	Concentration Factor	Remarks	References
Wire grass (*Aristida* spp.)	N.D.[a]			The sensitivity for Tl was between 0.01 and 0.1%.	21
Spanish forest plants	1 ppm (ash)				22
Yeast *Saccharomyces cerevisiae*				Qualitative detection spectroscopically by Richards and Troutman (1940). Yeast were grown on synthetic and natural media.	23
Beetroot, grape juice, chicory root, tobacco, beechwood, and seaweed	detected (ash)			Böttger (1864) cited.	24
Kale	0.14 ppm (dry)			Crystal violet spectrometry.	25
Wines (German and Portugese, 1941-1957)	56-684 ng/ℓ (~ 0.06-0.7 ppb) 13 samples 0.0551-0.0978 ppm (ash) 3 samples			Voltammetric determination. Three vineyard soils contained 0.17-0.40 ppm Tl.	24,26
Euphorbia virgata (willow spurge)	0.027 g/ton (ash) 0.022 g/ton (ash) 0.03-0.3 (ash) avg. 0.084 g/ton (ash)	Avg. 38 soils, 0.023 g/ton	Avg. 1.74 0.18-1.3	C_{bg} = 1. The biogeochemical coefficient, C_{bg}, given for each of the next 9 species from Ref. 27 is the ratio of the arithmetic mean concn. of Tl in the ash of plants to the same in *Euphorbia*. The avg. concn. factor for these plants and 38 soils is calcd., 0.040 (ash)/0.023 = 1.74. Dvornikov et al. (1973).	27,28
Salvia nemorosa	0.04 g/ton (ash N.D.-0.8 (ash)		0.3-1.3	C_{bg} = 1.81	27 28
Achillea millefolium (yarrow)	0.01 g/ton (ash) 0.04 g/ton (ash)			C_{bg} = 0.39 and 1.81	27 28
Achillea setacea	0.04-0.9 g/ton (ash)		0.18-16	C_{bg} = 1.1 and 1.81	27 28
Peucedanum ruthenicum also called *Lomatium daucifolium*, parsley family	0.04 g/ton (ash) 0.03 g/ton (ash) 0.04-0.5 g/ton (ash)		0.8-2		27,28
Artemisia absinthum (wormwood)	0.03 g/ton (ash) 0.02-0.6 g/ton (ash)		0.08-2	C_{bg} = 1.4	27,28

TABLE VI-1 (continued)

Plant Species, Organ	Average Thallium Concentration	Soil Concentration	Concentration Factor	Remarks	References
Artemisia campestris	0.057 g/ton (ash) 0.04-0.8 g/ton (ash)		 0.14-20	C_{bg} = 2.6	27,28
Artemisia Marchalliana	0.04-0.7 g/ton (ash) 4 samples		Avg. 0.11 ([ash]/[soil])	Dvornikov et al. (1976). This reference's plants are from the Donets Basin hydrothermal deposits. Values for 9 species from a Black Hills,[b/] mercury deposit are included in the ranges. The avg. concn. factor does not include a Black Hills value.	28
Artemisia inodora	0.02-0.3 g/ton (ash) 3 samples		2.2 (1 sample)		28
Echium vulgare (borage family)	0.1-0.3 g/ton (ash) 5 samples		0.18-5		28
Asperula humifosa	0.1-1.0 g/ton (ash) 3 samples		0.9-3.3		28
Verbascum ovalifolium (mullein)	0.01-0.7 g/ton (ash) 4 samples		0.3-1.8		28
Anthemis tinctoria (golden marguerite)	< 0.1-0.5 g/ton (ash) 3 samples		< 0.09-6.7		28
Tanacetum vulgare (Common tansy)	N.D.-0.2 3 samples		0.2-2		28
Juniperus turcomanica (the Turkmenian SSR endemic juniper species) needles	220 ppm				29
Grasses (composite of 20 samples)	0.10 ppm (dry)			Spectrochemical analysis, ± 7.67% (variation coefficient).	30
Onosma tauricum, ssp. dalmaticum	detected (ash) (0.00x %)			Alsar region. Two samples of water contained 0.0x % and 0.00x %.	9
Linaria triphylla	3,000.0 ppm (ash) 3,800.0 ppm (ash)			Alsar ore region, Macedonia (Yugoslavia). As and Tl among characteristic elements. Plants poisonous to cattle. Spectr. colorimetric detn.	9,10
Linaria vulgaris (butter-and-eggs, a toadflax)	0.1-0.4 g/ton (ash) 4 samples		0.18-2.2	Donets Basin soil.	28
Festuca sulcata (fescue grass)	0.2 (ash) 0.2-0.6 (ash)		1.7-5.6	C_{bg} = 9.09	27 28
Plantago major (common plaintain)	0.02 (ash)			C_{bg} = 0.9	27

TABLE VI-1 (continued)

Plant Species, Organ	Average Thallium Concentration	Soil Concentration	Concentration Factor	Remarks	References
Convonvulus arvensis (field bindweed)	0.02 g/ton (ash)			$C_{bg} = 0.9$	27
Ranunculus acer (buttercup)	0.02 g/ton (ash)			$C_{bg} = 0.9$	27
Dianthus sp., Caryophyllaceae flowers	520.0 ppm (ash)			Spectral analysis.	31
Galium sp., Rubiaceae (bedstraw) flowers	1,700.0 ppm (ash)				31
Centaurea sp., Compositae (star thistle)					
fruit	7.5 ppm (ash)				31
leaves and stalk	10.5 ppm (ash)				31
Eryngium sp.					31
leaves	0.03 ppm (ash)				
flowers	0.10 ppm (ash)				
Campanula sp., Campanulaceae (bluebell family) flowers, leaves, and stalk	599.0 ppm (ash)				31
Spinach and rye sample	1 ppm			Berg (1925) cited.	9
Moss (gold-bearing areas of the Aldan Region, USSR)	$\leq$ 0.5 ppm (ash)			Salje (1965) cited.	9
Larch	0.5-5.0 ppm (ash)			Salje and Astaykina (1966) cited.	9
Birch leaves	0.5 ppm (ash)			Salje and Astaykina (1966) cited.	9
Horsetail	0.7 ppm (ash)			Salje and Astaykina (1966) cited.	9
Moss	0.5-10.0 ppm (ash?)			Salje and Astaykina (1966) cited.	9
Lavatera sp., Malvaceae					
leaves and stalk	12.5 ppm (ash)			Zyka[9] cites all of Petkovic's values 10 times higher than those in CA or in the typewritten copy of Petkovic's original article provided us by the Gecloški Zavod Biblioteka in Yugoslavia.	31
flowers	4.5 ppm (ash)				
Echinops sp, Compositae (globe thistle), leaves	1.5 ppm (ash)				31

TABLE VI-1 (continued)

Plant Species, Organ	Average Thallium Concentration	Soil Concentration	Concentration Factor	Remarks	References
Melilotus indica (sweet clover)					
whole	0.5 ppm (ash)	0.05 ppm (siltstone soil A_1)	10	USGS anal.	8
whole	0.5 ppm	0.2 ppm (clayey soil A_1)	2.5	USGS anal.	8
whole	0.4 ppm	0.04 ppm (sandy loam A_1)	10	USGS anal.	8
"*Papaver Escholtszia*" whole	0.8 ppm	0.45 ppm (sandstone C)	1.8	USGS anal. The species may be *Eschscholtzia Californica*.	8
Head lettuce	0.021 ppm (dry wt)			Colorimetric detn. after spectral anal. indicated the presence of Tl.	26
Red cabbage	0.040 ppm (dry wt)				26
Green cabbage	0.125 ppm (dry wt)				26
Leek	0.075 ppm (dry wt)				26
Endive	0.080 ppm (dry wt)				26
Clover	0.008-0.010 ppm (dry wt)				26
Meadow hay	0.020-0.025 ppm (dry wt)				26
Beet leaf	0.025-0.030 ppm (dry wt)				26
Potato plant tops	0.025-0.030 ppm (dry wt)				26

TABLE VI-1 (concluded)

Plant Species, Organ	Average Thallium Concentration	Soil Concentration	Concentration Factor	Remarks	References
Potassium bitartrate (cream of tartar)	0.044-0.104 ppm			Obtained from sediments of wine manufacture, known as argols or wine lees.	24
Bread	0.0005-0.001 ppm (dry wt)				26
Cigarette tobaccos	0.050-0.110 ppm 3 samples				26
	0.024-0.100 ppm (air dried); avg. of 8, 0.062 ppm				26
Cigars	0.024-0.170 ppm (avg. 0.063 ppm)				26
Cigar stubs	0.057-0.170 ppm				26
Cigar tobacco ash	0.009-0.038 ppm			Eight mass balance determinations indicate an avg. 26% loss of thallium in the smoke.	26
Alpine and subalpine plants, mostly conifers, Rocky Mountains of Colorado, and subalpine plants, North Central Idaho.	N.D.-100 ppm (ash)			Detected in only 1% of several hundred Colorado plants more in Idaho. Values > 10 ppm (ash) considered high anomalous. Any detected amoutn (> 2 ppm) anomalous. Highest value in alpine fir needles in an area of mercury mineralization. Areas studied were within or adjacent to mineral belts. Only 7 samples contained ≥ ppm thallium in the ash. Analysis by 6-step semi-quantitative spectrography. (U.S.G.S. analysis, H. King).	7

a/ N.D. = Not detected.
b/ Probably not South Dakota Black Hills.

TABLE VI-2

THALLIUM CONTENT OF AQUATIC ANIMALS

Animal	Average Tl Concentration	Range	No. of Samples	Remarks	References
Ciona intestinalis (Ascidien), whole	-			Lives at depth 0 - 0.5 m. Analyses of Reference 11 spectroscopic.	11
Halichondria (sponge), whole	-			20 m deep	11
Cyanea capillata (jellyfish), whole	0.03 ppm			0 - 35 m deep	11
Metridium dianthus (sea anemone), whole	0.03 ppm			Lives near surface	11
Stichopus tremulus (Halothurien), without viscera	0.003 ppm			30 - 40 m deep	11 11
Brissopus lyrifera (Echinoidea), shell	0.001 ppm			30 - 35 m deep	11
Asterias rubens (starfish), without viscera	0.003 ppm			Lives in shallow water	11
Ctenolabrus rupestris (small fish), whole	-			1 - 20 m deep	11
Squalus acanthius (dogfish, a shark), without viscera	-			30 - 50 m deep	
				Concentration factor for sea organisms based on the above nine species and a seawater concentration of 0.11 ppb: >700	11
Fundulus heteroclitus an estuarine teleost (fish)	N.D.				32
Mytilus gallo - provincialis	-	1.60 - 2.09 ppm (ash)		Detected by atomic absorption spectrophotmetry. Samples collected from eastern Sicilian waters. Concentrations in samples from Syracuse waters (petrochem. industries) about the same as that from the port of Catania.	33
Vimba vimba seminal fluid	Detected in the ash.				34
Pacific salmon	-			^{208}Tl found in King salmon muscle and roe Chum muscle, and Silver muscle.	35
Clupanodon punctatus	0.19 ppm	0.00 - 1.29 ppm	30	Minamata District, Japan, five sites. Tl determination colorimetric (Rhodamine B).	12

TABLE VI-2 (concluded)

Animal	Average Tl Concentration	Range	No. of Samples	Remarks	References
Trachurus japonicus	0.22 ppm	0.00 - 1.54 ppm	24	Minamata District, Japan, five sites	12
Penaeus japonicus	0.43 ppm	0.05 - 2.93 ppm	24	Minamata District, Japan, four sites	12
Neptunus trituberculatus	0.05 ppm	0.00 - 0.24 ppm	11	Minamata District, Japan, two sites	12
Leiognathus nuchalis	0.04 ppm	0.02 - 0.05 ppm	3	Minamata District, Japan, one site	12
Platycephalus indicus	0.01 ppm	0.01 - 0.02 ppm	3	Minamata District, Japan, one site	12
Phinoplagusia japonica	0.16 ppm	0.05 - 0.29 ppm	3	Minamata District, Japan, one site	12
Nibea schlegel	0.10 ppm	0.02 - 0.40 ppm	8	Three sites	12
Upeneodies bensasi	0.33 ppm	0.10 - 0.77 ppm	3	One site	12
Monacanthus cirrhifer	0.17 ppm	0.10 - 0.29 ppm	3	One site	12
Lateolabrax japonicus	0.05 ppm	0.01 - 0.13 ppm	3	One site	12
Loligo japonica	0.22 ppm	0.0; - 1.00 ppm	8	Four sites	12
Trachinocephalus myopus	0.04 ppm	0.01 - 0.08 ppm	3	One site	12
Mugil cephalus	0.06 ppm	0.01 - 0.09 ppm	5	Two sites	12
Plotosus anguillari	0.09 ppm	0.07 - 0.11 ppm	3	One site	
Pagrosomus major	0.07 ppm	0.03 - 0.19 ppm	5	Three sites	
16 Minamata Bay organisms above	0.14 ppm	0 - 2.93 ppm	139	Minimata Bay bottom: Sandy mud and sand, average of 11 samples: 0.55 ppm. Range 0.14 - 1.13 ppm.	12
All Minamata fishes and shells examined:					
Viscera	0.34 ± 0.12 ppm				
Meat	0.08 ± 0.02 ppm				
Bone and Shell	0.11 ± 0.03 ppm				12
Mya arenaria (marine clam)		Dry weight basis			
Days 19 - 60	50 ppb (100 ppb)	4.11 - 4.69 ppm (7.39 - 5.74 ppm)		Clams were exposed for 88 days to 50 and 100 ppb Tl in aquarium experiments. Maximum concentration factor ~ 150 (dry weight). Authors give 18.6 (wet weight) and conclude there is no danger of thallium's accumulating in aquatic mollusks.	13
Day 74	50 ppb (100 ppb)	6.03 ppm (12.45 ppm)			
Day 88	50 ppb (100 ppb)	3.78 ppm (10.78 ppm)			
Day 111	50 ppb (100 ppb)	1.89 ppm (3.38 ppm)			
Day 118	50 ppb (100 ppb)	< 0.5 ppm (≤ 0.5 ppm)			
Mytilus edulis (marine mussel)					
Day 12	50 ppb (100 ppb)	4.26 ppm (6.33 ppm)		Mussels were exposed for 40 days to 50 and 100 ppb Tl. Maximum concentration factor ~ 85 (dry) Authors give 12.4 (wet)	13
Day 27	50 ppb (100 ppb)	2.91 ppm (6.33 ppm)			
Day 40	50 ppb (100 ppb)	2.17 ppm (6.33 ppm)			
Days 47 - 69	50 ppb (100 ppb)	< 0.5 ppm (6.33 ppm)			

TABLE VI-3

THALLIUM CONTENT OF TERRESTRIAL ANIMALS

Animal, Organ, Tissue, Etc.	Average Tl Concentration	Range	No. of Samples	Remarks	Refer
Dog					
Blood (poisoning)		5.8 - 347.0 ppm	4	Lethal doses 35 - 253 mg Tl/kg	3
22 tissues (poisoning)		0.0 - 1540 ppm		Highest value in spleen. Range for this organ 16 - 1540 ppm.	3
Cattle					
Hair	0.020 ppm			Geilmann et al., 1960, report an average for three analysis methods: spectroscopy, colorimetric, and voltammetric.	2
Hair (calf)	0.13 ppm				2
Horn	0.010 ppm				2
Hoof	0.016 ppm				2
Beef rib eye	N.D.			Sensitivity, 0.5 ppm Tl	3
Veal	N.D.			Sensitivity, 0.5 ppm Tl	3
Swine					
Pork	N.D.			Sensitivity, 0.5 ppm Tl	3
Hair	0.009 ppm				2
Hoof	0.011 ppm				2
Rat					
Blood (poisoning)	2.7 μg/ml or 2.8 μg/ml	1.1 - 5.8 or 1.0 - 6.1 μg/ml		Two analysis methods. Rats fed "relatively high levels in their drinking water" for 3 days.	3
Sheep					
Hair	0.009 ppm				2
Hoof	0.012 ppm				2
Horse					
Hair	0.007 ppm				2
Hoof	0.004 ppm				2
Potato beetle	0.018 ppm				2
Goat					
Hair	0.007 ppm				2
Hoof	0.009 ppm				2
Rabbits, hares					
Wild rabbit hair (? "wildkanin")	0.060 ppm				2
Field hare hair	0.017 ppm				2
Guinea pigs					
Liver (fatal poisoning)	13.3 - 190 ppm			Results of 12 papers, 1930 - 1968, tabulated.	39
Kidney (fatal poisoning)	2 - 190 ppm			Results of 12 papers, 1930 - 1968, tabulated.	39
Liver (fatal poisoning)	1.5 - 17.5 ppm		10	Oral doses, 20 - 80 mg/kg. Analyses at 3 - 14 days.	39
Kidney (fatal poisoning)	1.2 - 73.5 ppm		10	Oral doses, 20 - 80 mg/kg. Analyses at 3 - 14 days. The ratio of Tl in kidney and liver was lower for 18 animals dying of the poisoning than for 8 poisoned animals dying of acute asphyxia, possibly due to kidney function damage in the former group.	39

TABLE VI-4

"CONCENTRATION FACTORS"[a] FOR ^{208}Tl IN TISSUES OF OLYMPIC ELK AND DEER[40]

	Concentration Factor	Relative Concentration[b]
creas	0.14	4.7
rt	0.09	3
roid	-	-
er	0.23	7.6
ney	0.73	24
g	0.14	4.7
cle (shoulder)	0.03	1
:		
nternal	-	-
uscle	-	-
; bone	14	47
e	0.15	5
/		
cle		
houlder	0.04	1
eck	-	
ulder bone	22.7	57
e	0.37	9.3

sed on the ratio of ^{208}Tl concentration in the tissue to ^{208}Tl concentration in the stomach contents.
lculated for each species.
-yr-old female
-yr-old female; stomach contents mostly

TABLE VI-5 HUMAN TISSUE CONTENT OF THALLIUM

Animal, Organ, Tissue, etc.	Average Thallium Concentration[a/]	Range[a/]	No. of Samples	Remarks	References
Human Body	<0.086			Calculated from the generalization by Lisk that the human body burden is < 6 mg.	41
Human organs		0.001-0.006		Normals. Goenechea and Sellier, 1966.	42
Blood, serum		0		Normals.	15,43
Blood (poisoning)		3-6		Weinig and Schmidt, 1966, cited.	43
Blood	< 0.02 μg/ml < 0.05 μg/ml			Determined by polarography.	44
Blood, juveniles	0.003 μg/ml	0.0-0.08 μg/ml	320	Asymptomatic children, primarily 1-5 yr old. Newark, N.J. Detection frequency (i.e., 0.005 μg/ml): 20.6%.	45
Blood, juveniles(poisoning)		0.1-0.9 μg/ml	4	Atomic absorption.	50
Liver	0.47 dry wt [~ 0.14 wet wt]	< 0.4-0.9 dry wt	11	New Zealand subjects with no known metal exposure. Detected by atomic absorption spectrometry.	44
Liver (poisoning)	26.0			Suicide. Death on fifth day. Spectrophtometric analysis.	46
Tooth enamel	N.D.[b/]			Detection limit of spark source mass spectrometer is 0.04 ppm Tl.	47
Liver (Poisoning)	10			Pooth, 1967, cited.	43
Urine (poisoning)		0.8-26.9		Blanz, 1965, cited.	43
Urine (poisoning)	235		1	Suicide.	46
Urine (normal)		0.00007-0.0015	6	Mass spectrometry. Error < 10%.	15
Urine (normal)	0.011	0.011-0.012	3	Colorimetric determination. Males	48
Urine (subacutemyeloopticoneuropathy)	0.021 ppm	0.011-0.026	5	4 females, 1 male. Syndrome similar to poisoning by triorthocresyl phosphate. The latter has no ocular involvement.	48
Urine (normal)	N.D.		20	Detected by carbon rod atomization and atomic absorption.	49
Urine, juveniles (poisoning)		0.700-1.24 μg/ml	5	Atomic absorption.	50
Kidney (poisoning)	6.5		1	Suicide.	46
Kidney (poisoning)	8-28			Weinig and Schmidt, 1966, cited.	43
Kidney (normal)		0.0014-0.0041	6	Mass spectrometry	15
Brain (poisoning)	21.5		1	Suicide.	46
Brain (normal)	0.00042-0.00151			Several brain tissues from 2 individuals. High value in pineal gland.	15
Heart (poisoning)	8.5		1	Suicide	46
Heart muscle (normal)		0.00096-0.00195	6	Mass spectrometry	15
Spleen (poisoning)		6-12		Weinig and Schmidt, 1966, cited.	43
Fat (normal)		0.00015-0.00032	6	Mass spectrometry	15
Skeletal muscle (normal)	0.00705		6	Mass spectrometry	15
Lung (normal)		0.00036-0.0295	6	Mass spectrometry	15
Liver (normal)		0.00056-0.00285	6	Mass spectrometry	15
Colon (normal)		0.0006-0.0054	6	Mass spectrometry	15
Rectum (normal		0.009-0.0014	6	Mass spectrometry	15
Nails (normal)		0.0007-0.0049	6	Mass spectrometry	15
Fingernails (normal) males		0.040-2.4 ppm (dry wt)	7		26
Hair (normal)		0.0048-0.0158	6	Mass spectrometry	15
Hair (normal)		0.007-0.65 ppm (dry wt)	8	Colorimetric and spectrometry range 0.007-5.1 ppm. Highest value by a volumetric method.	26
Urine (normal)	N.D.-0.001 ppm (detection)		9	Spectrography. Range when detected excluding highest value 0.0001-0.0004 ppm. The cigar butt of the individual with 0.001 ppm in his urine contained 0.025 ppm Tl.	42
Urine (poisoning)	>0.35 μg/ml	0.00046-0.00069 ppm	2	Coulometric determination.	51
Skin (normal)		0.00146-0.00251 ppm	2	Mass spectrometry.	15
Bone (normal)		0.76-27 ppm	3	Mass spectrometry.	15
Bone (near pyrite roaster)	N.D.			3 fatal cases.	52 53

TABLE VI-6

THALLIUM CONCENTRATIONS IN DIFFERENT HUMAN ORGANS OF TWO INDIVIDUALS, ppb[157]

Organ	73-yr-old female, (death by heart failure)	16-yr-old male
Fat	0.15	--
Skeletal muscle	1.02	2.01
Heart muscle	1.95	2.11
Atrio-ventricular node	--	2.66
Achilles tendon	0.25	0.51
Ligamentum nuchae	0.54	--
Meniscus (intraarticular fibro-cartilage)	--	0.74
Rib cartilage	0.49	1.05
Epiglottis	--	1.72
Skin	0.46	0.69
Gingiva	0.91 (atrophied)	2.25
Gall Bladder Wall	0.40	0.79
Bladder (urinary)	0.61	1.16
Aorta (wall)	0.64	--
Cerebrum (gray)	0.56	1.25
Cerebrum (white)	0.97	1.24
Cerebellum	0.65	--
Hypophysis with dural sac	0.49	--
Hypophysis without dural sac	--	1.26
Medulla oblongata	0.55	--
Pineal gland	--	1.51
Corpus striatum (of brain)	--	1.27
Internal capsule (of brain)	0.60	--
Nucleus ruber (of brain)	--	1.23
Nucleus niger (of brain)	0.62	--
Choroid plexus (of brain)	0.42	0.76
Liver	0.76	1.42
Pancreas	0.81	1.54
Spleen	1.19	1.54
Kidney, medulla and cortex	2.69	--
Kidney, medulla	--	4.05
Kidney, cortex	--	3.68
Adrenals	0.92	1.32
Thyroid gland	--	1.37
Parotid gland	0.86	1.87
Lymph node	0.91	1.32
Breast (fatty degeneration)	0.32	--
Lung	1.80	0.72
Ovary (fatty degeneration)	0.47	--
Uterus, degenerated	0.61	--
Testicles	--	2.08
Epididymis	--	1.61
Prostate	--	1.37
Stomach wall	0.84	1.56
Duodenal wall	1.56	--
Middle small intestinal wall	0.89	1.54
Colon wall	4.04	5.40
Rectal wall	--	1.38
Spongiosa of the sternum	0.84	1.57
Cranial roof	1.44	--
Rib next to cartilage	1.99	--
Middle rib	2.15	--
Rib next to vertebral column	1.46	--
Thigh [bone] compacta	2.12	2.51
Tooth, whole	--	4.70
Toenail, whole	1.89	2.51
Hair, head	--	15.80
Hair, head, near root	9.69	--
Hair, head, far from root	7.71	--
Serum	0.26	0.33
Blood	0.33	0.59
Gall	0.35	0.53
Urine	--	0.26

BIBLIOGRAPHY. SECTION VI.

1. Polikarpov, G. G., A. Ya. Zesenko, and A. A. Lyubimov, "Dinamika Fiziko-Khimicheskogo Prevrashcheniya Radionuklidov Mnogovalentnykh Elementov v Srede i Nakoplenie ikh Gidrobiontami" ["Dynamics of Physicochemical Transformations of Radionuclides of Multivalent Elements in the Environment and Their Accumulation by Hydrobionts"] in Radiatsionnaya i Khimicheskaya Ekologiya Gidrobiontov, G. G. Polikarpov, Ed., Naukova Dumka (publisher), Kiev, USSR, 1972, pp. 5-42; Chem. Abstr., **78**, 240 (1973).

2. Polikarpov, G. G., "Radioecology of Artificial Nuclides," Proc. Symp. Hydrogeochem. Biogeochem., **1**, 356-371 (1970, published 1973).

3. Calspan Corporation, Heavy Metal Pollution From Spillage at Ore Smelter and Mills, preliminary copy of the final report version provided in March 1977, Calspan Corporation, Buffalo, N.Y., for the U.S. Environmental Protection Agency, Cincinnati, Ohio (in press).

4. Thompson, S. E., C. A. Burton, D. J. Quinn, and Y. C. Ng, Concentration Factors of Chemical Elements in Edible Aquatic Organisms, UCRL-50564, Lawrence Livermore Laboratory, University of California, Livermore, Calif, 1972 (revision of 1968 publication).

5. Matthews, A. D., and J. P. Riley, "The Determination of Thallium in Silicate Rocks, Marine Sediments and Sea Water," Anal. Chim. Acta, **48**(1) 25-34 (1969).

6. Magorian, T. R., K. G. Wood, J. G. Michalovic, S. L. Pek, and M. W. Van Lier, Water Pollution by Thallium and Related Metals, Calspan Corporation for the U.S. Environmental Protection Agency, Cincinnati, Ohio, 1974. [This is the draft version of the final report issued (see Reference 3) in 1977 and was obtained from the National Technical Information Service as PB-253 333.]

7. King, H., Branch of Exploration Research, U.S. Geological Survey, Golden, Colorado, personal communication (letter to E. E. Angino), February 1977.

8. U.S. Geological Survey, Branch of Exploration Research raw data, released by U.S.G.S. through Keith Robinson, Denver, Colorado, November 1976. [Most thallium analyses were performed by A. E. Hubert, W. H. Ficklin, and G. Crenshaw.]

9. Zyka, V., "Thallium in Plants From Alsar," Sb. Geol. Ved., Technol., Geochem., No. 10, 91-96 (1970).

10. Zyka, V., "Trace Elements in Plants From Alsar," Sb. Geol. Ved., Technol., Geochem., No. 12, 157-160 (1974).

11. Noddack, I., and W. Noddack, "Die Häufigkeiten der Schwermetalle in Meerestieren" ["The Frequency of Heavy Metals in Sea Organisms"], Arkiv. Zool., 32A(4), 1-35 (1939).

12. Hamaguchi, H., "Inorganic Constituents in Biological Materials. XIV. Thallium, Selenium, and Arsenic Contents of Minamata Districts," Nippon Kagaku Zasshi [J. Chem. Soc. Jap. Pure Chem. Sect.], 81, 920-927 (1960); Chem. Abstr., 54, 25337i (1960).

13. Zitko, V., and W. V. Carson, "Accumulation of Thallium in Clams and Mussels," Bull. Environ. Contam. Toxicol., 14(5), 530-533 (1975).

14. Johnson, C. A., "The Determination of Some Toxic Metals in Human Liver as a Guide to Normal Levels in New Zealand. Part I. Determination of Bi, Cd, Cr, Co, Cu, Pb, Mn, Ni, Ag, Tl and Zn," Anal. Chim. Acta, 81(1), 69-74 (1976).

15. Weinig, E., and P. Zink, "Ueber die Quantitative massenspektrometrische Bestimmung des normalen Thallium-Gehalts im menschlichen Organismus" ["Quantitative Mass Spectrometric Determination of Normal Thallium Content in Man"], Arch. Toxikol., 22(4), 255-274 (1967); Chem. Abstr., 67(1), 723w (1967).

16. Division of Radiological Health, Department of Health, Education, and Welfare, Ed., Radiologic Health Handbook, Revised ed., PB 121784R, Department of Health, Education, and Welfare, Washington, D.C., 1960.

17. Hamilton, E. I., and M. J. Minski, "Abundance of the Chemical Elements in Man's Diet and Possible Relations With Environmental Factors," Sci. Total Environ., 1, 375-394 (1972/1973).

18. Committee on Dietary Allowances and the Committee on Interpretation of the Recommended Dietary Allowances, Food and Nutrition Board, National Research Council, Recommended Dietary Allowances, 8th revised ed., National Academy of Sciences, Washington, D.C., 1974.

19. McChesney, E. W., "54. Pathways of Mineral Metabolism: Laboratory Mammals" in Biology Data Book, P. L. Altman and D. S. Dittmer, Eds., Federation of American Societies for Experimental Biology, Washington, D.C., 1964, pp. 192-196.

20. Tipton, I. H., and M. J. Cook, "Trace Elements in Human Tissue. Part II. Adult Subjects From the United States," Health Phys., 9, 103-145 (1963).

21. Rusoff, L. L., L. H. Rogers, and L. W. Gaddum, "Determination of Coppe and Estimation of Other Trace Elements by Spectrographic Methods in Wire Grasses From Salt Sick and Healthy Areas," J. Agric. Res., 55, 731-738 (1937).

22. Ugarte, Laiseca, J., "Forest Phytochemistry. III. Elements of Groups C and D," Inst. Forestal Invest. y Experiencias (Madrid), 21(21), 1-91 (1950); Chem. Abstr., 45, 10454h (1951).

23. Porter, J. R., Bacterial Chemistry and Physiology, Wiley, New York, N.Y., 1946.

24. Eschnauer, H., and R. Neeb, "Beiträge zur analytischen Chemie des Wein IX. Voltammetrische Bestimmung geringster Thallium-Gehalte im Wein" ["Analytical Chemistry of Wines. IX. Voltammetric Determination of Trace Amounts of Thallium in Wine"], Z. Lebensm.-Untersuch. u.-Forsch 112, 275-280 (1960).

25. Kothny, E. L., "Trace Determination of Mercury, Thallium and Gold With Crystal Violet," Analyst, 94(1116), 198-203 (1969).

26. Geilmann, W., K. Beyermann, K. H. Neeb, and R. Neeb, "Thallium ein regelmässig vorhandenes Spurenelement im tierischen und pflanzlichen Organismus" ["Thallium as a Trace Element for Animals and Plants"], Biochem. Z., 333, 62-70 (1960); Chem. Abstr., 55, 14528f (1961).

27. Dvornikov, A. G., L. B. Ovsyannikov, and O. G. Sidenko, "Biogeokhimichr Oreoli Rozsiyannya Khal'kofil'nikh Elementiv na Roduproyavakh Zolota Nagol'nogo Kryazhu (Donbas)" ["Biogeochemical Dispersion Haloes of Chalcophile Elements on Gold Ore Deposits of the Nagol'nyi Ridge (Donets Basin)"], Dopov. Akad. Nauk. Ukr. RSR, Ser. B, 35(6), 490-494 (1973).

28. Dvornikov, A. G., L. B. Ovsyannikova, and O. G. Sidenko, "Nekotorye Osobennosti Koeffitsientov Biologicheskogo Pogloshcheniya i Biogeokhimicheskikh Koeffitsientov na Gidrotermal'nykh Mestorozhdeniyakh Donbassa v Sbyazi S Prognozirovaniem Skrytogo Ptutnogo Orudeneniya" ["Some Characteristics of the Coefficients of Biological Absorption and the Biogeochemical Coefficients at Hydrothermal Deposits of the Donets Basin in Relation to Predicting Hidden Ore Mineralization"], Geokhimiya, No. 4, 626-633 (1976); Chem. Abstr., 85(10), 65891r (197

29. Karryev, M. O., "Comparative Characteristics of the Essential Oils From Central Asian Species of Juniperus," Isv. Akad. Nauk Turkm. SSR, Ser. Biol. Nauk, No. 1, 40-43 (1967); Chem. Abstr., 67, 93901r (1967).

30. Lakanen, E., A Method for the Determination of Inorganic Components of Plants, Maatalouden Tutkimuskeskus Maantutkimuslaitos Agrogeol. Julkaisuja, Helsinki, 1961, 26 pp.

31. Petkovic, M., "Talijum u Biljkama Alšara" ["Thallium in Vegetation in the Alshar Region"] in Ref. Savetovanja Savez Geol. Drustava SFR Jugoslav, 6th, 2, M. Strackov, Ed., Savez Geol. Drustava Ser. Jugoslavije, Ohrid, Yugoslavia, 1966, pp. 606-609; Chem. Abstr., 70, 75118y (1966).

32. Eisler, R., and G. LaRoche, "Elemental Composition of the Estuarine Teleost Fundulus heteroclitus," J. Exp. Mar. Biol. Ecol., 9, 29-42 (1972).

33. Castagna, A., and F. Sarro, "Primi Dati sulla Presenza di Alcuni Elementi Metallici in Mytilus galloprovincialis, lam. della Costa Orientale Sicula, Svelati con Spettrofotometria per Assorbimento Atomico" ["Occurrence of Some Metallic Elements in Mytilus galloprovincialis of the Eastern Coast of Sicily, Revealed by Atomic Absorption Spectroscopy"], Boll. Soc. Ital. Biol. Sper., 51(8), 477-483 (1975); Chem. Abstr., 84, 85068k (1976).

34. Kucherova, F. N., "Kationnyi Sostav Semennykh Kletok i Semennoi Zhidkosti Rybtsa Vimba vimba (L.)" ["Cation Composition of the Sperm Cells and Seminal Fluid of Vimba vimba (L.)"], Voprosy Ikhtiologii, 12(4), 728-732 (1972).

35. Jenkins, C. E., "Radionuclide Distribution in Pacific Salmon," Health Physics, 17(3), 507-512 (1969).

36. Pritschow, A. L., Distribution of Thallium in Tissues, Blood, Urine, and Feces, L. C. Card No. Mic 59-3366, University Microfilms, Ann Ann Arbor, Mich., 1959; Dissertation Abstr., 20, 1041 (1959).

37. Mitteldorf, A. J., and D. O. Landon, "Spectrochemical Determination of the Mineral-Element Content of Beef," Anal. Chem., 24, 469-472 (1952).

38. Singh, N. P., and M. M. Joselow, "Determination of Thallium in Whole Blood by Delves Cup Atomic Absorption Spectrophotometry," At. Absorpt. Newsletter, 14(2), 42-43 (1974).

39. Weinig, E., Jr., and W. Walz, Jr., "Die Thallium-Verteilung in Niere und Leber bei letalen Tl-Vergiftungen" ["The Distribution of Thallium in the Kidneys and Liver in Lethal Tl Poisoning"], Arch. Toxikol., 27(3/4), 217-225 (1971).

40. Jenkins, C. E., N. A. Wogman, and H. G. Rieck, "Radionuclide Distribution in Olympic National Park, Washington," Water, Air, Soil Pollut., 1(2), 181-204 (1972).

41. Lisk, D. J., "Trace Metals in Soils, Plants, and Animals" in Advances in Agronomy, Vol. 24, Academic Press, New York, N.Y., 1972, pp. 267-325.

42. Goenechea, S., and K. Sellier, "Über den natürlichen Thalliumgehalt des Menschlichen Körpers" ["Normal Thallium Content of the Human Body"], Deutsche Z. Gesamte Gerichtl. Med., 60(4), 135-141 (1967).

43. Geldmacher-von Mallinckrodt, M., and M. Pooth, "Gleichzeitige spektrographische Prüfung auf 25 Metalle und Metalloide in biologischem Material" ["Simultaneous Spectrographic Testing for 25 Metals and Metalloids in Biological Material"], Arch. Toxikol., 25(1), 5-18 (1969).

44. Bowen, H. J. M., The Elementary Composition of Mammalian Blood, U.K. Atomic Energy Authority Research Group Report AERE-R4196, Isotope Research Division (AERE), Wantage Research Laboratory, Berkshire, 1963.

45. Singh, N. P., J. D. Bogden, and M. M. Joselow, "Distribution of Thallium and Lead in Children's Blood," Arch. Environ. Health, 30(11), 557-558 (1975).

46. Tewari, S. N., S. P. Harpalani, and S. S. Tripathi, "Determination of Thallium in Autopsy Tissues and Body Fluids by Spectrophotometric Technique," Mikrochim. Acta, No. 1, 13-18 (1975).

47. Losee, F., T. W. Cutress, and R. Brown, "Trace Elements in Human Dental Enamel" in Trace Substances in Environmental Health-VII, D. D. Hemphill, Ed., Proceedings of University of Missouri's 7th Annual Conference on Trace Substances in Environmental Health, June 12, 13, and 14, 1973, Columbia, Mo., University of Missouri, Columbia, Mo., 1973, pp. 19-2[illegible]

48. Ikeda, Y., ["Toxicological Study on the Etiology of SMON"], Nippon Rinsho, 29(2), 766-772 (1971); abstract in TOXLINE thallium search.

49. Kubasik, N. P., and M. T. Volosin, "A Simplified Determination of Urinary Cadmium, Lead and Thallium, With Use of Carbon Rod Atomization and Atomic Absorption Spectrophotometry," Clin. Chem., 19(9), 954-958 (1973).

50. Berman, E., "Determination of Cadmium, Thallium and Mercury in Biological Materials by Atomic Absorption," At. Absorpt. Newsletter, 6(3), 57-60 (1967).

51. Christian, G. D., and W. C. Purdy, "Coulometric Determination of Thallium in Blood and Urine," Am. J. Clin. Pathol., 46(2), 185-188 (1966); Chem. Abstr., 65, 18976b (1966).

52. Thomas, F., A. Heyndrickx, and W. van Hecke, "Thallium-Gehalt in menschlichen Knochen" ["Thallium Content of Human Bones"], Deut. Z. Ges. Gerichtl. Med., 54(1), 91-95 (1963); Chem. Abstr., 60, 1029b (1964).

53. Machata, G., "Über den Thalliumgehalt in menschlichen Knochen" ["Thallium Content of Human Bones"], Deut. Z. Ges. Gerichtl. Med., 54(1), 95-98 (1963); Chem. Abstr., 60, 1029b (1964).

VII. PHYSIOLOGICAL EFFECTS

This section reviews the biochemistry (absorption and distribution, excretion, and biological interactions); human and animal toxicity, chiefly from the viewpoint of symptoms in the whole organism; and "special" toxicities to animal tissues, organs, organ systems, and the developing embryo and fetus. Many of the special toxicity studies were aimed at determining the effects on isolated organs or systems and explaining the mechanisms of thallium's effects.

A. Biochemistry

1. Absorption, transport, and distribution

a. Absorption and transport: Thallous salts are well absorbed regardless of route. Il'in (1938),[1/] contrary to earlier reports by Buschke, observed that rabbits given hypodermic, oral, cutaneous, or intravenous doses showed identical toxicity and pilotropic effects except that molting occurred sooner after hypodermic than oral administration.

Lie et al. (1960)[2/] determined the minor variations in radiothallium content of rats at days 1, 2, and 7 after being given thallous nitrate solution (pH 2 to 4 except for intravenous (i.v.) and subcutaneous (s.c.) routes, when it was neutralized to pH 6 with potassium hydroxide) of specific activity ~ 2,000 mCi/g, i.v., intraperitoneally (i.p.), intramuscularly (i.m.), s.c., endotracheally, or orally. The results indicate complete absorption by any route.

The mean percentage body burden by all routes at 2 days was muscle 35.07%, bone 20.61%, gastrointestinal tract 11.90%, skin 9.34%, kidney 8.32%, liver 4.35%, testes 3.00%, respiratory system 0.89%, salivary glands 0.59%, hair 0.25%, spleen 0.35%, brain 0.49%, and heart 0.41%. The actual thallium concentration was highest in kidneys and salivary glands. By 7 days, the mean percentage body burden in the hair was 1.56%.

Thallous salts taken orally are absorbed from the mucous membranes of the mouth and intestines. Appearance of effects generally requires 1 to 2 days. They appear only rarely as soon as 8 to 10 hr after ingestion (Prick et al.[3/]).

Munch and Silver (1931)[4/] concluded that thallium sulfate must be absorbed from the gastrointestinal tract and localized in the tissues to exert its fatal effect on poisoned rats. Death usually occurred on the 2nd or 3rd day. Ten times the lethal dose killed one to five rats on the first day, whereas 40 times the lethal dose killed four of five rats within 1 day.

Hallopeau (1898, cited by Heyroth[5]).killed guinea pigs within 48 hr by applying a 50% thallium salt ointment to their skins. Toxic symptoms in people from external use of thallium for epilation have been reported by several researchers.[5]

Karsybekova (1956)[6] noted that thallium absorption from thallium acetate depilatory plasters (for ringworm treatment) was extremely variable among individuals. Children excrete thallium faster than adults with the highest excretion in the first 4 weeks after application of the plaster.

Sanotskii (1961)[7] found that the rate of penetration of a 2% thallous carbonate solution rubbed on the skin of mice was 0.0025 $mg/cm^3/min$.

Wadachi et al. (1966)[8] determined that $^{204}Tl^+$ is cationic in its mode and character of contaminability of fresh pig skin.

Radioactive K^+ and Tl^+, when injected into the mealworm Tenebrio molitor, were actively transported from the hemocoel to the rectal complex to the same extent (Koefoed, 1975).[9]

The immature egg-cell membrane of the starfish Nordora punctiformis is permeable to monovalent ions in the order Tl^+ greater than K^+ greater than Rb^+ greater than NH_4^+ greater than Na^+, Cs^+ (Hagiwara and Takahashi, 1974).[10]

Goldfish intestinal tissue from both fresh and saltwater-adapted fish took up Tl^+ at a faster rate than K^+, Rb^+, Cs^+, Na^+, or Li^+ (Ellory et al., 1973).[11]

The same carrier system appears to transport K^+, Rb^+, Cs^+, and Tl^+ ($^{204}Tl^+$) in cultured rabbit lenses (Kinsey et al., 1971).[12]

Gehring and Hammond (1964)[13] found a relation between Tl^+ and K^+ in the uptake of Tl^+ by rabbit erythrocytes. The rate of uptake had an initial fast component followed by a distinctly slower one, but both slopes of disappearance from the external medium followed first-order kinetics.

Cavieres and Ellory (1974)[14] reported that Tl^+ substitutes for K^+ at the external K^+ sites of the sodium pump in human red cells. Lishko et al. (1973)[15] arrived at a similar conclusion.

b. Distribution

(1) Poisoned animals and humans: Rauws (1974)[16] in c puter modeling of thallium pharmacokinetics states that only a small fracti

of the body burden of thallium is in the plasma and extracellular fluid, and that almost all of the thallium is in the cells. It requires only about 2 hr to transport all the thallium ions into the cells, which is extremely short compared with the elimination half-life of about 4 days.

Andre et al. (1960, cited by Oehme, 1972[17/]) found little radiothallium in the blood 1 hr after absorption; but much had accumulated in the bones and renal medulla, with lower concentrations in gastric and intestinal mucosa, pancreas, salivary glands, hair, eye lens, and tongue epithelium. After 10 days, the concentration in the renal medulla had increased and the central nervous system concentration (not observed at 1 hr postabsorption) was higher than in the liver. Bone concentrations had decreased more rapidly than in the muscle. Thallium was not found in the blood at 10 days.

Gehring and Hammond (1967)[18/] followed the first-order plasma disappearance of ^{204}Tl and ^{42}K and their uptake by tissue. Although their ionic movements are related, once thallium is inside the cell, it is less readily released than potassium. Rats on higher potassium diets showed a higher LD_{50} for thallium.

In the first hours of thallium poisoning, the kidney concentration is much higher than in other tissues; but Reinhardt and Zink (1974)[19/] found that in the final stage of fatal thallium poisoning (s.c. injection of the LD_{50} of thallium acetate) in white mice, the tissue concentrations equalize.

Truhaut et al. (1957)[20,21/] injected 4-day-old rats with 50 γ thallous sulfate containing 1 μCi ^{204}Tl and injected them 5 days later with 250 γ. The highest amounts of thallium were found in the thyroid gland and endocrine organs in animals sacrificed 4 or 24 days after treatment.

On a 0.003% thallous acetate diet for 60 days, 80% of the male weanling rats and 60% of the females died within the 4th to 8th week. The tissue distribution and amounts were about the same as in rats fed 0.0035% thallic oxide: kidneys greater than bone and liver (Downs et al., 1960[22/]).

In subacute and chronic thallium poisoning in the rat, the concentration in the adrenal gland is relatively high. Retention in the adrenal gland of chronically poisoned rabbits is low.[23/]

Frey and Schlechter (1939)[24/] injected rabbits s.c. with lethal doses of thallium acetate. As early as 12 min afterward, thallium was found in the blood and gastric juice, concentrations in the latter never attaining the values in the blood. After receiving 111 mg thallium acetate per kilogram body weight, a 2.7-kg pregnant rabbit showed 1.4 mg%

Tl in the blood, 1.0 mg% in the gastric juice, and 1.2 mg% in the milk. Thallium was found in the placenta as well as in the kidney, liver, brain, and heart of the young.

The distribution of sublethal (half the LD_{50}) doses of radiothallium in rabbits at 24 hr, 43 hr, and 144 hr after administration given in Table VII-1 shows highest concentrations in the kidneys at all times.

Truhaut et al. (1957)[21] injected rabbits intraperitoneally with 25 mg/kg radioactive thallium sulfate having a specific activity of 100 mCi/mg Tl. The animals were killed 48 hr later. The blood cell plasma ^{204}Tl ratio was 0.66. The highest concentration of radiothallium was in the thyroid, followed in descending order by kidney, testicles, salivary glands, muscle, lung, liver, spleen, eye, brain, adrenals, pancreas, ovary, and skin. Rabbits each given a total of 35 mg Tl_2SO_4 within 22 days showed thallium in all organs, especially the adrenals, spleen, pancreas, kidney, and testicles. Thallium accumulated at the level of the hair follicles in the skin.

Truhaut (1959, 1960)[23] found thallium spectrographically in all organs of poisoned animals, with the highest concentrations in the kidneys, salivary glands, hypophysis, testes, ovaries, skeletal muscle, and eyes. Brain, spinal cord, bones, and adipose tissue showed low concentrations. He generalized that except in the dog concentrations are usually low in the liver, heart, spleen, and lungs.

Gettler and Weiss (1943, cited in Reference 25) found that thallium in poisoned dogs was highest on a per gram basis in kidney, pancreas, spleen, and muscle; but found no selective accumulation in bone (perhaps because thallium phosphate is about 50 times more water soluble than calcium phosphate).

Dogs given 55 to 500 mg thallium acetate by stomach tube showed the following thallium distribution after 24 hr: liver ~ 16%, intestines 8.5%, stomach 7%, heart 2%, kidneys 1.7%, lungs 1.5%, brain 1.2%, and spleen 0.7% of the dose given.[26]

Emara and Soliman (1950,[26] cited in Reference 3) gave 8.5 mg thallous acetate per kilogram to dogs, sacrificing them on days 1, 7, 11, or 15. Table VII-2 shows the tissue distribution of thallium.

On the basis of percent of dose per organ, during the first 20 hr after injection of goats with ^{201}Tl (100-700 μCi), the liver contained more thallium than the heart, kidney, lung, and spleen. However, the highest concentration of thallium per gram of tissue was in the heart during the first 10 min. At 25 min to 243 days, it was highest in the kidney (Bradley-Moore et al., 1975[27]).

A 93-kg calf given 4.48 mCi $^{204}TlNO_3$ orally and killed 48 hr later showed the following values for thallium recovery in the tissues (expressed as a ratio of the tissue concentration to the plasma concentration taken as 1.000): kidney 154, heart 31.0, pancreas 29.6, salivary gland 26.5, lung 22.5, liver 21.7, spleen 19.1, rib bone 18.2, thyroid 17.2, muscle 17.1, testes 14.3, thymus 11.1, compact bone 10.0, brain 5.44, cartilage 5.34, fat 4.55, marrow 1.55, bile 1.54, and red blood cells 1.27 (Potter et al., 1971[28/]).

Jenkins (1972)[29/] reported levels of fallout radiothallium in tissues of Olympic Peninsula elk and deer and Alaskan caribou (see Table VII-3).

Munch (1970)[30/] tabulated literature data on the distribution of thallium in blood, kidney, liver, muscle, spleen, and urine in humans and animals poisoned by thallium. Table VII-4 is adapted from a similar tabulation for human poisonings by Prick et al.[3/]

Tewari et al. (1975)[31/] found the following levels of thallium by a spectrophotometric method in the tissues of a suicidally poisoned young man (death on the 5th day): liver 26 ppm, kidney 6.3 ppm, heart 8.5 ppm, brain 2.15 ppm, and urine 233 ppm.

According to data in patients contaminated by intravenous injection of Thorotrast 18 years or more previously, 620% of the ^{208}Tl (probably only shows the distribution of its precursor ^{212}Bi) is in the sector including liver, spleen, and much of the lung, blood, bone marrow, and skeleton (Malatova and Dvorak, 1971[32/]).

(2) <u>Placenta and/or fetus</u>: Chick embryos that received 25 μCi ^{204}Tl (0.114 mg $TlNO_3$) on the chorio-allantoic membrane when 8 days old and sacrificed 10 days later showed the largest amounts of thallium in the yolk, blood-contaminated albumin, shell, and leg and foot with attached muscle. Slight achondroplasia was observed (Barclay et al., 1952[25/]).

Thallium was localized at the cellular level in developing chicken embryos by injecting 600 mCi/mg) and 1 mg thallous sulfate (a dose known to produce shortening and bowing of long bones and other deformities) into the yolk sac on the 11th day of incubation. Marrow spaces, vascular spaces, and surfaces of the bone trabeculae were the chief sites of ^{204}Tl localization. Differences in label distribution were not apparent at 1, 71, or 192 hr after injection (Ford et al., 1968[33/]).

Gibson et al. (1967)[34/] reported that the biological half-life of thallium (12.5 μCi $^{204}Tl_2SO_4$ injected subcutaneously before caesarean section 1 to 24 hr) in the pregnant Swiss-Webster mouse or Sprague-Dawley rat is 24 hr. At equilibrium (8th hr in the mouse or the 16th hr in the rat), the fetal/maternal tissue ratios were 0.84 in the rat and 0.46 in the mouse.

Increasing the thallium sulfate levels (25 μCi/ml of $^{204}Tl_2SC_4$) infused (0.2 to 6.4 mg/min/kg*) into 20-day pregnant rats increased the whole blood levels of thallium. The whole fetuses contained about 1/15 the thallium concentration in the maternal blood. Similar experiments with normal nonpregnant female rats showed that the ratio of thallium concentration in erythrocytes to that in plasma was only 1.5 to 2.0, whereas the potassium ratio was 22. At all dosages, about one-third of the total blood thallium remained in the plasma and was available for possible placental transfer (Gibson and Becker, 1970[35/]).

Pregnant guinea pigs and rabbits were given thallous acetate by stomach tube in doses of 1 to 2.5 mg/kg for up to 61 days (total doses 16 to 54 mg/kg). Thallium was always detected in the placenta of the guinea pigs but not the rabbits. It was almost always found in the fetuses. One rabbit aborted; one litter of guinea pigs had cataracts (Testoni, 1936[36/]).

Fitzek and Henning (1976)[37/] reported the distribution of thallium in the tissues of a poisoned pregnant house cat, who died 6 days after aborting four fetuses. The thallium concentration in the fetuses was 11.4 ppm (eviscerated fetus 11.7 ppm, abdominal organs 5.0 ppm, and heart/lungs 51.9 ppm) and in the placenta 0.4 ppm. The uterus of the cat contained 18.0 ppm; the highest maternal thallium concentration was 67.7 ppm in the kidney.

Neal et al. (1935, cited by Richeson[38/]) reported that a 38-year-old woman who was 5 months pregnant at the time of a fatal thallium intoxication, on autopsy, showed thallium in her tissues; but none was in the kidney or liver of the fetus. A woman who had become intoxicated by thallium in the 4th month of pregnancy gave birth to an apparently normal baby while still paralyzed. At delivery, the maternal and fetal urine, placenta, membranes, and amnionic fluid were free of thallium (English 1954, cited by Richeson). A woman who had ingested 1.2 g thallium at term (and survived) gave birth to a baby that died within 5 days. Thallium was in the baby's tissues (Horstmann, 1949, cited by Richeson). Van Oettingen (195? cited by Richeson) reported thallium in the urine and meconium of babies whose mothers had been poisoned by thallium in the 7th month of pregnancy.

2. Excretion

a. Radiothallium: Durbin et al. (1957, cited in Reference 2) reported data indicating the biological half-time of radiothallium is only 5.2 days.

---

* ". . . maternal death [resulted] in some animals (6.4 mg/kg/min for 64 min) . . ."

Lie et al. (1960)[2] gave rats radiothallium nitrate (specific activity 2,000 mCi/g, 20 to 400 μCi/rat) by various routes (i.v., i.p., i.m., s.c., endotracheally, and orally). For the first 21 days, the exponential body clearance rate had a half-time of 3.3 days regardless of administration route. Only about 1% of the administered dose remained at 21 days.

The ratios of fecal to urinary excretion also showed minor differences (1-3:1 at 2 days; 3-7:1 at 9-21 days) depending on the route of administration. Urinary excretion decreased faster.

Gehring and Hammond (1967)[18] found that increases of dietary potassium in rats and dogs enhanced tissue mobilization of ^{204}Tl and thallium elimination via the urine but not the feces.

Barclay et al. (1952)[25] injected seven male albino rats i.v. with 5 μCi ^{204}Tl (23 μg $TlNO_3$). Three of them immediately received a 10-mg injection (a lethal dose of Tl_2SO_4). All were sacrificed at 48 hr. More ^{204}Tl was excreted in the feces (9%) than in the urine (6%) by the animals receiving only ^{204}Tl. Slightly more ^{204}Tl was excreted in the urine than in the feces by the lethally poisoned rats. Most of the ^{204}Tl was recovered in the carcass (muscle and skeleton), but the relative concentration of these tissues was not high.

Potter et al. (1971)[28] followed the radiothallium output in feces, milk, and urine of a 502-kg dairy cow for 168 hr after oral administration of 8.96 mCi $^{204}TlNO_3$. The recoveries were 80, 16.4, and 3.4%, respectively, of the administered dose. The biological half-time "for an early component" was calculated as 36 hr. The half-residence time of ^{204}Tl in the cow was 56 hr.

Discrepancies in the half-time in the animal body and the slow excretion reported by various investigators may be due to the differences in behavior between radiotracer and chemically detectable levels. In the studies described here, the fecal:urinary excretion ratios were above 1 when only tracer levels were used.

Katayama (1963)[39] followed the rate of excretion of orally administered ^{204}Tl by rats for 7 days (see Table VII-5).

Lund (1956)[40] found thallium in the urine and feces of a rat 1 hr after i.p. administration of 10 mg Tl_2SO_4 (~ 10 μCi ^{204}Tl). After killing 24 hr later, the percent of the dose in some of the organs and excreta was: kidneys 3.85%, urine 1.03%, intestinal contents 3.52%, muscles 41.1%, and skeleton 17.6%.

The elimination curves in other rats given the same i.p. dose had "the exponential shape characteristic of substances eliminated in unit time as a certain fraction of the total amount of substance present in the organism at any given time." Within 25 days, 26.4% of the dose had been eliminated in the urine and 51.4% in the feces; 12% could not be accounted for; and 10.3% was retained in the tissues.[40/]

Thyresson (1951)[41/] observed that rats given radiothallium nitrate (activity: 0.5 mCi/g) s.c. in acute or chronic dosing schedules (e.g., 20 μg $TlNO_3$ per gram body weight* in an acute test) excreted more thallium in the feces than in the urine.

Rats given 2.8 mg Tl_2SO_4 (~ 28 μCi)/kg s.c. had equal concentrations of thallium in the blood serum and gastric secretions at 7 hr, but the urine concentration was 30 times that of the serum. In ligated bile duct experiments, it was shown that thallium is not appreciably excreted by the liver with the bile. All sections of the intestinal tract excreted concentrations of about the same order of magnitude of thallium. Experiments with rabbits given radioactive thallous sulfate orally indicated that thallium is excreted through the kidneys by glomerular filtration. Half of the filtered amount is reabsorbed in the tubuli. The passage of thallium and potassium through the kidney is closely associated. Plasma thallium is rapidly eliminated by the kidneys; however, most of the thallium is held strongly in the organs in the intracellular phase together with potassium (Lund, 1956[40/]).

Rabbits injected i.p. with 25 mg/kg radiothallium(I) sulfate of specific activity 100 mCi/mg Tl showed 58% of the thallium in the urine at 24 hr and 17% in the feces; at 48 hr, 13% in the urine and 3% in the feces (Truhaut et al., 1957[21/]).

A middle-aged human female dying from osteogenic sarcoma metastatic to the lungs was dosed orally with 500 μCi ^{204}Tl activity followed by 45 mg doses of thallium sulfate every 3 days until 225 mg had been given. She died on the 24th day with 45% of the administered dose still in her body. An average 3.2% of the body burden was excreted in each 24-hr period. (Joseph (1939) had reported that labeled potassium is excreted at the rate of 6 to 7% per day for 66 hr.) Tissues with highest levels of thallium were hair, kidney, heart muscle, primary bone tumor, and spleen (Barclay et al., 1952[25/]).

b. Stable thallium: Prick et al.[3/] summarize that thallium is excreted in perspiration, tears, secretions of the mucous membrane of the bronchial and nasal tubes, bile, milk, urine, and feces.

---

* Approximately 20 mg/kg.

Thallium is excreted very slowly like lead. It has been found in the urine months after poisoning and in the vomitus and feces 35 days after ingestion.

Oehme (1972)[17] gives the biological half-life of thallium as 3 to 8 days. According to Oehme, urinary excretion of thallium is about half fecal excretion, persisting for up to 3 months after administration. The long retention period is ascribed to the sparing solubility of thallium (as the chloride[42]).

Fabre (cited in Reference 3) found the following amounts in urine on the 10 consecutive days that a 13-kg dog lived after ingesting 576 mg thallous acetate (446 mg Tl): 0, 0, 10, 10, 12, 17, 16, 17, 19, and 15 mg (total urinary excretion 28% of the administered dose).

Rats given 0.38 mg thallous acetate/100 g by gastric cannula three times weekly for 15 days (total ingested 8.62 to 8.97 mg) excreted 5.70 to 9.20% of the dose in the urine (Paolaggi et al., 1972[43]).

Shaw (1934)[44] fed a dog 25 mg thallium (as Tl_2SO_4) per kilogram. The dose was not fatal. During the first 3 days, the dog excreted 32% of the dose in the urine. By the end of 36 days, 61.6% of the dosage had been eliminated in the urine. At the then rate of elimination (0.22% of the dosage), 80 more days would have been required to completely remove the thallium.

Ehrhardt (1927)[45] found that lactating rats fatally poisoned by thallium secrete enough thallium into their milk to inhibit the offsprings' growth and development and to produce alopecia.

Mrozikiewicz and Widy (1960)[46] gave six female rats 35 mg thallium sulfate per kilogram after parturition. The young fed on their milk showed characteristic darkening in the roots of the tactile hairs 24 hr later, showing that thallium had passed into the milk of the poisoned females.

A 19-year-old soldier who had ingested thallium-containing rat poison still showed thallium in the urine and feces 5 months after exposure (Arnold et al., 1964[47]).

Whether more thallium is excreted in the urine than feces is debatable. Prick et al.[3] in their 1955 review concluded that more is excreted in the urine. Truhaut (1960)[23] found that urinary excretion is much more important than fecal excretion in thallium poisoning (contrary to the results of Thyresson (1951)[41] for rats but in agreement with results of Shaw for dogs and DeFrey, Schlecter, and Testoni for rabbits). Fecal excretion appears to be favored when only radiothallium is used in tracer doses.

Testoni (1936)[36/] reported that at 30 days after administration to laboratory mammals, 72% of the thallium given had been excreted via the urine, and 14% via the feces. At 1.5 months, thallium was not found in the feces; but at 3 months, it was still in the urine.

Stevens et al. (1974)[48/] state that fecal excretion in man is less than urinary excretion until late in the intoxication. In a case of acute thallium poisoning of a human, on the 14th day 0.56 mg% (5.6 ppm) thallium was found in the urine and 47.2 mg% in the feces (Prix and Dvorackova 1963[49/]).

Jacobs (1962)[50/] found up to 200 μg thallium in the urine of 21 exposed individuals by a colorimetric method, but none in the urine of unexposed people.

Paolaggi et al. (1972)[43/] could not detect (limit less than 10 μg/liter) thallium in the urine of normal humans by voltammetry with anode redissolution.

Goenechea and Sellier (1967)[51/] reported the values* for thallium in the urine and feces of 10 normal humans (Germans) given in Table VII-6. (The cigar stub of individual No. 7 with the highest thallium content in urine contained 25 ppb thallium.) The average daily excretion of thallium in the urine was about 0.45 μg; in the feces, perhaps about 0.2 μg (fecal concentration about five times urinary (~ 0.3 ppb); thus, ~ 1.5 ppb x (100 g water + 25 to 50 g dry matter)). Data of Table VII-7 indicate higher urinary thallium excretion (~ 1.2 μg/day). An average of the data from both tables gives about 0.75 μg thallium in the urine (0.5 ppb) and 0.3 to 0.4 μg in the feces (assumption of 2.5 ppb). Thus, normal humans (German) excrete about 1 μg thallium per day in feces and urine. Since Standard man excretes 2.5 liter water daily,[52/] a significant amount of thallium may also be excreted in sweat (0.5 liter).

3. Biological interactions

a. Effects on enzymes: Thallium alters the keratinization process and is toxic to epidermal cells, possibly by enzyme inhibitions, although cytochrome c, cytochrome oxidase, cozymase, and diaphorase were not affected at concentrations having maximum effect on tissue respiration (Schwartzman and Kirschbaum, 1962[53/]).

Glucose-6-phosphate dehydrogenase activity in the red blood cells of rats acutely poisoned by thallium nitrate decreased to 20%, while the activities of glyceraldehyde-3-phosphate dehydrogenase and lactate dehydrogenase remained the same (Erich and Waller, 1967[54/]).

* Determined by an extraction-spectrographic method.

The relative enzyme activating efficiency of monovalent cations for the glycerol dehydrogenase of Aerobacter aerogenes is NH_4^+ greater than Tl^+ greater than K^+ greater than Rb^+ and is independent of pH. However, changing the pH from 6.1 to 7.5 to 9.0 decreased the affinity for NH_4^+ and Tl^+ but increased the affinity for K^+ and Rb^+ (McGregor et al., 1974[55/]).

Oxygen uptake by liver, heart, and kidney homogenates is inhibited by thallium salts. Succinic acid suppresses the inhibition. Thallium (~ 0.61 M) inhibits the enzymes that dehydrogenate succinic acid and higher fatty acids.[23/]

Phosphatases play a major role in bone formation, which is disturbed in chronic thallium poisoning. Although up to 0.1 M thallium does not inhibit acid phosphatase, 0.001 M thallium inhibits alkaline phosphatase. Cysteine offers no protection.[23/]

Thallium (0.004 M) inhibits pseudocholinesterase in serum, but true cholinesterase is less sensitive, becoming depressed at concentrations producing parasympathomimetic effects in vivo. Cysteine, glutathione, and disodium thiomalate are not completely protective.[23/]

Tl^+ as well as K^+, NH_4^+, and Cs^+ activates mitochondrial pyruvate carboxylase from chicken liver (Barden and Scrutton, 1974[56/]).

Tl^+ binds many ligands at least 10 times more strongly than K^+. Manners et al. (1970)[57/] reported that a 10-fold increase in binding resulted in thallium's being 10 times more effective in activating pyruvate kinase and vitamin B_{12}-dependent diol dehydratase than K^+. Williams[58/] summarized data for several enzymes activated more strongly by Tl^+ than K^+ (Table VII-8).

Kayne (1971)[59/] reported that the kinetically determined affinity of Tl^+ rabbit muscle pyruvate kinase is about 50 times that for K^+ with maximum activation at 2 mM Tl^+ (~ 400 ppm). The maximum velocity of the Tl^+-activated system is about 60% that of the K^+-activated one. High Tl^+ concentrations (greater than 10 mM) inhibited the reaction, indicating likely formation of the Tl^+-ADP complex. Four binding sites for $^{204}Tl^+$ appeared to correspond to the number of active sites.

Yeast aldehyde dehydrogenase is activated by Tl^+ in concentrations up to 0.0038 M (~ 780 ppm) but at higher concentrations, inhibition offsets activation (Bostian et al., 1975[60/]).

L-Threonine (deaminating) dehydratase of seven species of marine planktonic algae was activated 100 to 161% by 0.05 M thallous nitrate (1% Tl) while K^+ and NH_4^+ (required for in vivo expression of the enzyme) caused a three- to fivefold activation (Antia et al., 1972[61]).

Britten and Blank (1968)[62] noted that Tl^+ replaces K^+ in activation of the Na^+, K^+-activated (-dependent, or -sensitive) adenosine triphosphatase (ATPase) of rabbit kidney and that its affinity for the K^+-activating site was 10 times that of K^+. Na^+, K^+-activated ATPase is a membrane-bound enzyme that is involved in electrolyte transport across the membrane. The enzyme, requiring sulfhydryl groups for activity, is inhibited by metals that bind -SH, such as lead and mercury.[62]

A Na^+-dependent phosphorylation of Na^+, K^+-activated ATPase followed by a K^+-dependent dephosphorylation appears to be the reaction sequence of the enzyme. When ATP binds to a single site on kidney cell ATPase, binding is inhibited by K^+ and the inhibition is antagonized by Na^+. The sequence of apparent ion affinity is $Rb^+ = K^+$ greater than Tl^+ greater than NH_4^+ greater than Cs^+ for driving ATP from the binding site. The sequence for the dephosphorylation step in ATPase is Tl^+ greater than Rb^+ greater than K^+ greater than Cs^+ greater than NH_4^+ greater than Li^+ (Albers, 1967; and Hegyvary and Post, 1971, cited by Lee[63]).

Robinson (1970)[64] reported the activation of the Na^+, K^+-dependent ATPase of rat brain by NH_4^+ and Tl^+.

Natochin and Skul'skii (1972)[65] found that Tl^+ cannot replace K^+ in the K-Na pump system in frog skin. Tl^+ is tightly bound in the membrane and inhibits the K-Na pump action. The Na^+, K^+-dependent ATPase of frog skin is activated to the same extent by one-tenth as much Tl^+ as K^+. The affinity of Tl^+ for the activating site was about 10 times that of K^+. The inhibition of Na^+ transport in frog skin by 0.5 to 1.0 mM thallous carbonate is similar to that caused by ouabain (Maslova et al., 1971[66]).

Adding Tl^+ to the incubation medium of Ehrlich or rat ovary ascites tumor cells increased the activity of the (Mg^{2+}, Na^+, K^+)- and Na^+, K^+)-dependent ATPase (Ivaschenko and Balmukhanov, 1974[67]). Thallium (5 mM) stimulated glycolysis and oxygen uptake by K^+-deficient tumor cell cultures (Ehrlich ascites, rat ovary ascites, sarcoma 37, lymphosarcoma NK, and Zajdela ascites hepatoma) more than K^+, Rb^+, Cs^+, ClO_4^-, or I^- (Ivaschenko et al., 1973[68]).

Tl^+ can replace K^+ in activation of the ATPase of fragmented red cell membranes. The concentration of K^+ required to activate ATPase in the presence of Na^+ was 10 times that of Tl^+ (Skul'skii et al., 1973[69]). The results of Lishko et al. (1973)[15] were similar. Tl^+ (0.1 to 1.0 mM, 20 to 200 ppm) increased the membrane ATPase activity of human erythrocytes and the efflux of Na^+ from the cells. Again, Tl^+ had 10 times the affinity for

ATPase that K^+ did. In human erythrocytes, 0.1 to 1.0 mM Tl^+ in a K^+-free medium, inhibits the ouabain-sensitive self-exchange of Na^+ but activated oubain-sensitive ^{22}Na outward transport and transport-related ATPase. The ouabain-sensitive ^{22}Na efflux and the (Na^+ + Tl^+)-ATPase were inhibited by 5 to 10 mM Tl^+. Higher Na^+ concentration inside the cell protected the system from Tl^+ inhibition at the higher concentrations (Skul'skii et al., 1975[70/]).

Tl^+ inhibits the Na^+, K^+-dependent ATPase activity of guinea pig hearts. The inhibition by Tl^+ is dose-dependent at concentrations of Na^+ and K^+ similar to those in intracellular fluid. Tl^+ has a dose-dependent positive inotropic effect on electrically stimulated left atrial strips that is not sensitive to pretreatment with propanolol. Although the onset of the Tl^+-induced positive inotropic effect does not depend on the number of contractions as with cardiac glycosides, the same degree of Na^+-pump inhibition is apparently produced by concentrations of Tl^+ and cardiac glycosides that produce similar inotropic effects (Ku et al., 1975[71/]).

At concentrations affecting oxidative phosphorylation, Tl^+ specifically inhibits both the influx and efflux of K^+ across the mitochondrial membrane. An energy-independent process was involved in the binding by mitochondria of 10 to 15 nmole $^{204}Tl^+$ per milligram protein (Barrerra and Gomex-Puyou, 1975[72/]).

Thallium apparently uncouples mitochondrial oxidative phosphorylation in energized mitochondria, with mitochondrial membranes able to differentiate Tl^+ from K^+. It stimulates succinate oxidation in rat liver mitochondria and is actively accumulated in mitochondria during ascorbate oxidation (Melnick et al., 1976[73/]).

Inturrisi (1969)[74/] reported that thallium can produce a decrease in Na^+-induced phosphorylation of microsomes and increase hydrolysis of ATPase. He ascribed the Tl^+-induced activation of the latter to a K^+-like dephosphorylation mechanism.

The results of magnetic relaxation and kinetic studies of the interaction of Tl^+ with Na^+, K^+-activated ATPase indicate that the transport site for monovalent cations is located in the active site. The monovalent cation shares a common ligand with Mn^{2+}, presumably the bound phosphate (Grisham et al., 1974[75/]).

Williams (1971)[58/] points out that with neutral ligands, the binding of potassium and thallium is similar, but anions bind Tl^+ about 10 times more strongly. He speculates that K^+ and Tl^+ in enzymes bind at monoanion phosphate and carboxylate centers whose "hole" size closely matches their ion radii of ~ 1.4 Å. Proton nuclear magnetic resonance studies conclusively show that Tl^+ and K^+ are bound in polyphosphate chelates in pyrophosphate, adenosine diphosphate, and adenosine triphosphate complexes. In ATP crystals, Tl^I is bound to all three phosphates. Stability constants for some pertinent Tl^I complexes are given in Table VII-9.

Lindegren (1971)[32,76] and Lindegren and Lindegren (1973)[77] report that thallium from a nutrient broth whose glucose has been exhausted is oxidized in the cristate mitochondria of yeast cells, the Tl_2O_3 staining the resting cells dark gray. The Tl_2O_3 crystals are discharged from the mitochondria along with fat into the nuclear vacuole. Since the cytoplasm is not coagulated, thallium must directly enter the mitochondria of up to 30% of the yeast cells, depending on conditions. The protoplast discharges the Tl_2O_3 crystal into a pinocytotic vesicle between the cell wall and the plasma membrane.

Although large amounts of Tl_2O_3 deposited in the mitochondria of Fleischmann bakers' yeast (a Saccharomyces species) when 125 ppm thallium was added to the nutrient broth, the cells proliferated as rapidly as the controls when planted on nutrient agar. Fleischmann bakers' yeast was the only yeast culture that oxidized thallium(I) (Lindegren and Lindegren, 1973[78]

The sequence of activation of the aminoacyl transferase by monovalent cations in the transfer of amino acids from aminoacyl tRNA to polypeptides is NH_4^+ greater than K^+ greater than Na^+, Rb^+, Tl^+ greater than Li^+, Cs^+ (Levine et al., 1966; cited by Lee[63]).

At 1 mM concentration (200 ppm), Tl^+ protects ribosome function (capacity for protein formation is lost in K^+-deficient medium and ribosomes dissociate) during K^+ deficiency. Tl^+ substitutes for K^+ in the polyuridine-directed binding of phenylalanyl-tRNA to ribosomes and also in the puromycin-induced release of nascent polypeptides (Hultin and Näslund, 1974[79]).

Lethal doses of thallium nitrate in mice reduced in vivo and in vitro (40 to 50 mM Tl^+) amino acid incorporation into protein and caused disaggregation of liver polyribosomes. The isolated ribosomes showed reduced activity in polyuridine-dependent poly(phenylalanine) synthesis as well, an unusual effect for a hepatotoxin. Hultin and Naslund (1974)[80] concluded that the strongly binding, heavy thallous ions probably block some translocation cycle reaction that is dependent upon recurrent translocations of K^+.

Thallium could influence amine-synthesizing or amine-degrading enzymes or transport mechanisms. Tl^+ ions did not affect the ATPase important for amine transport of the amine-storing granula of bovine adrenal medulla and splenic nerve. However, 5×10^{-6} M Tl^{3+} (1 ppm) inhibited the granula ATPase of the splenic nerve, and 3×10^{-5} M Tl^{3+} (6 ppm) inhibited that of the adrenal medulla (see also Burger and Starke, 1969).[81] It has not been shown in humans whether in vivo oxidation of Tl(I) to Tl(III) is responsible for altered catecholamine metabolism.[82]*

* Catecholamines are a group of sympathomimetic compounds, such as dopamine, norepinephrine, and epinephrine, derived from catechol (1,2-dihydroxybenzene) and an amine.

Starke et al. (1969)[83] injected rats with a single dose of 10 mg/kg* thallium as Tl_2SO_4 and found within 21 days neither blood pressure rise nor increased renal catechol excretion or alteration of the amine content of the heart, brain, or adrenals. During repeated injections of 2.5 mg/kg, the norepinephrine excretion increased on the average by 71 to 93 ng/mg creatinine.

b. Effects on mitosis, meiosis, and chromosomes: Thallium can inhibit or alter cell division in higher plants and animals.

Levan (1945; cited in Reference 84) noted that several metal salts and especially those of thallium act upon the meristems of *Allium cepa* (onion). Avanzi (1956)[85] enumerated the mitotic aberrations in *A. cepa* root tips induced by thallium acetate solutions (365 to 1,460 ppm Tl). Thallium acetate, acting for 2 to 12 hr, induced prophasic separations and prevented the functioning of the individual chromosomes' centromeres. A 20% solution of thallous ion (as thallium acetate) affected all mitotic phases in *A. cepa*, but a solution containing ~ 200 ppm Tl^+ decreased anaphases and teleophases more than prophases and metaphases (Nanetti and Marras, 1957;[86] and Nanetti, 1956[87]). Beltrami (1958)[88] reported that 20 to 2,000 ppm Tl^+ (as TlCl) inhibited mitosis in the vegetative apexes of *A. cepa* more and faster in dilute solutions. Truhaut (1964)[84] observed chromosomal abnormalities and antimitotic activity in *A. cepa* roots exposed to 0.001 *M* (~ 200 ppm) to 10 *M* thallium as the sulfate or acetate. His experiment on the antimitotic effect of thallium in regenerating rat kidneys gave equivocal results.

Treating larvae and nymphs of *Anopheles maculipennis atroparvus* with thallous acetate solutions (78 and 155 ppm) retarded mitosis and meiosis and caused death. Reproductive capacity compared with the controls was unimpaired, indicating the majority of gonad cells reached maturity. At high doses, thallous acetate disturbed or abolished spindle activity, disturbed or inactivated centromeric activity, and caused aberrant movement of at least one chromosome (Frizzi and Lecis, 1965[89]).

Verne and Sannie (1933 snd 1935, cited in Reference 84) reduced mitoses and growth in cultured chicken embryo heart fibroblasts by thallium chloride in Ringer's solution at dilutions *M*/10,800 (20 ppm) to *M*/1,200 (170 ppm).

In mammals, thallium accumulates in hair follicles and testes, tissues characterized by marked mitotic activity, and shows definite antimitotic activity antagonized by glutathione and cysteine.[23]

---

* LD_{50} = 13.2 mg/kg.

Thallium ion leads those of a group of very active metals (cadmium, copper, osmium, mercury, silver, titanium, tantalum, gold, platinum, chromium, and cobalt) in the ability to break chromosomes of pea plants. For thallium, the threshold value range (at which visible chromosome breakages in Pisum rootlets occur) is 0.2 to 2 ppm. The action of the metal ions is attributed to complex formation with phosphate, amino radicals, etc., causing the chromosome to lose its contractability and ability to enter normal division (Von Rosen, 1954[90/]). Von Rosen (1957)[91/] designated thallium (nitrate) as a typical radiomimetic metal causing mutagenesis in Pisum abyssinicum in pot experiments. Mutations were all chlorophyll mutations. Abnormalities arising from such metals are less frequent than those from irradiation because of a poisoning effect on vitality.

c. Antagonistic or synergistic effects: Hollo and Zlatarov (cited by Ganther, 1974[92/]) reported that feeding rats 6 to 9 mg sodium selenate daily for several days after a 30 mg/kg dose of thallium acetate prevented death. Most of the rats given only thallium died.

Rusiecki and Brzezinski (1966)[93/] found that selenate increased fecal excretion of thallium in poisoned animals, but thallium concentrations in the bones, liver, and kidney were increased two- to threefold. Perhaps, selenium combines with thallium to produce a less toxic compound.

Thallium does not protect against chronic selenosis induced by selenite. Levander and Argrett (1969)[94/] fed male Holtzmann rats 10 ppm selenium as sodium selenite or selenate in the diet for 7 weeks along with 10 ppm thallous acetate in their drinking water. Six of the seven initial rats survived with moderately severe gross liver damage. (The rats given no "protective" metal had severe damage.) Thallium caused a marked increase in selenium content of kidney tissue of selenate-fed rats. In acute studies, the rats were given s.c. doses of 30 mg/kg thallous acetate, quite close to the LD50. Thallium inhibited pulmonary selenium excretion in rats given subacute doses of sodium selenite.

In chronic thallium poisoning, dietary cystine, methionine, and betaine offer rats some protection. Thallium may interfere with the metabolism of sulfur-containing compounds, but it does not block free sulfhydryl groups in the skin. Compounds such as amino alkylisothiouronium, diphenylthiocarbazone, monothioglycerol, and mercaptopropane give varying protection, possibly by shielding protein sulfhydryl groups by forming reversible disulfide bonds with thallium (Stavinoha et al., 1959; cited by Oehme[17/]). Cysteine offers plants some protection. The cytological effect but not the toxic effect of 1,000 ppm Tl acetate on onion roots was inhibited by simultaneously treating the roots with 500 to 1,000 ppm cysteine (Avanzi, 1957[95/]).

Buschke and Konheim (1939)[96] reported that dihydrotachysterol counteracted the rachitic changes and growth inhibition caused by thallium. On the other hand, Selye et al. (1961)[97] studied the aggravation of symptoms of dihydrotachysterol overdosage by thallium. Combining a single dose of thallium acetate that produced no detectable organ lesions in the rat with a nonnephrotoxic dose (0.5 mg oral) of dihydrotachysterol caused severe nephrocalcinosis limited to the corticomedullary junction line and greatly aggravated calcification in the aorta or traumatized skin regions.

Selye et al. (1962)[98] found that concurrent administration to rats of sodium acetate, sodium citrate, sodium dihydrogen phosphate, and disodium hydrogen phosphate with thallium acetate induce renal calcification. With the latter two salts, the effect was intense and sometimes fatal. Sodium chloride did not have this effect.

Rygh (1949)[99] and (1950)[100] reported that zinc and thallium or barium and thallium produce decalcification of osseous tissue. Rats, which can produce their own l-ascorbic acid, developed scurvy when given vanadium, strontium, thallium, barium, or zinc along with a diet lacking methylnornarcotine.

Further discussion regarding compounds for which evidence of a protective effect in poisoning has been reported is in Subsection VII.B.1.a.(5) on therapy and antidotes and VII.B.2.f. on reproductive and teratogenic effects.

d. Other reactions: As of November 1976, proposals by Taylor and Shore to the U.S. Energy Research and Development Administration, Biomedical and Environmental Research Division, to study in vivo biochemical methylation of thallium and its effects on cellular toxicity[101] have not been funded.[102] Speciation of thallium and chronic administration were to have been the important aspects of the study.

In reactions with other group B vitamins, thallous acetate gave a yellow precipitate with riboflavin, and thallic chloride gave a dark brown precipitate at pH 8 to 9 with riboflavin, a light yellow precipitate at pH 3 to 4 with nicotinic acid, and a reddish violet color with B_6 azo dye. No reaction was observed with vitamin B_6 itself (Aliev and Gasanov, 1964[102]).

Tl(III) forms a 4:1 complex with bovine serum albumin (BSA). The degrees of dissociation of the metal-BSA complexes indicate that Tl(III) is bound less tightly than manganese(II) or copper(II) but more than cadmium, mercury, and lead divalent ions (Sudmeier and Pesek, 1971[104]).

Sundararajan and Whitney (1969)[105] studied the binding of Tl(I) to whole casein at pH 7.0, ionic strength 0.4, 2°C. About 4 moles thallium are bound per mole of casein.

Witschi (1965)[106] concluded that thallium (1 to 2 μCi ^{204}Tl) was not bound by human erythrocytes or plasma in vitro. After standing with the red cells for 1 hr, the ^{204}Tl could be removed by plasma, potassium chloride at pH 6.9, and phosphate at pH 6.0 and 7.4 and was distributed evenly between the cells and the washes.

Hatem (1961)[107] suggested that formation of the unstable complex of Tl(I) and histamine, represented as $[TlHi]^+$ (dissociation constant $\sim 10^{-3}$), was an initial process in the provocation of cancer by thallium poisoning. Champy and Hatem (1957)[108] suggested that the specific toxicity of thallium for neurons, especially sympathetic neurons, was due to their enrichment in histamine.

Cohen (1962; cited by Lee[63]) showed that Tl(I) has a very high affinity for the potassium-specific binding sites of Chlorella. It also competes with K^+ for binding sites on macromolecular substances in barley root tissues (Bange and Van Iren, 1970; cited by Lee[63]).

B. Human and Animal Toxicity

1. General toxicity

a. Humans

(1) Accidental and intentional exposures

(a) Symptoms and diagnosis: Humans have been poisoned by thallium compounds from therapeutic, cosmetic, accidental, and homicidal exposures.

A story that probably best illustrates the insidiousness of acute thallium poisoning is found in Prick et al. (1955).[3] A woman had attempted murder 13 times by putting water containing a dissolved Celio pellet (2% Tl_2SO_4) in the victims' food or beverage. Seven had died by the time a correct diagnosis was made and the murderess exposed. Among the widel[y] differing diagnoses were: encephalitis, brain tumor, bulbar paralysis, apopl[exy], tabes, alcoholic neuritis, endocrine disorders, typhus, pneumonia, and epilep[sy]. Human thallium poisoning victims have also been diagnosed as having trichloroethylene poisoning and arsenical polyneuritis (Polson and Tattersall, 1969[42]

Acute oral toxic doses of thallium usually produce severe hemorrhagic gastroenteritis within 24 hr, sometimes quickly followed by delirium and convulsions, then severe depression and coma. Respiratory depression with pneumonia or respiratory paralysis is the usual cause of death. Cardiac disturbance and dehydration are other causes of death. Death has also been attributed to progressive impairmant of the brain and vagus nerve. Symptoms do not usually appear before 12 hr and sometimes not before 48 hr. Neurological symptoms do not appear for 2 to 5 days. Other poisoned individuals may show stomatitis, mild gastroenteritis, emaciation, achylia gastrica, a bluish gum line, skin eruptions, rhinorrhea, conjuntivitis, blepharitis, strabismus, mydriasis, acroparesthesia, muscle and subcutaneous tissue pain, polyneuritis, blindness due to optic nerve atrophy, liver and kidney damage, incontinence, heart degeneration, tachycardia, conditions resembling angina pectoris, lymphocytosis, eosinophilia, hyperglycemia, glycosuria, acetonuria, amenorrhea, nail changes, encephalopathy, insomnia, drowsiness, restlessness, choreic movements, epileptic fits, auditory disorders, excitement, dementia, delirium, and convulsions.[3-5,42,109,110]

Thallium may be distinguished from other poisons causing polyneuritis by progressive sensory disability that begins in the soles of the feet and/or big toes, spreads to the trunk, and is less severe in the upper limibs. Motor phenomena, occurring later, include progressive tiredness, weakness, and finally paralysis of the legs. Pains occur in the joints and trunk. The polyneuritis may resemble Landry's syndrome. The diaphragm may be paralyzed after involvement of the vagus and phrenic nerves. Damage to the central nervous system produces convulsions or myoclonic or choreiform movements. Ptosis, strabismus, and optic atrophy may be observed (Polson and Tattersall, 1969[42]).

Horizontal white bands or white cross lines on the nails may be a late sign but are not necessarily characteristic of thallium poisoning.[42]

Masoud et al. (1973)[110] consider polyneuritis, alopecia with eyebrow involvement, and optic atrophy as diagnostic of chronic thallium poisoning. Polson and Tattersall (1969)[42] agree, but state that alopecia and polyneuritis usually do not occur for at least 10 days. Symonds (1953, cited by Polson and Tattersall), however, was unable to detect thallium in specimens from three patients coincidentally having polyneuritis, alopecia, and optic atrophy. Sleep disturbances may be characteristic with patients dozing by day but unable to sleep at night even when sedated with opiates.

Alopecia does not always occur even after severe poisoning. Bank et al. (1972)[82] suggest that "since clinical diagnosis depends on this inconstant sign, a significant number of unsuspected cases of thallium poisoning probably occur."

Prick et al. (1955)[3] regarded acute toxicity from thallium poisoning as illness ending in death after 7 to 10 days. Poisonings in which an acute phase passes into a "much more quietly developing syndrome, which sometimes lasted 6 months or longer and showed new symptoms during this period" were termed chronic. The subacute form was classed as having symptoms lasting no more than 6 weeks to 3 or 4 months.

Gettler and Weiss (1943; cited by Pritschow[111]) regarded cases that end in death within about 30 hr of a large dose of thallium as acute; in the subacute poisoning, the patient lives up to 3 or 4 weeks. The only characteristic sign of chronic poisoning is the hair loss.

Pritschow (1959)[111] used the following classification: "Acute toxicity" can be produced from single or multiple (short intervals) doses or exposure with short- or long-latent period, but the symptoms are always severe. The types of "subacute" toxicity are like those of acute toxicity except that the symptoms are mild. Chronic poisoning arises from multiple doses over a long interval, shows a long-latent period, and produces either severe or mild symptoms.

(b) Case histories: Clinical use dates from about 1898 when thallous acetate was prescribed as an "anti-hydroticum" for "night sweats" in phthisis (tuberculosis).[3] The dose was 100 mg at bedtime for no more than four successive nights. Leg pain, abdominal pain, and alopecia were reported in some patients.[4]

Thallium acetate was a widely used depilatory in ringworm treatment for almost 50 years. The therapeutic dose of 8.0 mg/kg body weight produced toxic symptoms in about 40% of the patients.* Numerous fatal therapeutic thallium poisonings were reported up until 1949. It was sometimes not appreciated by the prescriber that the dose could not be safely repeated within 3 months.[110] Prick et al. (1955)[3] report that only 6 of 114 children treated with 8 mg thallium/kg as an epilatory experienced muscle pain, diarrhea, stomatitis, joint pain, albuminuria, and folliculitis. Another 500 patients were treated without "serious symptoms," but 25% had leg pain, gastrointestinal disorders, conjunctivitis, and lichinoid rash. In other cases, an eruptic fever similar to measles or scarletina, sciatic neuritis, and "intestinal colics with obstipation" were observed.

Curry (1972)[112] remarks that thallium has figured in forensic toxicology since 1928 and is "in many geographical areas still a major weapon for the murderer."

The only mass human thallium poisoning in the United States due to its use as a rodenticide was the 1932 episode in California** involving tortillas made of Thalgrain (thallium-treated barley flour) stolen from a warehouse by a Mexican laborer. Thirty-one members of two families were exposed, 14 were hospitalized, and 6 people died from the acute poisoning

---

* In animals, 6 to 8 mg Tl_2SO_4/kg causes hair loss without other symptoms.[3]

** Another "outbreak of thallium poisoning" occurred in California in March 1931. The victims were treated for heavy metal poisoning, but thallium poisoning was not diagnosed at the time. Five adults completely recovered, but a 15-month-old child died.[113]

Twenty-two people developed alopecia and leg and feet pains (Munch et al., 1933[113]). Heyroth (1947)[5] described the preliminary report of 11 of these patients as due to a separate incident.

Reed et al. (1963, cited in Reference 114) stated that hundreds of cases of thallium poisoning from accidentally ingested pesticides had occurred in Texas since 1932. Over 130 children had been poisoned between 1954 and 1959.

Munch (1934)[115] reviewed the literature prior to January 1934 in an effort to determine the frequency of thallium poisoning. He tabulated injuries from industrial exposure, from clinical use as a depilatory given both orally and by injection; and from suicidal, homicidal, or accidental ingestion of pesticide preparations. Clinical use had poisoned 692 people, with 31 deaths reported.* There were 53 suicidal and homicidal poisonings with 10 deaths, and 21 accidental rodenticide poisonings with 5 deaths. Thus, 778 humans had been poisoned by thallium; 46 (6%) died. Of these poisonings, 447 (5.6%) occurred among 8,006 children treated with thallium acetate depilatory; 8 died.

Heyroth (1947)[5] reviewed the literature on thallium poisoning. Several more poisonings occurred, mostly due to ingestion of thallium salts, in the 14 years after Munch's summary.

Prick et al. (1955) in the book Thallium Poisoning[3] described numerous cases of acute thallium poisoning in humans.

Cavanagh et al. (1974)[116] have recently produced an extensive and critical review of the effects of thallium salts in humans and mammals, with particular emphasis on the changes found in the nervous system.

Adults are more susceptible to thallium poisoning than prepubertal children. The minimum lethal dose for an adult is about 800 mg (about 12 mg/kg).[3] Munch and Silver (1931)[4] stated that 8 mg/kg in adults had been toxic and that the maximal dose for any patient should be 300 mg.

Doses of about 14 mg/kg and above prove fatal to man in 2 to 3 days, death being from gastroenteritis, shock, and dehydration. Doses of about 7 to 14 mg/kg in one or two doses allow survival for a week or two with severe gastroenteritis. The peripheral neuritis usually developing during the second week produces progressive and fatal paralysis (Cavanagh, 1973[117]).

---

* At least six deaths occurred when thallium was used in the recommended dosage of 8 mg/kg.

Few human case histories qualify as due to chronic exposure other than occupational exposures. Gefel et al. (1970)[118] described a prolonged case of thallium poisoning in a man who either had repeatedly ingested it because of a mental disturbance or a criminal act by someone else. He eventually developed all the usual symptoms of full-blown thallium poisoning and died during his fifth hospital stay.

Patterson (1975)[119] reported the case of a 37-year old man of subnormal intelligence who exhibited choreiform movements in the limbs and thallium in the urine. Symptoms improved during treatment with diphenylhydantoin and bethanechol chloride and later levodopa. Patterson termed his symptoms a chronic thallotoxicosis and hypothesized that the man had been ingesting a thallium rodenticide obtained 10 to 15 years previously while employed as a stock clerk at a rural southern store. The first (gastr intestinal) symptoms of thallium poisoning appeared years after he worked there.

In both of these "chronic" cases the doses and dose rate of ingested thallium were unknown. A person exhibiting thallium toxici whether from chronic or acute dosing usually suffers disabling symptoms.

(2) Occupational exposures

(a) Symptoms: Prick et al. (1955)[3] generalized that chronic occupational thallium poisoning involves hair falling out, atrophic skin reactions with seborrheic symptoms, intestinal spasms, and psychic excitement.

Sessions and Goren (1947)[109] described symptoms from over-exposure in industry: appetite loss, dryness of the mouth, burning of the tongue, gingivitis, stomatitis, nausea, vomiting, diarrhea followed by constipation, fatigue, limb pain, alopecia, severe eye affection, kidney inflammation, albuminuria, polyneuritis, lymphocytosis, eosinophilia, and erythemas and other rashes. They recommended that in periodic examination of plant workers exposed to thallium, attention should be particularly directed toward discovering kidney dysfunction, albuminuria, casts, changes in the eye background, decreased visual acuity, blood changes, pain in the limbs, and falling out of hair. Workers should change clothes at work, wear protective gloves when handling thallium compounds, and use respirators in operations producing dust or fumes.

Initial industrial intoxication is characterized by excitement and insomnia. After exposure for a few weeks or months, leg join pain, muscular cramps and weakness, loss of the Achilles tendon reflex, and sometimes polyneuritis occur. The hair falls out after a few months. Saliv tion, anorexia, vomiting, diarrhea, weight loss, depression, fatigue, hyster laughter, cyanosis, tachycardia, bradycardia, and albuminuria are also noted

Meyer (1928, cited by Heyroth[5]) observed lymphocytosis in the blood of workmen exposed to thallium. About 40 to 58% of the leukocytes were lymphocytes. Sometimes, 70% of the cells were eosinophiles.

(b) Case histories: Munch (1934)[115] reviewed the injuries following industrial exposure to thallium compounds, citing Buschke and Langer (1927), Meyer (1928), Rube and Hendricks (1927),[120] Teleky (1928),[121] and his own unpublished data on two American dealers handling thallium salts and dusts, where no thallotoxicosis was evident. Of a total of 10 factories handling pyrites (1 plant, 6 affected workers), dust (2 plants, 4 poisonings), and thallium salts (7 plants, 2 poisonings at one), a total of 12 workers were affected, with no deaths. Nine recovered; at the Prussian plant handling thallium-containing dusts where four poisonings occurred, one worker lost his sight and two suffered chronic albuminaria.

Teleky (1928)[121] described the cases of workers inhaling the dust of metallic thallium, thallium oxide, and thallium sulfate. All but 3 of 14 men suffered the usual mild to severe symptoms of toxicity: sleep disturbances, stimulation, leg pains, cataracts, optic nerve atrophy, albuminuria, alopecia, blood changes. Neither the amounts taken in by the men nor the workplace air concentrations were reported.

Meyer (1928, cited by Richeson[38]) had reported on three of the poisoning cases, which had occurred during thallium recovery from flue dust at a sulfuric acid plant. One man was permanently blind after 3 months' exposure. All of the workers showed lymphocytosis and eosinophilia. These are probably three of the cases described above by Munch.

Rigner (1951, cited by Richeson[38]) reported poisonings of two Swedish thallium rodenticide workers. Initial symptoms in one of them resembled those of a heart attack.

Glömme and Sjöström (1955)[122] observed four cases of occupational poisoning initially exhibiting sensory changes in the extremities, minor and uncertain signs of polyneuritis, and alopecia. The highest value for thallium in the urine was 380 μg/liter. Due to improper handling, the main route of poisoning was probably the skin. Each case had occurred at a different Swedish plant manufacturing rodenticides.

Between October 1, 1950, and February 14, 1958, 15 men had been employed in a small Ohio operation separating industrial diamonds from abrasives. One of the steps involved the use of a solution of organic thallium salts to separate the diamonds from certain minerals on the basis of specific gravity. The thallium salts solution was reconcentrated by distillation in a fume hood. However, other steps with the solutions allowed the dried salts to accumulate on the outside of the vessels. Eleven of the 15 developed thallium poisoning symptoms between 3 weeks and 3 months after their first exposure. Gastrointestinal symptoms sometimes mimicked chronic appendicitis. High blood pressure, limb aches and pains, albuminuria, and irritability were common symptoms. Polyneuritic and neurological disturbances were mild or absent. Hair loss was not frequent. Thallium was not detected in the air of a new plant, so exposure was ascribed only to skin contact in two cases (Richeson, 1958[38]).

Egen (1955)[123] described the clinical course of thallium poisoning in a 45-year-old lithopone plant worker in Germany who had handled about 6 kg thallium/day (almost 1.75 tons/year) for 24 years while wearing protective clothing and mask. Landry-type ascending polyneuritis, speech and sight disturbances, and hair loss finally appeared.

At an industrial plant where concentrated solutions of thallium salts were used for heavy-medium separations, complaints indicative of thallium intoxication had arisen despite precautions--heating at low temperatures, washing of thallium-containing residuals, handwashing before eating or smoking, and careful housekeeping to prevent dust accumulation. After the complaints, a fume hood was installed and wet mopping substituted for sweeping and dusting. There was wide variation in the air contamination during working hours, with highest thallium-in-air concentrations occurring during centrifugation and screening of residual solids and concentrations of thallium solutions. Significant improvement was noted after new recommendations were followed: all concentrating operations and cooling of hot solutions were done in a fume hood, screen covers were used during screening, floors and tables were wet mopped, and dust scattering during handling was avoided (Hill and Murphy, 1959[124]).

Budrin and Meshcherskaya (1975)[125] described a fatal poisoning of a 25-year-old woman who accidentally swallowed a reagent containing thallium formate-malonate during a titration. The dose received was not estimated.

Tikhova (1964)[126] (1967)[127] stated that during sulfatizing roasting of thallium-containing dusts from lead production, workers were also exposed to sulfuric acid vapors and fine dusts (containing up to 50% lead) in concentrations attaining 30 to 62.5 mg/m^3 at the granulator. In processes involving thallium precipitation with sodium sulfide, the workers were also exposed to hydrogen sulfide (0.028 to 0.1 mg/liter). Arsine evolves during thallium cementation at 0.0031 to 0.0058 mg/liter.

During melting and pouring of thallium, thallium oxides attained concentrations of 0.026 to 0.1 mg/m^3 in the air. A maximum concentration of 0.12 to 0.18 mg/m^3 thallium occurred at the end of melting. Volatile thallium chloride also evolved. During pouring, the entire surface of the thallium became oxidized and total thallium concentrations in the air were 13.5 to 17.4 mg/m^3.

The amount of thallium at the end of work (4 to
hr) on the palms of a worker engaged in producing thallium sulfate and
composing the bichromates was 300 to 350 μg compared with 5 to 20 μg at
e beginning. While thallium was being dissolved, the average air concen-
ation was 0.117 mg/m^3; and during packing the concentration was 0.274
$/m^3$. The wash from the hands of the workers rose from 5 to 20 μg at the
art of work to 350 to 480 μg at its end.

Production of thallium bromide-thallium iodide (KRS-5)
ystals comprises cutting the crystals with a disk saw, machining on a
the, and wet grinding. During cutting and machining, the thallium concen-
ation in the air was 0.0036 to 0.0072 mg/m^3. By the exhaust hood during
inding, the concentration was 0.0036 mg/m^3. Accumulation on the workers'
nds did not exceed 5 μg.[126,127]

The laboratories of E. C. Taylor of Princeton
iversity and A. McKillop of the University of East Anglia in England for
e past decade have been primary centers of research on the use of thal-
um in organic synthesis. Worker precautions include the wearing of rub-
r gloves at all times, conducting operations in a fume hood, and good
usekeeping practices. After a person shows signs of thallium poisoning,
rther contact with the metal is avoided until urine tests for thallium
e negative.[128]

(3) Accepted levels of exposure: In the United States,
e threshold limit value (TLV) for thallium soluble compounds in workplace
r is 0.1 mg/m^3.[129,130] The value, according to the American Conference
Governmental Industrial Hygienists,[114] is largely based on analogy with
her highly toxic metals, since satisfactory data do not exist from which
derive a threshold limit. The TLV has been adopted as an OSHA limit.
e U.S. standards are applied in Argentina, Great Britain, Norway, and
ru. The Federal Republic of Germany also has set 0.1 mg/m^3 as the limit.
e USSR, as of 1972, stipulated 0.01 mg/m^3 as the ceiling value for thal-
um in the work environment. (This ceiling had been in effect since at
ast 1964.[131]) East Germany, Sweden, and Czechoslovakia do not have a
allium standard.[132]

TRW Systems Group[133] in 1973 suggested provisional limits
air of 0.001 mg/m^3 and in potable water and soil of 0.005 mg/liter (5 ppb).
value of 5 ppb for soil scarcely seems realistic when background levels can
almost a thousand times higher.

Dawson (1974)[134] suggested 0.1 mg/liter (0.1 ppm) as the
sired maximum ambient environmental level for aquatic life, based on a com-
rison of marine water limits of the National Academy of Sciences (0.05 mg/
ter[135]) and proposed Environmental Protection Agency Numerical Criteria
r Water Quality (1/20 LC_{50} = 0.1 mg/liter[135]). Dawson also suggested a
mit of 10 mg thallium/liter irrigation water (10 ppm) presumably based on
e very limited information in McKee and Wolf (1963).[136]

As of 1974, there were no EPA, NAS, or NTAC criteria fo thallium in irrigation or livestock waters; freshwater (aquatic life, wild-life, and public supply); or recreational waters.[135]

(4) Epidemiology: The incidences of poisoning among populations exposed to thallium therapeutically or occupationally are described above. We have found no suggestions of environmental thallium bei the cause of human illness. In the Alsar Region of Macedonia, Yugoslavia, soil thallium levels are so high naturally that grazing animals have been poisoned by the pasture plants. An examination of human medical records i this region could be helpful in understanding the potential human health hazards of thallium in the environment.

Williams and Riegert (1971)[137] described an epidemic of alopecia areata in 12 men at a Canadian oil refinery employing 190 peop The outbreak occurred between July 1962 and September 1963, in eleven of t 60 maintenance workers and an office accountant. Thallium was ruled out b cause the alopecia was not diffuse and there were no signs of systemic poi ing. There had also been no use of rat poison in the refinery. There was no recurrence within the following 5 years. However, urine samples were analyzed polarographically for thallium. Half of the initial urine specimens contained 8 to 15 μg Tl/liter. Some of the unaffected workers had similar levels. After 3 months, some workers were excreting more thallium some less, and some the same amount. The several hundred chemicals used a produced at the plant could not all be analyzed; but thallium was found in epoxy paint (0.55 ppm), coal tar epoxy (0.12 ppm), wiping rags (a trace), dust on the locker room floor (0.01 to 0.22 ppm, with the higher levels ne the lockers of the maintenance workers).

(5) Therapy and antidotes: Heyroth (1947)[5] summariz early literature on therapy of thallium poisoning. Stevens et al. (1974)[4] and Munch (1970)[30] have reviewed all the kinds of treatments and antidote that have not proved especially helpful in improving the prognosis of thal-lotoxicosis. The former review is the more extensive. It also discusses hemodialysis as a treatment (thought by the authors to be helpful only wit the first 24 hr after acute intoxication) and Prussian blue as an effectiv antidote, first recommended ca. 1970. Table VII-10 summarizes most of thi literature.

Prussian blue (PB) is potassium ferrihexacyanoferrate($[KFe^{III}Fe^{II}(CN)_6 \cdot \underline{n}\ H_2O]$, soluble colloidal form. It acts an an ion exchan for univalent ions, its affinity increasing with increasing ionic radius i the order Li^+ less than Na^+ less than K^+ less than Cs^+. PB had been used treat ^{137}Cs poisoning, but Heydlauf (1969)[138] was the first to use insolu PB to remove thallium from the tissues of acutely intoxicated rats. Kamer et al. (1971)[139] confirmed that the colloidal form of PB had a higher bin ing capacity. They were the first to treat humans with PB, which is appar not toxic itself and does not release CN^-. Tl^+ is exchanged for both stoichiometrically and nonstoichiometrically bound K^+ ions, thereby preventing reabsorption from the intestine and increasing the inflow of thallium into the gut by increasing the difference between intra- and extraluminal free thallium concentration.[48/] The complex is hardly absorbed and is stable; so is the thallium complex.

In clinical trials, Stevens et al. (1974)[48/] obtained very good results with PB treating patients, who had taken up to 1 g thallium, shortly after the intoxication. None developed clinical symptoms of thallium poisoning. Good results were obtained even when treatments were started late.

Rauws (1974)[16/] reported that in rats the half-life of radiothallium of about 4 days was reduced to about 2 days by Prussian blue pretreatment. He concluded that reabsorption of thallium is inhibited only about 70% by PB.

In the most recent reference found on the topic, Manninen et al. (1976)[140/] reported that PB increased the survival of rats treated with lethal doses of thallium from 0 to 50%.

b. Animals: The following working definitions for thallium dosing have been used to subdivide the animal studies on thallium toxicity:

Acute	One dose
Subacute	Dosing for 2 days to 4 weeks
Subchronic	Dosing for 6 to 13 weeks
Chronic	Dosing for more than 13 weeks

(1) Mammals

(a) Acute studies: Table VII-11 summarizes the published information regarding acute toxicity of thallium compounds to mammals where an LD_{50} or LD_{Lo} has been determined. For inorganic thallous compounds, the LD_{50}'s for all routes of administration fall in the range of about 15.8 to 71 mg/kg; for thallic compounds, 5.66 to 72 mg/kg.

Il'in et al. (1938)[1/] found certain trivalent thallium compounds are among its most toxic forms. Buschke and Peiser (1925) reported that organiothallium(III) compounds, such as $(CH_3)_2TlBr$ are about 10 times as toxic to mice as thallous acetate, the lethal dose to mice of $(CH_3)_2TlBr$ being 5 to 6 mg/kg.[4/]

Heyroth (1947)[5] reviewed the medical literature on thallium poisoning, citing 282 references; and Munch and Silver (1931)[4] reviewed the pharmacology of thallium, citing 148 references. Both refer t animals and human cases. The symptoms produced are much the same as in hum poisoning, so their descriptions will not be reproduced here in any depth.

The nervous system, digestive tract, and circulato system of animals are damaged by acute thallium poisoning as shown by restlessness, tremors, ataxic gait, convulsive movements, then partial paralysi of the legs, weight loss, bloody diarrhea or constipation, dypsnea, and respiratory failure. Only two reports (earlier than 1947) mention cardiac depression. Polydipsia and polyuria or anuria are sometimes noted.[5]

Truhaut (1960)[23] reported that the principal injured organs of rats poisoned by thallium were the skin (general hyperplasi hair follicle atrophy), testis (spermatogenesis inhibition), and kidney (vacuolar degeneration of the convoluted tubules). Diffuse inflammatory lesions, often hemorrhagic, occurred in other organs. An acutely intoxicated rabbit showed degenerative lesions in the white matter of the spinal cord.

Munch and Silver (1931)[4] found that the "minimum lethal dose" of thallium sulfate (the lowest dose killing all or practicall all the test animals within 5 days) was about 25 mg Tl/kg, both for rats given thallium orally and rabbits given thallium i.v. By comparison, the minimum lethal dose for strychnine is 20 to 25 mg/kg; for arsenious oxide, 100 mg/kg; red squill powder, 250 mg/kg; and barium carbonate, 750 mg/kg.

Chusid and Kopeloff (1962)[141] bilaterally implant sharp thallium fragments (sterilized for at least 24 hr in 70% alcohol) in brain of a Macaca mulatta monkey. The monkey "appeared toxic from the time of operation and died 6 days later," a much quicker exertion of toxicity th given by the other fatally toxic implants (nickel, antimony, and cadmium); "no gross residuals of thallium were recognizable" at the implantation site

Truhant (1964)[84] removed the left kidneys of four rats and 80 hr later injected them s.c. with 20 mg thallium metal/kg. Thre of the animals died within 24 hr.

Levkovich (1938)[142] studied the effects of extern temperature, diet, and stress on the action of sublethal oral doses of thal acetate and thallous aluminum sulfate, $TlAl(SO_4)_2 \cdot 12H_2O$, to sheep and Angor rabbits (304 animals). Lower temperature (-3°C compared with 11 or 24°C) caused a higher weight loss (the toxic progress "much slower and less sharp" but mortality was reduced and toxic diarrhea did not occur. A hunger diet after dosing caused hair loss in 46.7% of the animals, diarrhea in 96.7% an death in 10.0%; whereas, on a normal diet, diarrhea and death did not occur and hair loss was only 3.3%. Sheep under pasture conditions showed no hair loss or death and only a 1.7% weight loss. Large stalls allowing freedom of movement also favored thallium resistance; but when confined in pens with dry food, 66.7% of the sheep showed hair loss, strong toxic effects, and 15.1% average weight loss.

Case (1974)[143/] described symptoms of thallium poisoning in sheep exposed due to careless distribution or storage of baits: inappetence, excessive salivation, depression, stasis of the rumen, progressive weakness, with death occurring within 1 day to 2 weeks. Surviving sheep may or may not shed nearly all of their wool. The pathological findings include intense hemorrhagic gastroenteritis, toxic hepatosis, parenchymatous nephritis; and hyperemia of the brain, kidneys, and spleen.

B. Munch et al. (1974)[144/] reported findings in Denmark in the period 1963 to 1970 of thallium poisonings in 48 red foxes and 5 badgers taken for examination because they were dead or obviously diseased. Central nervous system and gastrointestinal symptoms were often noted. Fox stomachs were empty, but the remains of mice in badger stomachs indicated they had been secondarily poisoned.

Clausen and Karlog (1974)[145/] examined 60 wild martens and badgers over a 2-year period in Denmark. Twenty-two percent were actually poisoned by thallium, while another 26% showed thallium in their tissues. Poisoning was slow developing and mostly seen in well-nourished animals.

Mention of accidental thallium poisoning in dogs and/or cats is rather frequent in recent literature; e.g., DeSloovere et al. (1971),[146/] Carpenter (1971),[147/] Atkins and Johnson (1975),[148/] and Breukink and van der Lee (1976).[149/] An older bibliography of canine poisoning is given by Vismara (1961).[150/] Gettler and Weiss (1943, cited by Browning[151/]) estimated the lethal s.c. or i.v. dose for dogs as 12 to 15 mg/kg.

Two dogs infused through the right carotid artery with 1.00 and 1.75 mg Tl^+/kg/min (as Tl_2SO_4) lived 4 hr 13 min, and 2 hr 20 min, respectively, receiving 245 to 253 mg Tl^+/kg. Autopsy of both animals showed edema and acute passive congestion of the lungs. Urinary excretion of Tl^+ began within 20 to 30 min of infusion (Pritschow, 1959[111/]).

When 80 male Angora rabbits were given thallium acetate hypodermically, cutaneously, or i.v. (5% solutions), or given thallium acetate powder per os in bread boli, the pattern of rather mild symptoms*

---

* General depression, loss of appetite, diarrhea, weight changes, hypersalivation, and increased nasal secretion. Only one rabbit showed disturbed movement coordination.

experienced was very similar regardless of route. Sixty-four rabbits molted totally; only four died. The doses received were not estimated, but in preliminary experiments of 10 rabbits receiving 20 to 165 mg doses cutaneously, 6 died; of 6 receiving 20 to 25 mg by hypodermic, 4 died; of 9 receiving 11 to 25 mg i.v., 2 died; of 5 receiving 20 to 25 mg into the stomach with oil, 2 died; of 11 receiving 20 to 30 mg into the rectum, 2 died; and of 5 receiving 25 mg in boli, 1 died (Il,in, 1938[1/]).

In rats fatally poisoned by 30 mg/kg thallium sulfate administered by stomach tube (average survival time 71.2 hr), serum glutamic-pyruvic transaminase and serum glutamic-oxalacetic transaminase activities increased. For the former transaminase, the activity increased from 30 to 40 units in the 12th hr after poisoning; to 70 units, 24th hr; and to 97 units, 48th hr. For the latter transaminase, same times, the increases were from 45 to 63, 94, and 105 units, respectively. Adrenalectomy increased the average survival time to 73.8 hr. Since it did not prevent thallium deposits in hair roots, the effect of thallium on the hair is apparently not through the adrenals (Mrozikiewicz and Widy, 1967[152/]).

Selye and Mécs (1974)[153/] compared the effects upon thallium toxicity in rats of bile-duct ligature, partial hepatectomy, partial nephrectomy, dosing with a catatoxic* steroid (PCN) or dosing with a syntoxic** steroid (triamcinolone). The Sprague-Dawley rats (10/group) were injected s.c. with either 16 mg thallium chloride/100 g body weight (the approximate weight of each rat) for the steroid-treated rats, or 10.6 mg/100 g for the surgically treated rats in 0.5 ml peanut oil. Only triamcinolone had any protective effect against mortality (reckoned on the fourth day) or on nephrocalcinosis estimated on the day of death.

Sanotskii (1961)[7/] determined that the subcutaneous LD50 for thallous carbonate in adult albino mice is about 27 mg/kg. That same dose given to juvenile, aged, and pregnant mice gave mortalities of about 37, 70, and 39%, respectively. The progeny of the pregnant mice were stillborn.

Sanotskii noted in histological examination lobar pneumonia in some animals after endotracheal administration. Acute toxicity was not closely related to the solubility of the thallium compound. Even one-time s.c. doses of 0.00024 mg Tl_2CO_3/kg in rats disturbed hair regeneration. (This would correspond to ~ 15 μg in a 70-kg human.) Compare the excretion of thallium in Canadian workers suffering alopecia areata in Subsect VII.B.1.a.(4).

---

* Catatoxic - assisting biodegradation of a toxicant.

** Syntoxic - improving tissue tolerance of a pathogen without attacking it

(b) Subacute studies: Two dogs fed 5 mg/kg/day chronic experiments" died on the seventh and eighth days. The dis-
ıtion pattern of thallium in the tissues observed by Pritschow (1959)[111]
ed with that found by Gettler and Weiss, Fridi, Thyresson, Durbin et al.,
Heyndrickx. Pritschow did not detect thallium in the feces of these ani-
. These chronically poisoned animals contained up to 107 ppm thallium
rt) in their tissues; two acutely poisoned dogs by infusion contained
o 1,540 ppm (spleen). (The thallium determination was based upon pre-
tation of $(TlI)_2 \cdot BiI_3$ from the acid-ashed organic material, oxidation
bromine, and iodometric titration.)

Swain and Bateman (1910, cited by Heyroth[5]) reported
200 mg thallium distributed in doses over 13 days was as toxic to a dog
similar amount given within 4 days.

Daily oral doses of 0.2 mg thallous acetate for sev-
weeks severely poisoned rats (Buschke and Peiser, 1922; cited by
oth[5]).

Rats given 6.5 mg thallium/kg s.c. every other day
weeks showed decreased red blood count, catalase activity, prothrombin
and phosphatase activity and increased blood sugar. Some of the effects
e intoxication were alleviated by feeding 2.5 mg vitamin B_{12}/kg (Malachovskis,
[54]).

In subacute tests, male albino rats poisoned by in-
stric* doses of 0.6 or 1.25 mg/kg thallous sulfate daily for 15 days re-
catalase activity 48%; thallous carbonate (2.5 or 5 mg/kg for 4 to 11
, 51%. The relative content of the enzyme per million erythrocytes was
ed an average 41.6%. Excretion of hippuric acid at the end of the
ous sulfate poisoning period, after s.c. dosing with 400 mg sodium ben-
over 6 hr, was 53% less than that of the controls. Significant changes
e ratio of the protein fractions of the blood serum appeared toward the
day. The relative content of albumins decreased 38.2% from the starting
; the relative content of globulins increased correspondingly by 26.3%;
mount of total protein was constant (Tikhova, 1964[126]).

Downs et al. (1960)[22] fed male weanling rats diets
ining 0.0002% (about 1.6 ppm thallium) and 0.001% thallous acetate for
th. There was no effect on growth and all survived (five in each group).
005 to 0.5% thallous acetate in the diet, mortality was 60 to 100% within
ys.

---

ven incorrectly in Chem. Abstr., 71(12) 53248j (1969) as i.p. The
Russian versions state that the dose was introduced into the stomach.
The word for i.p. "vnutribryushinnyi" was not used.

Sanotskii (1961)[7] exposed five albino rats to 0.0005 to 0.001 mg/liter (average 0.005 mg/liter) of mixed TlI-TlBr dust 8 to 10 times within a month. At the end of the month, two of the poisoned rats and one of the controls had died from unrelated causes. The length of time the remaining poisoned rats could swim in 17°C water was somewhat less than that of the controls. When 20 rats were exposed to a 1:1 mixture of TlI and TlBr for 2 to 3 weeks at ~ 0.0005 to 0.0007 mg/liter, toxic effects were not observed although thallium accumulated in their bodies. The rats were then exposed to 0.0012 to 0.0152 mg/liter and acute intoxication was observed. Poisoning was not seen when five rats were exposed daily except Sundays for 52 days to 0.00038 mg/liter.

(c) <u>Subchronic studies</u>: Toya et al. (1961)[155] studied the effects of 8 ppm thallium in the drinking water for 80 days on endogenous copper and zinc in rat tissues. Copper levels were less than 30% of the controls in the cerebrum, lung, liver, muscle, and blood. Zinc decreased in the cerebellum.

Downs et al. (1960)[22] observed some sex differences in response to feeding 0.003% thallous acetate diets to weanling rats for 60 days. Males showed a moderate growth depression. Alopecia occurred. Within the fourth to the eighth week, 80% of the males and 60% of the females died. See also Section (d) chronic studies.

Dzialek (1965)[156] dosed 1- to 18-month-old rats with 0.1 to 0.11 mg thallium acetate/day for more than 40 days. The younger rats were more resistant, mostly showing depilatory effects. Five- to six-month-old rats lost weight, and the old rats frequently showed decreased mobility, discoordination of movements, and blindness.

Sappino (1936)[157] reported that the water content of the thymus of poisoned white rats* was reduced especially in the females, whereas the males showed reduced water content in the thyroid, but the females did not. Guinea pigs showed a slight reduction in water content of the thymus but not in the thyroid. Only the rats showed any organ weight changes: both the thymus and the thyroid weights decreased.

(d) <u>Chronic studies</u>: Herman and Bensch (1964)[158] injected 5 to 50 mg/kg thallium acetate s.c. into Sprague-Dawley rats at weekly intervals (duration not given in the reference, which is an abstract). The acute histologic changes comprised enteritis (mainly in the colon) and renal tubular changes (nearly spherical cytoplasmic protrusions into the tubular lumina, especially in the proximal tubules). Chronic histologic cha were loss of mitochondrial cristae; accumulation of electron-dense material within the mitochondria, especially in the liver; and intranuclear inclusion bodies and degenerating cells in the brain.

---

* About 0.75 to 1.5 mg/kg s.c. for 22 to 50 days.

Chronically poisoned rats (Herman and Bensch, 1967[159]) who received 10 to 20 mg thallium acetate/kg followed by weekly s.c. injections of 5 mg/kg (one or two injections may have been reduced to 2.5 mg/kg or withheld because of signs of toxicity) and were killed at 4 to 26 weeks after the first injection, showed alopecia occasionally with black speckling of periorbital hairs, sometimes persisting up to 4 months; irritability during handling about 3 to 8 weeks after the initial injection; diarrhea, periorbital redness, and conjunctivitis in two of the 15 rats; poor hair luster; cyanosis of the tail or loss of a distal tail segment in three rats; hind limb dragging in two rats for less than a week at 1 and 7 weeks postinjection; and poor weight gain. No cataracts were observed. Their organ pathology is discussed in Subsection VII.B.2.

Downs et al. (1960)[22] had shown in subacute tests that 60 to 100% of male weanling rats fed at least 0.005% thallous acetate in the diet died within 10 days. They fed weanling rats (ten per group, five each sex) a diet containing 0.0005% (total dose ~ 10 mg/rat) and 0.0015% thallous acetate for 15 weeks. The rats grew as well as the controls, but alopecia was observed at the higher dose (equivalent to about 1 to 3 mg/kg/day). On a 0.005% diet, deaths began in the second week and most were dead within 2 weeks. There were no survivors.

When a mixed sex group of weanling rats was fed 0.002% thallic oxide in the diet for 15 weeks, hair follicle atrophy and alopecia were observed (total dose ~ 40 mg/rat). Males showed a body weight depression at the fourth week as high as that of females fed a 0.0035% diet. All males survived. Two females died. Females grew as well as the controls.[22]

Fed 0.0035% thallic oxide, the weanling rats showed alopecia. The body weight of the males was markedly depressed; the weight depression was less marked for the females. Only one male rat survived, most deaths occurring after the first month. Again, two female deaths occurred that were probably not related to thallium. At 0.005 to 0.05% thallic oxide, severe growth depression and increased mortality were observed.[22]

Tikhova (1964,[126] 1967[127]) gave daily doses of 0.35 mg Tl_2SO_4/kg (oral or s.c.) or 0.25 mg Tl_2CO_3/kg (oral) to rabbits (1% of the absolute lethal dose) for 6 months (total: ~ 100 mg/rabbit). Towards the end of this period (5 months) behavioral changes were noted: aggressiveness, retardation in some, and paralysis of the rear limibs. Poisoning by thallous sulfate reduced the alkaline phosphatase 62% (by thallous carbonate, 72%). In the chronic experiment in the course of 6 months, the relative decrease of the content of albumins in the blood serum was 30% less than the starting level and was 27.6% and 25.5% less in additional oral dosing tests with the sulfate and carbonate. The hyperglobulinemia was principally caused by the β-globulins. The development of hypoalbuminemia and, correspondingly, hyperglobulinemia led to a change of the ratio of the albumin-globulin coefficient. The content of SH groups in the serum decreased from 70 to 100 μmole/100 ml in the controls to 35 to 40 μmole/100 ml in the poisoned animals.

(2) Birds: Ward (1932)[160] determined that the minimum lethal dose (MLD) i.p. in female mallard ducks is about 25 mg/kg. Wild geese had been killed by eating thallium-poisoned barley. The oral MLD from this source is about 50 mg/kg. Feathers on the back just above the uropygial gland were lost after ingesting sublethal doses. Liver and kidneys contained 6 to 11% of the administered dose following deaths within 1.25 to 13 days.

Linsdale (1931, cited in Reference 44) had attributed the deaths of many California game birds to thallium poisoning, but laboratory examinations were negative.

Shaw (1934)[44] determined an approximate oral MLD of about 30 mg/kg (thallium from thallous sulfate as the dry salt) in domestic mallards and wild white geese. When given in solution or coated on grain, the MLD was about 15 mg/kg. Geese appeared more susceptible than ducks. Starved quail were most susceptible, with an MLD of at most 12 mg/kg. The deposition in the tissues of geese was dependent on dose and time that elapsed before death. The relative concentrations found decreased approximately in the order: thigh greater than bone, breast, or liver greater than intestines or brain greater than kidney greater than heart much greater than fat. (The concentration factor from the dosage to the thigh muscle was ~ 1 to 2.) The tissues still retained 33 to 71% of the original dosage after 13 days. Shaw concluded that secondary poisoning in a human from eating a poisoned game bird was improbable.*

Cromartie et al. (1975)[161] report that of 37 bald eagles found sick or dying in 1971 to 1972 in the United States, all contained organochlorine pesticides or metabolites and nine had been poisoned by thallium as well. Most of the birds had died from illegal shooting.

* Continual eating of such birds might have been slightly more hazardous. For example, a bird ingesting 20 mg/kg did not die for 13 days. Its breast and thigh muscles contained 10.8 and 14.2 mg/kg at that time. A man eating from 0.25 to 0.5 lb of such fowl per day would ingest about 2 to 4 mg thallium daily. The value is certainly much lower than the recommended single therapeutic epilant dose for an adult human of 300 mg but about 1,000 times more than the estimated normal dietary intake of thallium.

Bean and Hudson (1976)[162] gave three fasted immature golden eagles (Aquila chrysaetos) single doses of thallium sulfate in a capsule inserted through glass tubing to the level of the proventiculus. Bird No. 1, given 60 mg/kg, survived with remission in symptoms at 4 days. Birds 2 and 3, given 120 mg/kg, died at 6.5 and 4.5 days after ingestion. Symptoms included loss of coordination, imbalance, and falling, reluctance to move, and finally immobility; anorexia; belligerence and fear-threat displays; and slow and labored breathing. At least 80% of the thallium ingested was recovered in the tissues and feces of Bird No. 3, indicating high absorption and slow excretion. Bird No. 3, which died sooner, had almost three times as much thallium in its kidneys as Bird No. 2; 3 ppm was detected in the feathers of Bird No. 2.

(3) Fish: Nehring (1962)[163] reported that 10 to 15 ppm thallium as the nitrate is lethal to rainbow trout (Salmo gairdneri), 60 ppm to perch (Perca fluviatilis), and 40 to 60 ppm to roach (Rutilis rutilis) (threshold values for the appearance of toxicity--presumably death--within 3 days). At 4 ppm, trout were killed after 14 days and the roach, after 8 days. Neither species was harmed at 2 ppm Tl^+ for 17 days. Perch withstood 15 ppm Tl^+ for 17 days without effect, but were killed after 14 days at 20 ppm. Perch were killed in 0.4 day at 500 ppm Tl^+. Poisoning symptoms included deleterious effects on blood pressure and the nervous system.

Zitko et al. (1975)[164] reported that 0.03 ppm is the LD_{50} for Atlantic salmon.

At 400 ppm of thallium as the chloride, a 30 to 50 g toadfish (Porichthys) died within 50 hr after it had shown signs of respiratory distress; and the solution killed another in 53 hr. At twice that concentration, a toadfish died in 20 hr. A rock cod (Cottidae) was killed in a few hours by respiratory distress at the lower concentration. Poisoned fish swam in unusual positions. Injection of a 20 mg/kg thallium solution into the lymphatic trunk killed a blue cod (Ophiodon elongatus) by the next day. Half that concentration killed a bluefish (Sebastodes) within 4 days; another died in 3 days after 16 mg/kg (Swain and Bateman, 1910; cited in Reference 4).

(4) Other animals: Effects on the reproduction of birds, amphibians, and mammals are discussed in Subsection VII.B.2.f. The effects of thallium on Paramecium bursaria are described in Subsection VII.C.2 because of the presence of an algal endosymbiont, which allows the host to withstand exposure up to 9 ppm thallium concentrations before greater than 50% suppression.[165,166]

Nehring (1962)[163] found that the threshold limit for the appearance of thallium toxicity (lethality) within 3 days was 2 to 4 ppm Tl^+ for Daphnia (water fleas) and 4 ppm for Gammarus (swimming amphipod crustaceans).

Dilling and Healey (1926)[167] found that 0.4 ppm thallium is fatal to tadpoles. Thallium was reported to retard tadpole growth and metamorphosis (Buschke and Peiser, 1923; cited by Heyroth[5]). However, thallium did not antogonize iodine or thyroid extract during tadpole metamorphosis (Huxley and Curtis, 1928; cited by Heyroth[5]). Thallium may have slightly accelerated metamorphosis.

2. Special toxicities to animal tissues, organs, and organ systems

a. Histopathology: Pritschow (1959)[111] reviewed the histological literature according to the changes--parenchymatous degeneration, fatty infiltration, hemorrhage, congestion, edema, inflammation, karyolysis, vacuolation, necrosis, and atrophy--in the various organs of thallium-poisone humans and animals. The digestive system, nervous system, and kidneys contai the predominant lesions.

Vismara (1961)[150] described the severe degenerative-necrotic and hemorrhagic changes in the liver and kidneys of four dogs poisoned by thallium. The skin of one animal showed serious changes; another, severe changes in the encephalon.

Herman and Bensch (1964[158], 1967[159]) performed extensive light and electron microscopic studies of Sprague-Dawley rats acutely, subacutely, and chronically poisoned by thallium. All but one of the eight rats of the acute group received 20 to 50 mg thallium acetate/kg s.c. The subacute group (four rats) received 2 to 3 injections of 10 to 15 mg/kg at intervals of 1 week. The chronic group (15 rats) received initially 10 to 20 mg/kg usually with weekly injections of 5 mg/kg thereafter. Members of the acute group were killed at intervals up to 5 days; the subacute group, 16 days; and the chronic group 4 to 26 weeks after the first injection. The acute animals showed diarrhea, marked weight loss, anorexia, and lethargy. The subacute group displayed similar symptoms with a later onset. The symptoms of the chronic group are given in Subsection VII.B.1.b.(1).(d).

The light microscope revealed in the acutely poisoned rats: renal eosinophilic casts, enteritis, and severe colitis. Foci of perivascular cuffing was often seen in the subacute group. The brain showed a rare focus of recent necrosis. The chronic group organs appeared unchanged with the light microscope.

All of the groups consistently showed changes in the kidney, liver, and intestine under the electron microscope. In the kidney, swollen tubule cells, dilated or partly ruptured endoplasmic reticulum; intraluminal casts in the tubules; and increased numbers of autophagic vacuoles, lysosomes, lipid droplets, and residual bodies in the tubular cells. Autophagic vacuoles, lysosomes, lipid droplets, and mitochondrial changes are not specific for thallium poisoning and are similar to cellular injury sequelae of many agents.

Mitochondria of the kidney, liver, and intestine most frequently showed the degenerative changes: swelling (thought to be due to thallium accumulation within them), partial or total loss of cristae and mitochondrial granules, increased size and density of the latter occasionally with electronlucent cores, and increased numbers and stacking of mitochondrial cristae. Liver cells of the subacute and chronic groups showed infrequent accumulations of electron-opaque material in the center of the mitochondria accompanied by displacement of the mitochondrial granules. The authors suggest that thallium may combine with the sulfhydryl groups of the mitochondria, thereby interfering with oxidative phosphorylation.

Tikhova (1964)[126] reported the following morphological changes in the organs of poisoned animals (presumably both of chronically poisoned rabbits or of subacutely poisoned rats): dystrophic changes of the liver with multiple-macrocellular infiltrates along the course of the hepatic ducts. Changes of the renal tissue comprised polyemia and turbid swelling of the kidney glomerulus. There was lymphoid infiltration of the mucous lining of the stomach. More profound dystrophic changes were found during histological studies of the subacutely poisoned animals.

Gettler and Weiss (1943, cited by Heyroth[5]) found similar degenerative changes in the organs of fatally poisoned humans, but they did not find characteristic lesions "in the more acute cases." They found only congestion of the internal organs, stomatitis, and punctate hemorrhages along the digestive tract.

Severe parenchymatous changes were found in three boys fatally poisoned by thallium, especially in the boy who had survived for 3 days. Fatty degeneration of the heart was especially prominent in the latter. All three showed "tabby cat" striation of the left ventricle (both ventricles in one boy). The liver lobules showed fatty degeneration and nephrosis.[42]

In the series of California poisonings, seven autopsies revealed stomatitis, fatty liver and central liver necrosis, lung edema, meningeal congestion, renal damage, gastroenteritis, and widespread degeneration of the nerve cells and axons in the brain.[42] The only endocrine gland affected was the adrenal with marked hyperemia, small medullary hemorrhages, and areas of necrosis and nuclear degeneration.[5]

b. Oncological effects: Buschke and Peiser (1923)[168] gave 100 rats 0.2 mg thallium salt per os. Papillomas of the forestomach occurred (pathology in 81% of the stomachs).

The forestomach of chronically poisoned rats showed inflammatory proliferative lesions such as hyperkeratosis, papillomas, and epithelial cysts and tufts extending into the muscularis mucosa in areas lined by flat epithelial cells. The lesions were not seen in the main stomach or on other organs covered by flat epithelium. (Heyroth[5] cited four papers published 1927-1929).

Olivier (1929)[169] reported that long feeding of small doses (amounts or duration not given) of thallium acetate to rats produced inflammatory mucosal proliferation in the esophagus and cardiac end of the stomach.

Champy et al. (1958)[170] reported (with no experimental details) that chronic oral or cutaneous administration of thallium to mice that never developed spontaneous cancers caused degeneration, papillomas, precancerous lesions, and cancers of the female genital tract. The incidence of lesions and mortality were not given.

Thallic chloride, when injected i.p. at the tumor site into Sprague-Dawley rats carrying ascitic Walker 256 carcinoma, increased median survival time 769% and the number of long-term survivors 25%. It had no effect when injected at a different site and no effect on L1210 leukemia cells in vitro or in vivo (Hart and Adamson, 1971[171]). Thallium chloride also had no effect on several types of solid tumors. The optimal dose of thallic chloride for the ascitic Walker 256 carcinoma was 3.0 mg/kg; of thallic acetate, 3.9 mg/kg (Adamson et al., 1975[172]).

c. Effects on skin and hair: Skin disorders caused by thallium poisoning include scaly erythema, eruptions, keratinization of the epithelium, ecchymoses, and petechia.[5]

KRS-5 crystals (42 mole % TlBr + 58 mole % TlI, water solubility 0.2 mg/ml at 20°C) used for infrared transmission, may come into contact with human skin during ordinary handling of the large single crystals in the spectroscopy laboratory or when reflection techniques are used to study living skin. Oertel and Newmann (1972)[173] found no skin hypersensitivity among 20 guinea pigs (10 controls) to powdered KRS-5 exposed 6 hr once a week for 3 weeks to a thin layer of KRS-5.

Some investigators* believe the effect is due to action on the central or sympathetic nervous system. The theory that systemic hair loss from thallium poisoning is due to autonomic abnormality is favored by the absence of local skin changes and the impermanence of the alopecia. On the other hand, "sensory hairs" in animals are spared; and when autotransplants of denervated skin are exposed to thallium, the hair is lost.[5/]

Other investigators have described atrophy and/or hyperplasia in structures of the skin and hair follicles and believe that the direct action of thallium on these structures causes alopecia. For example, Dixon (1927) studied the mechanism of thallium depilation and the degeneration of hair follicles due to thallium. Hair growth was interrupted when the larger polygonal cells failed to become stratified cells of the hair.[5/]

Butcher (1964)[174/] found that after thallium administration to rats, the papilla of the hair follicle persists from generation to generation and the surviving cells of the outer root sheath become progenitors of the new hair bud.

Zaun et al. (1972)[175/] described the dystrophic changes in the hair of two humans poisoned by vitamin A and thallium.

Gerdts (1974)[176/] suggested a chemical test for thallium poisoning based on the typical pigment deposition in the roots of the hair, which occurs a few days after poisoning. Jindrichova and Rockova (1969)[177/] had proposed simple examination of the hair roots for these melanin deposits as an aid in diagnosis.

d. Effects on the eye and nervous system: LeJeune (1972)[178/] cites six papers in briefly reviewing the ocular effects of thallium poisoning. The most frequent sign is retrobulbar neuritis with reduction of vision and central scotoma. Edema of the papilla nervi optici is more rare. Partial recovery of vision is possible but slow.

Several reports appeared before 1947 discussing cataract formation in animals chronically poisoned by thallium. Lid inflammation, intraocular hemorrhage, retrobulbar neuritis, and optic nerve atrophy had also been reported. Castelas (1941) and Donski (1932) reported the cataractogenic

---

* Among them was A. Buschke, "a dermatologist who published over 40 speculative papers that were meager in experimental details." He advocated an endocrine dysfunction mechanism for the action of thallium, but thallium's influence on the endocrine glands does not exceed that of other general cellular poisons.[5/]

effect of thallium in doses sufficient to cause hair loss and other toxic symptoms.[5/]

Thallous ion (as ^{204}Tl) is stored in ocular structures by at least two different mechanisms:

* Storage on melanin pigment gives very high thallium levels in the iris, choroid, and ciliary body of pigmented animals.

* Active transport. Storage based on the similarity of Tl^+ and K^+ causes high thallium levels in lens, retina, and optic nerve of both pigmented and albino animals.

The mechanisms were studied chiefly in "Dutch Midget" rabbits with black and white fur and dark brown irises, who were injected i.m. with a single dose of 15 mg $^{204}Tl_2SO_4$ (activity 33 µCi/mg Tl) or three divided doses of 2 mg/kg at 48-hr intervals (Potts and Au, 1971[179/]).

Nervous tissue damage in animals poisoned by thallium include degenerative changes in the Purkinje cells, the medial geniculate body, and the nuclei of the oculomotor nerves of the brain; thinning of the neurilemma in the peripheral nerves, brain, and spinal cord; deformation and displacement of ganglion cell nuclei; paleness, vacuolation, and lack of Nissl granules in the cortical cells; congestion of the meninges; and thickening of the pia.[5/]

Tharanne (1957)[180/] described the organ lesions in mice that died after about 15 days of being given 0.1 mg thallium acetate daily in their drinking water. The sympathetic nervous system was especially affected.

Hayashibara (1964)[181/] performed an electroencephalographic study of thallium acetate-poisoned rabbits. Thallium alopecia was thought to be accompanied by some dysfunction in the diencephalon.

Pentschew (1970)[182/] pointed out that brain toxins such as lead, manganese, and thallium affect the central nervous system without actually being there, operating through mediators. For example, rats are very susceptible to the riboflavin deficiency induced by thallium. Optic nerve and tract were invariably affected.

Karkos (1971)[183] described the damage to the brain cells of a 13-year-old girl who died from taking 0.5 tablespoonful of a thallium-containing rat poison. The changes included generalized and fatty degeneration of ganglion cells, damage of the axons, myelin sheath degeneration, and acute swelling of oligodendroglia.

Tackmann and Lehmann (1971)[184] concluded that both morphologically and electrophysiologically, the nerve damage pattern in guinea pigs (given single doses of 15 or 18 mg thallium sulfate by injection) apparently corresponds to pure axonal degeneration.

Bank et al. (1972)[82] found axonal degeneration and segmental demyelination in the sural nerve by electron microscopic study of tissues of thallium-poisoned patients.

Spencer et al. (1973)[185] showed the sensitivity in vitro of nerve cells, especially the axons, to 5 to 10 ppb thallous acetate or sulfate. Two hours after exposure, the mitochondria in peripheral nerve-fiber axons had begun to swell. Progressive swelling of the matrix space of these mitochondria transformed the organelle into an axonal vacuole, which coalesced with adjacent axonal vacuoles to produce massive internal axon compartments. Electron microscope mass spectrometry showed that the membranes of these axon compartments had an affinity for thallium. The fiber swelling retracted the myelin from the nodes of Ranvier; but degeneration did not occur, and impulses could still be propagated. Similar, less severe changes were seen with cultured dorsal root ganglion neurons and central nerve fibers.

Cavanagh (1973)[117] reviewed the peripheral neuropathy caused by thallium poisoning. His own findings implied that long nerves (in the leg and extraocular muscles) are affected earlier than short nerves and that larger diameter fibers of sensory nerves are attacked selectively. Chromatolysis of motor nerve cells and degeneration of the gracile tracts in the dorsal spinal cord are the only central nervous system changes. The gastrointestinal symptoms are most likely due to the effect of thallium on potassium exchange and transport across the intestinal epithelium as well as denervation in the autonomic nervous system. Denervation of the glossopharyngeal and vagus nerves was thought to produce the tachycardia and hypertension. Thallium may interfere with protein synthesis in the hair follicle by binding cysteine. Its effect upon proteins in mitochondrial membranes may be similar.

Cavanagh et al. (1974)[116/] concluded from reports of cases in the literature that all cranial nerves except I and VIII (which are tested with difficulty) are affected by thallium poisoning. Reed et al. (1963) had noted optic neuropathy in 7% of 61 children. Cranial nerve involvement usually occurs before optic neuropathy. Nonsensory-type ataxia may be noted in about three-fourths or more of a group of intoxicated children. The frequently abnormal electroencephalogram is characterized by nonspecific slow wave activity. Such changes may be persistent and sometimes permanent. T-wave changes in the electrocardiogram could be related to the similarities of thallium to potassium.

Cavanagh et al. (1974)[116/] noted a distal "dying back" type of pattern in sensory and motor nerves consistent with the clinical symptoms of three men homicidally poisoned with thallium acetate. More recently Kennedy and Cavanagh (1976)[186/] have again found a case exhibiting the dista nature of the nerve fiber degeneration in a man poisoned by thallium* and th especial sensitivity of the long fibers.

Some other recent reports of neuropathy have implicated thallium. Ikeda (1971)[187/] found slightly increased zinc levels and doubled thallium levels (relative to levels in normal urine) in the urine of five patients exhibiting subacute myelopticoneuropathy, whose etiology was unknown. A research chemist exposed to fumes of gallium, indium, and thallium halides, when the glass vacuum-distillation apparatus being used to separate them by fractional distillation broke, experienced irritant coughing for sev eral hours. Two weeks later he developed nauseous vertigo while traveling b train. Diagnosis was vestibulitis of unknown origin (Utidjian, 1973[188/]).

e. Effects on the heart and blood pressure: Heart injury by thallium poisoning has been evinced by sinus tachycardia, bradycardia, angin pectoris, and reduced or increased blood pressure. Injury to the vagus by thallium produces electrocardiographic changes.[5/]

Thallium, like potassium, depresses the heart directly and by action on the parasympathetic nervous system. Hypotension induced by thallium masks a stimulating action on the sympathetic nervous system. The hypo tension disappears on vagotomy; hypertension, occurring simultaneously with renal vasoconstriction, appears by blocking the parasympathetics with atropine. Truhaut (1960)[23/] hypothesized that the effects of thallium were caused by destruction (oxidation) of epinephrine. Thallium reproduces the

---

* A pest-control officer who had access to thallium sulfate as a 0.2% gel for killing moles.

effects of potassium, but with greater efficiency, on isolated rat and frog hearts (Hughes et al., 1976[189] and Rusznyak et al., 1968[190,191]).

Merguet et al. (1969)[192] assumed that the hypertension, tachycardia, and increased excretion of catecholamines in humans poisoned by thallium was due to increased sympathetic tone. Renal impairment or aldosteronism did not appear to be likely causes of the hypertension. In five cases of human poisoning urinary excretion of catecholamine was markedly increased and in two cases urinary excretion of porphobilinogen and total porphyrins temporarily increased. Propanolol (a β-receptor blocking agent) treatment appeared to be promising. The α-receptor blocking agent phentolamine also effectively reduced blood pressure but not pulse rate. (See also Bock et al., 1968[193] and Merguet et al., 1968[194].)

Hantschmann (1970)[195] briefly reviewed blood pressure changes in human thallium poisoning. He used phentolamine and propanolol alone and together to reduce blood pressure. The results were contrary to a neurally caused hypertension due to paralysis of the depressor nerves. Apparently, thallium directly or indirectly stimulates the sympathetic nerves so that pyrocatecholamines are released at the peripheral endings of the sympathetic nerves.

Schutt and Kersten (1976)[196] assumed that thallium directly damages the myocardium in acute thallium poisoning. The EKG's of two poisoned dogs were studied. Tachycardia, rhythm- and repolarization-disturbances, and occasional heart murmurs were observed.

Lemeijer and van Zweiten (1976)[197] found an acute hypotensive effect of thallium poisoning contrasted to the hypertensive effect noted at longer terms. Intravenous doses of 3 to 100 mg thallous sulfate/kg to anesthetized rats rapidly produced a dose-dependent hypotensive effect and decrease in heart rate, the maximum fall in blood pressure occurring within 3 to 5 min. A direct influence of Tl^+ (10^{-5} to 10^{-3} $\underline{M}$) on the sinus node presumably caused a concentration-dependent decrease in frequency of spontaneously beating guinea pig isolated atria. The effect was not altered by atropine, cocaine, or change in potassium level in the medium. Tl^+ caused relaxation of vascular smooth muscle both in isolated aortic strips of the rabbit and in isolated perfused rabbit ear.

f. Reproductive and teratogenic effects: Thallium poisoning deleteriously affects sexual behavior, reproductive organs, and egg and fetal development and survival. See also Subsection VII.A.3.b. for the effects of thallium on meiosis and chromosomes.

Sexual activity is usually lessened in chronically poisoned animals. Three papers prior to 1947 claimed thallium inhibition of ovarian hormone production in rats and mice. Others found no effect on ovarian function or histological appearance. Still others did find lesions in the sex glands. Testes may atrophy and be free from sperm.[5]

Ovarial hormone production in mice was said to be inhibited by feeding them thallium for 3 weeks (no other dosing details given). Maturing ova and large follicles were present, but no "estrual crisis" occurred (Zondek and Aschheim, 1927[198]).

In experiments confined to young and castrated rodents, the results of Truhaut (1960)[23] contradicted those of Buschke et al., who had ascribed the inhibitory effect of thallium on the estrous cycle to its action on the hypophysis and ovary. Truhaut showed that the action on estrone occurred in the vaginal epithelium.

On the other hand, Arashima (1933)[199] concluded that the alterations of sex functions of female rats exposed to thallium chloride were those associated with nutritive changes only. For 20 days, sexually immature female rats were injected s.c. with 25 to 50 μg thallium chloride ($TlCl_2$ in the abstract) in solution and larger rats were given 100 μg. Daily doses of 50 μg were then increased up to 350 μg. Changes included weight loss, delay in estrus, hypertrophic and degenerated kidneys, and alopecia. Ovaries, uterus, and corpora lutea were unaffected.

Champy et al. (1958)[170] stated that chronic oral or cutaneous (Tl_2S) administration of thallium caused degeneration of the female genital tract similar to that found in castrates or animals with denervated uteruses (no experimental details given except that 60 sibling mice had been used that had never developed spontaneous cancers). Papillomas, precancerous lesions, and cancers were found in the genital tract (mortality and cancer incidence not reported).

DiGaudio and Hirshfield (1973)[200] exposed mating types A and B of _Paramecium bursaria_ to 3 to 7 ppm thallous nitrate or acetate. Thallium delayed clumping and pairing and shortened the time during which paramecia paired. Nonconjugationing paramecia exposed to 5 or 7 ppm thallium died within 10 days, but 3 ppm did not decrease the population until after 2 weeks exposure.

Landauer (1931)[201] reported that chicks fathered by chronically thallium-poisoned cocks suffered higher mortality during the first 3 weeks after hatching than did chicks fathered by normal cocks. The chicks died suddenly, before which they had seemed normal. Chicks surviving 3 weeks were quite strong. If the poisoned cocks fertilized a second group of eggs, the chicks died sooner after hatching.

Dilling and Healey (1926)[167] reported that dilutions of *N*/5000 and *N*/1000 thallium nitrate (48 and 240 ppm) did not delay tadpole development in the egg, but even *N*/500,000 dilutions (0.48 ppm) killed tadpoles on emergence from the egg. Thallium, copper, and zinc were more toxic than lead to the emerged tadpoles.

The inhibition of *Paracentrotus lividus** egg development by Tl^+ could be counterbalanced by K^+ but not by Li^+ or Zn^{2+} (Lallier, 1968[202]). At 200 ppm thallium as TlCl in seawater, the segmentation of Stage IV of the blastula of *P. lividus* and *Arbacia lixula** was delayed. Partially hatched blastulas developed in 20 ppm thallium, gastrulas at 10 ppm, larva at 4 ppm, and small pluteus at 2 ppm. Dilute solutions of KCl, RbCl (but not LiCl or CsCl), sodium thiomalate, and the tetrasodium salt of EDTA protected against TlCl toxicity. Segmentation of *A. lixula* was prevented by 20 ppm thallium as $TlCl_3$. The thiomalate and EDTA salt (but not KCl) counteracted $TlCl_3$ (Lallier, 1969[203]).

Karnofsky et al. (1950) and Ridgway and Karnofsky (1952) (cited in References 33, 35, and/or 204) produced achondroplasia in embryonic chicks injected with 0.4 to 0.7 mg thallium sulfate/egg between incubation days 4 to 10. Injection at day 8 on the chorioallantoic membrane or at day 4 in the yolk sac gave a 90 to 100% incidence of achondroplasia. Defective cartilage growth led to micromelia (bowing and shortening of the extremities) and mandibular defects. Thallium sulfate produced early fetal death and reduced fetal size as well. The younger the embryo at injection and the higher the dose, the more severe the effect by thallium.

Bertran (1963)[205] found that injecting 5 mg thallous nitrate into the vitelline sac of chicken eggs that had been incubated at 38°C for 3 to 5 days diminished the head size of 67% of the chicks that it did not kill, but had no effect on feather development.

Červeňanský et al. (1964)[206] induced phocomelia and micromelia in chick embryos with thallium, but dosing details were not given.

Ford et al. (1968)[33] found necrosis and defective maturation of hypertrophic chondrocytes by injecting 1.2 mg thallium sulfate into embryonic chicks on the 11th day of incubation. A 30% decrease in mucopolysaccharide content in cartilage occurred without a simultaneous loss of collagen, suggesting interference in mucopolysaccharide synthesis.

---

* Sea urchin species.

Embryos injected with 1.0 mg thallous sulfate on the 9th day and examined on the 19th day showed growth retardation (26 g compared to 40 g for the median weight of the controls), mild parrot-beak deformity, microphthalmia, microcephaly, reduced size of all skeletal elements, and 50% mortality (2% due to infection).

At the 1.2 mg dosage, embryos on the 19th day that had been injected on the 11th day weighed 34 g. Parrot beaks, microphthalmia, and microencephaly were absent, but the other deformities were present. The size reduction of skeletal elements was not so severe.

Thallium-treated chicks showed much less intracellular and extracellular alkaline phosphatase activity than the controls in the hypertrophic zone (zone nearest the diaphyses, which consists of maturing hypertrophied chondrocytes). There was no consistent difference in the water or hydroxyproline content between treated chicks and controls. Total hexosamine content and type differed.

Hall (1972)[204] injected thallium sulfate (0.6 mg dissolved in 0.5 ml saline) onto the chorioallantoic membrane on the 7th day of chicken egg incubation, which coincided with the onset of embryonic osteogenesis. Long bone growth was reduced and areas of necrotic cartilage formed. Collagen, calcium, and lipid accumulated in the necrotic chondrocytes, but there were subnormal amounts of acid mucopolysaccharide, which was probably the cause of the achondroplasia. The thallium-produced syndrome is similar to certain genetic defects.

Thallium disturbs the histogenesis and morphogenesis of the cartilaginous model of the bones in the critical teratogenic period of organogenesis in the chicken. The disturbance on chondrogenesis was chiefly on interstitial rather than perichondrial growth. The growth plate of the bone showed the greatest changes, with the primary changes in the chondrogenesis layer and the secondary changes in the transformation layer and in the metaplastic rebuilding of cartilage to bone (Škrovina et al., 1973[207]).

Roth (1975)[208] studied the osteo-vaculo-neural macro-relatio in teratogenic micromelia and suggested that the cause is insufficient prima growth of the nervous trunks. Roentgenograms and stained tissues of chick em bryos (whose deformations were induced by 1.5 mg semicarbazide or 0.5 mg thal lium nitrate injected into the yolk sac at 5 to 7 days of incubation) support the contention that the length of bone growth is determined by the space avai able along the nervous trunks.

Gibson et al. (1967)[34] administered thallium sulfate to pregnant mice, rabbits, and rats. Only slight toxic or teratogenic effects were noted in the embryos or fetuses.[35]

Dose levels of 2.5 mg thallous sulfate/kg at gestation days 8, 9, and 10 (early) or days 12, 13, and 14 (late) given i.p. to pregnant Sprague-Dawley rats did not affect the concentrations of potassium in plasma or erythrocytes of control rats or rats fed a low-potassium diet. This level did not affect significantly the fetal resorption rate and low fetal weight observed in low-potassium-diet rats; but if the dose was received in early gestation, the phalanges failed to ossify. A dose of 10 mg Tl_2SO_4/kg during late gestation killed all maternal rats fed the low-potassium diet. Even the lower doses produced typical toxic symptoms in all rats (Gibson and Becker, 1970[35]).

If given during late gestation, 2.5 mg/kg thallium sulfate produced hydronephrosis (16/34) and absence or nonossification of vertebral bodies (21/35) in the fetuses (8/14 and 14/15, respectively, for rats on a low-potassium diet). The defects were also produced when given early in gestation, but the frequency was not statistically significant.[35]

Nogami and Terashima (1973)[209] found that treating 6- and 9-day-old rats with 20 and 40 µg/g thallous sulfate reduced $^{35}SO_4$ incorporation into cartilage mucopolysaccharides (MPS) to 50% that in the controls; $^{35}SO_4$ incorporation into MPS in fetal epiphyseal cartilage, sternal cartilage, and liver was similarly reduced. Apparently, Tl_2SO_4 generally inhibited MPS synthesis. Columnar cartilage of the long bone was hypoplastic and the calcifying zone was defective.

The teratogenic effect of thallium on chicken embryos is not modified or alleviated by simultaneous administration of manganese, biotin, riboflavin, or nicotinamide, deficiencies of which cause similar gross skeletal changes (Karnofsky et al., 1950; cited in Reference 33).

Four days of incubation with thallium salts produces teratogenic effects in developing chicken embryos: micromelia, parrot beak, and shortness of the tail with different stocks of white leghorns showing wide variation in response. Glucose, sorbitol, sorbose, and arabitol given with thallous acetate reduce the toxicity and teratogenicity of thallium. Giving imidazole reduced mortality and incidence of malformations, especially of the legs. Presumably, thallium interferes with imidazole groups of certain enzymes (Landauer, 1960[210]).

Cortisone acetate decreases collagen and hexosamine-containing acid mucopolysaccharide synthesis and potentiates the achondroplasia occurring in chick embryos from thallium sulfate. Having opposite effects on collagen and acid MPS, vitamin C alleviates the thallium-induced syndrome (Hall, 1972[211]).

g. Effects on blood, bones, and endocrine glands: Chronically poisoned rats show a transitory hyperglycemia before blood sugar lowering. Hyperglycemia follows intravenous injection of a thallium salt. Subcutaneous injections given repeatedly increase blood levels of potassium and cholesterol and decrease calcium. Erythropenia and leukocytosis are only occasionally observed. Characteristic changes in the formed elements of the blood were not observed in the acute cases in California, but Flamm (1926) noted a 21.7% increase in leukocytes, a 39% increase in lymphocytes, and an 80% increase in eosinophiles in the blood of children 16 days after receiving an epilating dose of thallium.[5/]

Various theories have been espoused to explain thallium toxicity. Thallium has been though to exert its primary effect at the cellular level (Truffi), on the endocrine glands (Buschke et al.) and on the autonomic nervous system (Peter, Dixon, and others).[23/]

According to Sessions and Goren (1947),[109/] thallium apparently attacks the autonomic nervous system, especially the sympathetic, in chronic poisoning to cause alopecia; cataract; growth inhibition; inhibition of sexual development; skeletal disorders; and diseases of the thyroid, parathyroid, sex, and adrenal glands.

Truhaut (1960)[23/] did not find specific lesions in the endocrine glands of thallium-poisoned mammals. In the terminal phase of poisoning, the basal metabolism rate is slightly lowered. In Heyroth's review (1947),[5/] five papers are cited in which thyroid function or tissue was affected by thallium poisoning, and five papers are cited in which they were not. Heyroth felt that the endocrine dysfunctions reported for thallium poisoning were no worse than might occur with any general cellular poison.

According to Buschke and coworkers, growth retardation by thallium in warm-blooded animals is due to failure of the body to store calcium. The effect resembles rickets.[4/] The bones of young rats poisoned by thallium are soft, curved, poorly calcified, and show fusiform osteoid tissue deposits. The marrow is fibrous. Sixty-eight percent of rats given 0.2 mg thallous acetate daily developed bony lesions within 4 to 6 weeks.[5/]

Thallium decreased the ash content of rat teeth, but did not change the phosphorus content markedly. Small doses of thallium increased calcium content; large doses decreased it (Urabe, 1936; cited by Heyroth[5/]). Phosphorus retention may be impaired during the first stage of poisoning, but calcium stores decrease gradually with advancing intoxication (Rominger et al., 1931; cited by Heyroth[5/]).

C. Effects on Microorganisms and Plants

1. Microorganisms: Gottbrecht (1880), Schulz (1886), and Richet (1917) reported yeast growth stimulation by thallium. Richards (1932)[212] found that about 0.08 to 8 ppm in a sugar solution culture promoted higher growth of *Saccharomyces cerevisiae*, but concentrations 80 to 8,000 ppm retarded growth 95 to 98.5%. *Bacillus* lactic acid[213] was reported to be stimulated by 0.1 ppm thallium in milk. Khare et al. (1966)[214] found that about 0.6 to 10 ppm thallium (as thallous sulfate) increased growth of the nitrite-forming bacterium *Nitrosomonas* for 48 to 336 hr, with maximum growth at 1.3 ppm thallium.

Table VII-12 tabulates much of the literature information on the effects of thallium on bacteria, fungi, and mycoplasmas. The range of thallium concentrations having no toxic effects is 1.0 to ~ 1,600 ppm. Some soil organisms, in particular, are unaffected by 10 to 90 ppm, yet others show toxic effects at 30 to 90 ppm.[215]

Although yeast and some bacteria still show some growth at up to 8,000 ppm thallium, much lower concentrations can be toxic to some microorganisms. For example, 0.82 to 8.2 ppm thallous ion retards or completely stops nitrate formation by *Nitrobacter agilis* in soil suspension.[216] Thallium, 1 to 2 ppm (as thallous sulfate), inhibits spore germination of *Aspergillus niger* in nutrient solution although the effect is sharply mitigated by the presence of finely divided organic matter serving to adsorb Tl^+.

McCool (1933)[217] concluded that the injurious effects of thallium on crop plants were not due to the absence of nitrates in the soil. Nitrate formation in loam soil to which greater than 1,900 ppm thallium (2,400 ppm Tl_2SO_4) had been added was retarded only 5.8%.

2. Aquatic plants

a. Accumulation: The draft Calspan report[218] reports some in-house aquarium experiments by Wood with Lake Erie *Ulothrix* algae exposed to 0.005 to 0.5 ppm thallium ion for 1 to 23.5 hr. The original thallium concentrations when the algae were collected were usually 0 to 0.42 ppm with one high value: 12.7 ppm. The concentration rates ranged from 4 ppm/hr at 0.5 ppm to 220 ppm/hr at 0.005 ppm. Possibly, the higher concentrations killed or inhibited the organisms; whereas, the lowest concentrations stimulated their growth. These experiments were not described in the final report version we have seen.[219] The latter reports values indicating concentration factors of 9.3 to greater than 430 for algae growing in contaminated waters. The examined algae contained 0.043 to less than 0.85 ppm.

Lazorenko and Polikarpov (1972)[220] found that ^{204}Tl is not accumulated by alginic acid. (However, ^{137}Cs and ^{65}Zn were also not accumulated by alginic acid; yet their concentration factors in brown algae ranged from 2 to 186.)

Skul'skii et al. (1972)[221,222] reported that lowering temperature and light intensity or adding inhibitors reduced Tl^+ uptake by the Black Sea alga Ulva rigida. The effect was less on Rb^+ uptake, but there was no effect on Cs^+ uptake. Higher levels of energy metabolism enhanced the differences in uptake. Tl^+ and K^+ metabolism was similar, but NaF inhibited only Tl^+ uptake. Transport of the ions is coupled to a metabolically dependent flux across the plasmalemma. Coupling intensity decreases in the sequence Tl^+ greater than Rb^+ greater than Cs^+.

From the results presented in Table VII-13, Polikarpov et al.[223,224] concluded that accumulation of Tl^+ by Ulva rigida is a result of active transport with Tl^+ participating in enzymic processes acting together with energy-exchange processes. Results were comparable with those of West and Pitman (1967) for potassium.

Solt et al. (1971)[225] also reported that thallium (^{204}Tl) uptake by Chlorella fusca was energy dependent. Unlabeled, thallium inhibited the light-dependent uptake of ^{86}Rb more than the ^{204}Tl uptake was inhibited by Rb. Chlorella also takes up Tl^+ via the potassium pathway (Broda, 1974[226]).

b. Toxicity: DiGaudio (1975)[166] found that the threshold concentration (i.e., the concentration at which slight effects were noticeable) was 3 ppm Tl^+ as the nitrate or acetate for two algal species Chlorella vulgaria and Anacystis nidulans and the protozoan Paramecium bursaria. The limiting concentration (severe inhibition--greater than 50% suppression occurs) for the algae is 7 ppm and the ciliate, 9 ppm. Effects studied were growth inhibition, respiration, and extracellular polysaccharic production. Under photosynthetic conditions oxygen evolution and carbon dioxide consumption occur in the Paramecium until 9 ppm thallium. Presumably its respiration is affected by its endosymbiont.

Overnell (1975)[227] reported that 15-min incubation in the dark with about 1.5×10^{-5} M (~ 3 ppm Tl) Tl^+, Hg^{2+}, and Cu^{2+} gave 50% inhibition of light-induced oxygen evolution by the freshwater alga Chlamydomo reinhardii in whole-cell experiments. The effect was not altered by washing which indicated a rapid irreversible uptake, possibly by the K^+ or Rb^+ pump. A 2×10^{-4} M Tl solution inhibited the Hill reaction in whole cell experimen by 60%; in broken-cell experiments, no inhibition was observed. A modified Mehler reaction was not inhibited by 2×10^{-5} M Tl^+. The reaction in broken cells was not affected either. Overnell concluded that the principal effect of Tl^+ is on NADP reduction or the dark reactions.

Overnell (1975)[228] reported that Tl^+ (2 to 20 ppm) and Hg^{2+} affect light-induced oxygen evolution more than they do potassium leakage (an indication of cell damage) from the cells of the marine alga *Dunaliella tertiolecta*. (Oxygen evolution in *Phaeodactylum tricornutum* was more sensitive to Hg^{2+} but less sensitive to Tl^+.) Apparently, according to Overnell, photosynthesis is inhibited by Tl^+ at the chloroplast in concentrations noninjurious to the cell membrane.

3. *Terrestrial plants*: Table VII-14 summarizes the literature data on thallium's effects on terrestrial plants. An extensive early literature exists on the possibility of thallium's being the cause of frenching of tobacco.[229] Concentrations as low as 0.04 ppm in water induce chlorosis in tobacco seedlings. Tobacco plants show toxic effects when grown in soil containing 0.25 ppm thallium. Some plants (e.g., corn and/or clover[213,230]) in soil are said to be stimulated by $\leqslant$ 10 ppm thallium, yet others[217] have found ~ 2 ppm thallium in soil injurious to corn. Seeds of various crop plants (e.g., oats) show some germination at 70 to 2,500 ppm thallium,[231] but plants themselves are often badly injured or killed by concentrations of ~ 7 ppm.[217] Graminae, including oats and grasses, seem more resistant. Higher concentrations (e.g., 408 ppm) reduce oxygen uptake in excised roots.[232]

The shoots of oats grown in clay loam are reduced about 17% in length with 450 to 550 ppm Tl_2SO_6 in the moistening solution.[233] Twelve-day-old crop plants in loam soil are completely inhibited by 1,100 ppm in the irrigation water.[213] Although the final Calspan Corporation report[219] indicates that some thallium-contaminated waters are used for irrigation (e.g., in the Hayden/Ray, Arizona, vicinity), raw wastewaters seldom contain as much as 0.03 ppm thallium.

TABLE VII-1

DISTRIBUTION OF SUBLETHAL DOSES OF THALLIUM[a]
IN 2-kg RABBITS[234]

	9.7 mg Tl/kg After 24 hr Concn. (ppm)	9.6 mg Tl/kg After 43 hr Concn. (ppm)	8.5 mg Tl/kg After 144 hr Concn. (ppm)
Kidney	65.40	24.70	39.50
Liver	7.13	3.22	2.00
Femur	6.84	3.78	3.30
Lumbar vertebra	6.42	4.32	2.87

a/ Radioactive thallium (salt unspecified) giving a total radiation dose o[f] 1 mCi. Route not given.

TABLE VII-2

DISTRIBUTION OF THALLIUM IN DOGS GIVEN 8.5 mg/kg[3,26]

	mg Tl/100 g							
	Day 1		Day 7		Day 11		Day 15	
	mg/100 g	% of Dose in Total Organ	mg/100 g	% of Dose in Total Organ	mg/100 g	% of Dose in Total Organ	mg/100 g	% of Dose in Total Organ
Liver	3.4	15.9	3.2	15.3	2.1	13.5	2.1	10.8
Kidney	3.3	1.7	3.3	2.5	3.3	2.3	3.8	2.7
Spleen	1.5	0.7	1.5	1.2	1.8	1.3	2.1	1.3
Heart and heart's blood	1.8	2.0	1.7	1.5	1.0	1.1	0.6	0.6
Lung	0.9	1.5	1.3	1.3	1.1	1.2	0.9	1.0
Brain	0.8	1.2	1.1	1.2	1.2	1.4	1.8	1.4
Intestines and contents	2.3	8.5	1.3	5.9	0.6	3.9	0.5	3.3
Stomach and contents	3.6	7.2	3.8	5.6	1.5	3.6	1.3	2.7

TABLE VII-3

RADIOTHALLIUM IN OLYMPIC PENINSULA ELK AND DEER AND ALASKAN CARIBOU[29]

	$^{208}Tl(^{232}Th)$ Disintegrations/min/kg	Relative Concentrations
Elk muscle	0.6	1.5
Elk liver	5	12.5
Elk kidney	16	40
Elk bone	310	780
Deer muscle	0.4	1
Deer bone	227	570
Caribou muscle	1.6	4
Caribou liver	4.1	10
Caribou kidney	3.2	8
Caribou bone	98	245

THALLIUM CONCENTRATION IN HUMAN POISONING, mg/100 g TISSUE[37]

	Adult (13 days)[a]	Adult (weeks)	Adult (weeks)	Adult (?)	Adult (?)	Adult (2 days)	2 Children (1.5 days and 1 day)	Child, 6-1/2 (7 days)	Child, 10 Years (6 days)
Liver	0.89 [15.6][b]	0.1	Traces	7.2		3.4	Total in organ and intestines 5-9% of the total Tl taken.	0.5	0.5
Kidney	0.36 [1.4]	Traces	Traces	9.8		4.2		0.8	0.8
Spleen	0.13 [0.2]	-	Traces	7.0	10.6			1.0	1.0
Heart				4.0	4.0				
Lung				6.7	6.7				
Brain		0.3	Traces	2.4		3.1		0.6	-
Muscle		-	Traces	8.7				0.1	0.4
Bone				9.0		1.8			
Large intestine		0.5	0.3					0.9	1.6
Intestine, unspecified						70			
Small intestine		0.4	0.1					1.0	1.0
Stomach						12			
Blood						5.5			

a/ Numbers in parentheses indicate time elapsed from intake to death.
b/ Numbers in brackets indicate mg Tl in total organ.

TABLE VII-5

RATE OF EXCRETION OF ORALLY ADMINISTERED ^{204}Tl BY THE RAT, PERCENT OF DOSE[39]

Elimination Product	Day 1	2	3	4	5	6	7	Total
Urine	3.00	1.96	1.97	1.16	0.94	0.95	0.43	10.41
Feces	7.15	9.42	9.15	7.02	6.44	5.67	3.39	48.24
Total	10.15	11.38	11.12	8.18	7.38	6.62	3.82	58.65

TABLE VII-6

THALLIUM IN URINE AND FECES OF NORMAL HUMANS[51]

Case	Urine (ppb)	Feces (ppb)
1	0.1 0.4	0.6 3.0
2	0.1	a/
3	0.2	a/
4	0.1	a/
5	0.1	a/
6	0.3	0.1
7	1.0	a/
8	0.8	a/
9	0.02 ppb[b]	< 0.02 ppb[b]
10	0.02 ppb[b]	0.1

a/ Not looked for.
b/ Detection limit for the extraction-spectrometric method used.

TABLE VII-7

THALLIUM CONCENTRATION IN MORNING URINE AND URINE SALTS IN HUMANS WITH NORMAL NUTRITION[235]

	Urine (ppb)	Average in Urine Salts (ppb)	Estimated Amount of Tl Excreted in Urine (μg/day)
Vegetarian			
40-year-old female	1.34		2.0
43-year-old female	1.69	18	2.5
12-year-old male	0.92		1.4
Smoker			
28-year-old male	1.42		2.1
25-year-old male	0.40	17	0.6
24-year-old male	0.69		1.0
Nonsmoker			
26-year-old male	0.13		2.0
60-year-old female	0.39	6	0.6
48-year-old female	0.53		0.8

a/ Assume 1,500 g urine/day.

TABLE VII-8

THE EFFECT OF THALLIUM ON SEVERAL ENZYMES AND OTHER BIOLOGICAL SYSTEMS[58]

System	Order of Efficiency	Reference
Diol-dehydratase	$Tl^+ > NH_4^+ > K^+ > Rb^+ > Cs^+ > Na^+ > Li^+$	Foster and Williams (unpublished as of 1971). See also Manners et al. (1970)[57]
Pyruvate kinase	$Tl^+ > K^+ > Rb^+ > Cs^+ > Na^+ > Li^+$	Radder, Waller, and Williams (unpublished as of 1971)
Phosphatases	$Tl^+ > K^+ > Rb^+ > Cs^+ > NH_4^+ > Na^+, Li^+$	Inturussi (1969)
Na/K ATPases (K-function)	$Tl^+ > K^+ > Rb^+ > Cs^+ > Na^+ > Li^+$	Britten and Blank (1968)[62]
Erythrocyte transferases	Tl^+ moves with K^+	Gehring and Hammond (1967)[18]
Muscle excitation	$Tl^+ > K^+ > Na^+, Li^+$	Mullins and Moore (1960)

TABLE VII-9

STABILITY CONSTANTS OF SOME THALLIUM(I) COMPLEXES AT IONIC STRENGTH 0.075[58]

Ligand	$Log_{10}\ K_{Tl}$	$Log_{10}\ K_K$
PO_4^{3-}	2.25	-
HPO_4^{2-}	0.75	-
$P_2O_7^{4-}$	3.05	1.5
$HP_2O_7^{3-}$	2.35	-
Ribose-5-phosphate^{2-}	0.90	-
Adenosine diphosphate^{3-}	1.20	-
Adenosine triphosphate^{4-}	2.00	1.0
Ethylenediamine tetraacetate^{4-}	5.8	1.0
Nitrilotriacetate^{3-}	4.4	1.0

TABLE VII-10

ANTIDOTES USED IN EXPERIMENTAL AND CLINICAL TREATMENT OF THALLIUM POISONING

Substance	Favorable	Unfavorable or Not Effective	Species	Remarks
NaI		Reference 236 (1965)	Guinea pigs	Soly. of TlI (6.4 mg/ℓ) great enough to allow absorption of fatal amounts of Tl.48/
	Reference 30: 17 reports	Reference 30: 4 reports		The unsuccessful results may have been due to inadequate dosage or delay in treatment
		Reference 236 (1965)	Guinea pigs	
$Na_2S_2O_3$	Reference 30: 13 reports	Reference 30: 4 reports		i.v. dosage
		Reference 48: 4 reports[a]		
KI	Reference 30: 5 reports			In addition to stimulating thalliuresis, insol. TlI is produced.
KCl	Reference 30: 6 reports	Reference 236 (1965)	Guinea pigs	1.5-5 g daily favored orally or i.v. for 5 days to 3 weeks. Sometimes excitement increases.
	Reference 48: 2 reports	Reference 237 (1971)	Guinea pigs	
NaCl		Reference 140 (1976)	Rats	Improved short-term survival but total mortality unchanged.
Dithiocarb (Diethyldithio-carbamate)	Reference 238 (1967)	Reference 239 (1971)	Rats	Forms a Tl chelate <u>in vitro</u>. Causes Tl redistribution in tissue besides increased Tl excretion.
		Reference 240 (1970)	Humans	
		Reference 241 (1970)	Humans	
		Reference 242 (1969)	Rats	
	Reference 237 (1971)		Albino guinea pigs	Given s.c., it lowered lethal dose 50% to 65%.
		Reference 243 (1971)	One human	
		Reference 48: 2 reports	Humans	Thalliuresis increased. In one report, a patient briefly developed hypotension and psychosis.
BAL (Dimercaprol)	Reference 30: 7 reports	Reference 244 (1966)	Humans	
		Reference 30: 17 reports		
	Reference 48: 4 reports	Reference 48: 9 reports		
EDTA, NaEDTA, or CaEDTA (versenate)		Reference 244 (1966)	Humans	
		Reference 245 (1962)	Rats	
	Reference 30: 4 reports	Reference 30: 3 reports		i.v.
		Reference 48: 3 reports	Animals	
	Reference 48: 2 reports	Reference 48: 3 reports	Humans	

TABLE VII-10 (continued)

Substance	Favorable	Unfavorable or Not Effective	Species	Remarks
Dithizone (diphenylthiocarbazone)	Reference 246 (1969)	Reference 244 (1966)	Humans	
			One human	
	Reference 236 (1965)		Guinea pigs	
	Reference 238 (1967)		Rats	
		Reference 240 (1970)	Humans	Prussian blue is recommended by the Netherlands National Poison Information Center.
	Reference 48: 1 report		Rats	
	Reference 48: 2 reports		Dogs	
	Reference 48: 1 report		7 Humans	Thalliuresis not increased. Favorable effect based on clinical grounds. Neurological symptoms increased when dithizone dosage increased from 20 to 40 mg/kg day. Others have reported similar ill effects.
		Reference 48: 2 reports	Humans?	
	Reference 247 (1974)		Rats	
	Reference 30: 3 reports	Reference 30: 5 reports		
Sodium selenate	Reference 93 (1966)		Rats	
Charcoal	Reference 30: 7 reports Reference 48: 3 reports		Animals	Twice daily oral doses 500 mg charcoal/kg for ≥ 5 days.
Milk or Ca salts	Reference 30: 10 reports			
Various cathartics	Reference 30: 7 reports			
Pilocarpine HCl	Reference 30: 7 reports			Obsolete. Thought increased sweating would increase Tl elimination.
S-amino acids				
cystine	Reference 30: 6 reports	Reference 30: 3 reports		Cystine increases thalliuresis in chronic, but not acute poisoning.
Cystine, cysteine, and/or methionine	Reference 48: 1 report	Reference 48: 2 reports	Rats	
cysteinamine	Reference 48: 2 reports	Reference 48: 1 report		
methione		Reference 30: 6 reports		Results equivocal.
Others				
cytochrome c		Reference 248 (1964)	Rats	Peritoneal injections extended survival time only briefly.
thioacetamide		Reference 236 (1965)	Guinea pigs	
antidotum metallorum, $NaHCO_3$, Dowex resins penicillamine, [illegible]		Reference 48: 6 reports	Animal and humans	

TABLE VII-10 (concluded)

Substance	Favorable	Unfavorable or Not Effective	Species	Remarks
Thioacetamide	Reference 48: 1 report	Reference 249 (1960) Reference 48: 1	Dogs	Thioacetamide itself toxic.
Rhyolitic pumice	Reference 250 (1974)		Rats	Not as effective as Prussian blue
Zirconium phosphate and	Reference 251 (1963) and		Rats	Tl content of liver, muscle, and
Zirconium polyphosphate	252 (1965)		Rats	kidneys lowered > 50%.

a/ Numbers of reports given in the table for Reference 48 (1974) represent only those additional reports that were not already cited in Reference 30 (1970).

TABLE VII-11

ACUTE TOXICITY OF THALLIUM AND ITS COMPOUNDS

Substance	Species	LD_{50} (or other), mg/kg	Route	Toxic Effects	Reference
Thallium	Rat	(TD_{Lo} = 0.8/day)	Oral	Neoplasm	168
Thallous acetate	Wistar albino rat	(LD_{Lo}[a/] = 29) 32	Oral		22
		(LD_{Lo} = 20) 23	Intraperitoneal (i.p.)		
	Guinea pig	(LD_{Lo} = 12)	oral		
		(LD_{Lo} = 7)	i.p.		
	Rabbit	(LD_{Lo} = 19)	oral		
		(LD_{Lo} = 13)	i.p.		
		(LD_{Lo} = 20)	Intravenous (i.v.)		
	Dog	(LD_{Lo} = 10)	oral		
Thallous bromide	Albino rat	(LD_{Lo} = 29) Absolute mortality 35	Intragastric		7
Thallous carbonate	Albino mouse	(LD_{Lo} = 18) ~ 27 (LD 100 = 32)	Subcutaneous (s.c.)		7
	Albino rat	~ 18	s.c., intragastric, or endotracheal		7
	Male mouse	(LD 100 = 25) 21	Intragastric		126 and 127
Thallous chloride	Male mouse	23.7	Intragastric		126 and 127
Thallic chloride	CDF, male mouse	6.90 (LD_{10} = 6.00)	i.p. daily for 10 days	Pneumonitis Alopecia at 20 days	171, 172, and 253
	Sprague-Dawley female rats	5.66 (LD_{10} = 4.85)	i.p. daily for 10 days	Pneumonitis Alopecia at 20 days	171, 172, and 253
Thallous chloride	Swiss albino male mice	20 (Tl) (within 30 days)	i.p.		254
Thallium iodide	Albino rat	(LD_{Lo} = 28) Absolute mortality 55	Intragastric		7
Thallium nitrate	Rat	(LD_{Lo} = 20)	s.c.		Abderhalden's Handbook cited in Reference 255
	Mouse	32.5	Oral		J. Faculty of Agri., Tuttori Univ. (Japan), 5, 15 (1969), cited in Reference 255
	Dog	(LD_{Lo} = 45)	Oral		Abderhalden's Handbook cited in Reference 255

TABLE VII-11 (concluded)

Substance	Species	LD_{50} (or other), mg/kg	Route	Toxic Effects	Reference
Thallous oxide	Mouse	71 (Tl)	i.p.		Lyublina (1965) cited in Reference 256.
Thallic oxide	Wistar albino rat	(LD_{Lo} = 20) 39	oral		22
		(LD_{Lo} = 62) 72	i.p.		
	Guinea pig	(LD_{Lo} = 5)	oral		
		(LD_{Lo} = 30)	i.p.		
	Rabbit	(LD_{Lo} = 30)	oral		
		(LD_{Lo} = 60)	i.p.		
		(LD_{Lo} = 39)	i.v.		
	Dog	(LD_{Lo} = 20)	oral		
Thallium(I) selenite	Rat	(LD_{Lo} = 50)	Oral		National Academy of Sciences, National Research Council, Chemical-Biological Coordination Center, Review (Washington). Cited in Reference 255.
Thallous sulfate	Rat	15.8	Oral		94
	Mouse	23.5	Intragastric		126 and 127
	Mouse	29	Oral		255
	Sheep	(8 mg/kg causes thallitoxicosis, 24 mg/kg is lethal dose) 143/			
	Adult French male mouse	46(Tl)	Oral		23
	Adult French male mouse	41(Tl)	s.c.		23
	Adult French male mouse	33(Tl)	i.v.		23
Thallous sulfide		MLD = 30	oral, i.v., i.p.		257
Thallous salt (?)	Mouse	23(Tl)	i.p.		Lyublina (1965) (Russian book) cited in Reference 256.

a/ Lowest single dose showing any lethality within 14 days.

TABLE VII-12

EFFECTS OF THALLIUM ON MICROORGANISMS

Organism	Medium	Concentration	Effect	References
Saccharomyces cerivisiae	20 g sucrose, 3 g $(NH_4)_2SO_4$, 2 g KH_2PO_4, 1.5 g asparagine (Eimer and Amend brand), 0.25 g each at $MgSO_4$ and $CaCl_2$, 1,000 ml distilled H_2O. Optimum pH 3.8 maintained by addition of $Ca(OH)_2$		Tl reduces sugar utilization (the retarding effect of excretion products then appeared later). The period of constant growth was longer, giving a greater crop of yeast.	258
Yeast	Sugar solution	800-2,000 ppm, 2,000 ppm	Stimulated Retarded Richet, 1917, cited	213
Saccharomyces cerevisiae (yeast)	Sugar solution, see above.	0.0001 mg/ml TlOAc (0.08 ppm Tl) 0.001 mg/ml TlOAc (0.8 ppm Tl) 0.01 mg/ml TlOAc (8 ppm Tl) 0.1 mg/ml TlOAc (80 ppm Tl) 1 mg/ml TlOAc (800 ppm Tl) 10 mg/ml TlOAc (0.8% Tl)	25% higher growth 59% higher growth 44% higher growth 95% less growth 96% less growth 98.5% less growth Decrease in budding occurs later in cultures contg. Tl. Gottbrecht, 1880, and Schulz, 1886, also reported yeast growth stimulation by Tl.	212
Pseudomonas aeruginosa		0.6 m M TlOAc (122 ppm Tl)	Kunze, 1972, cited. Prevents growth	259
Staphylococcus aureus S. faecalis		39 m M TlOAc (~ 8,000 ppm Tl)	Kunze, 1972, cited. Does not prevent growth	259

TABLE VII-12 (continued)

Organism	Medium	Concentration	Effect	References
Saccharomyces cerevisiae	Liquid culture	305 ppm Tl[a] (as Tl_2SO_4)	Growth rate inhibited 50%.	259
	Agar	20 ppm Tl[a]	Colony size reduced 50%.	
	Agar	41 ppm Tl[a]	Colony formation prevented.	
	Liq.	82 ppm Tl[a]	Toxicity of Tl increasingly alleviated by increasing the K concentration of the medium up to that of the standard medium.	
		~ 410 ppm Tl[a]	Increasing the K concentration slightly alleviated the Tl inhibition of O_2 consumption.	
		-	Tl rapidly bound to cell surfaces and also progressively accumulated by several energy-dependent transport systems (possibly reflecting > 1 K transport system). Affinity for Tl uptake higher than for K.	
Streptococci	1.0% peptone, 1.0% Lemco, 0.1% NaCl, 0.1% TlOAc, 0.5% glucose, pH 6.8.	780 ppm Tl	Excellent growth by 39 cultures of S. agalactiae, one culture of S. pyogenes and two of S. lactis; but one culture of S. agalactiae and two of S. pyogenes failed to grow. Browning and Gulbransen, 1918, and Kinnear, 1931, also reported TlOAc or $TlNO_3$ as selective agents for streptococci.	260
Proteus vulgaris and Aerobacter aerogenes	1.0% peptone, 1.0% Lemco, 0.1% NaCl, 0.1% TlOAc, 0.5% glucose, pH 6.8	780 ppm Tl	Slight growth, only occasionally for A. aerogenes.	260
Alcaligenes viscosus, Pseudomonas aeruginosa, P. fluorescens, Escherichia coli, Staphylococcus aureus, Bacillus subtilis	1.0% peptone, 1.0% Lemco, 0.1% NaCl, 0.1% TlOAc, 0.5% glucose, pH 6.8	780 ppm Tl	No growth	260

TABLE VII-12 (continued)

Organism	Medium	Concentration	Effect	Reference
Bacteria			P. Eisenberg, 1918, cited. Decreasing bacterial toxicity. Ag > Hg > Pt > Au > Co > Cd > Ni > Zr > Tl > Cu > Pb	261
Escherichia coli	Peptone water	0.005 *M* TlCl (~ 1,000 ppm Tl)	R. D. Hotchkiss, 1923, cited. No multiplication (3-day incubation period).	261
Escherichia coli	Peptone water	0.001 *M* TlCl (~ 200 ppm Tl)	Allowed multiplication. TlCl was comparable with $SnCl_4$ and $NiCl_2$. Chlorides of Zn, Cu, Fe, Co, Pb, Al, Ce, Cd, and Hg were more toxic.	261
Escherichia coli	Liquid culture	370 ppm Tl[a] (as Tl_2SO_4)	Growth rate inhibited 50%.	259
	Agar	~ 20 ppm Tl[a]	Colony size reduced 50%.	
	Agar	82 ppm Tl[a]	Colony formation prevented.	
	Liquid	200 ppm Tl[a]	Presence of K did not alleviate Tl toxicity as well as seen with *S. cerevisiae* and *B. megaterium*.	
		~ 410 ppm Tl[a]	Increasing the K concentration did not affect inhibition of O_2 consumption.	
		-	Tl rapidly bound to cell surfaces and also accumulated by an energy-dependent transport system with more affinity for K than Tl.	
E. coli (Tl-resistant strain)	Agar	~ 200 ppm[a]	Colony size reduced 50%.	259
	Agar	820 ppm[a]	Colony formation prevented	
		-	Surface binding of Tl equivalent to wild-type *E. coli*, but showed no energy-dependent uptake of Tl (yet K uptake unimpaired).	

TABLE VII-12 (continued)

Organism	Medium	Concentration	Effect	References
Escherichia coli	Water	800 ppm	Inhibits or stops growth. A. Buschke and F. Jacobson, 1922, cited. Tl inhibits bacilli in the presence of O_2.	213
Bacillus lactic acid	Milk	0.1 ppm 1.0 ppm 12.5 ppm 1,500 ppm	Growth stimulated. Growth normal. Growth retarded. Growth inhibited or stopped. Richet, 1917, cited.	213
Fomes amnosus (a fungus)	"Culture"	180 ppm	Growth inhibited or stopped. Bateman, 1933, cited. Tl_2SO_4 about half as toxic as $HgCl_2$, 3 times as toxic as $AgNO_3$.	213
Mold fungi	Agar	16 ppm 16-160 ppm	Growth retarded. Growth inhibited or stopped. Fargher et al., 1930, and Morris, 1926, cited.	213
Alternaria tenuis conidia (a fungus)	Unbuffered solution of $TlNO_3$. No spore germination stimulant. 50,000 spores/ml	2.9×10^{-4} M or ~ 60 ppm, pH 6.7	The ED_{50} (median effective dose). Germination inhibited.	262, 263
Bacillus megaterium	Liquid culture	0.015 mM Tl_2SO_4 (6.1 ppm Tl)[a]	Growth rate inhibited 50%.	259
	Agar	10 ppm Tl[a]	Colony size reduced 50%	
	Agar	30 ppm Tl[a]	Colony formation prevented.	
			Toxicity of Tl increasingly alleviated by increasing the K concentration of the medium up to that of the standard medium.	
Staphylococcus aureus and Staphylococcus epidermis, 42 coagulase-positive strains and 48 coagulase-negative	Agar	$\leq$ 500 ppm Tl	S. epidermis showed growth	264
		$\leq$ 8,000 ppm	S. aureus showed growth. Its Tl resistance depended on the inoculum size.	

TABLE VII-12 (continued)

Organism	Medium	Concentration	Effect	References
Thiobacillus ferrooxidans	K-free	Tl_2SO_4: > 0.0001 M (~ 41 ppm)	Inhibited growth.	265
	4.56 mM K	0.001 M	Nontoxic	
Azotobacter 3 strains	Solid	10^{-4} mole Tl salt in Petri dish streaked with an Azotobacter suspension. [20 or 40 mg Tl]	Highly toxic. Among 51 salts tested, Tl_2SO_4 was as toxic as Na_2CrO_4 and only slightly less toxic than the most effective growth inhibitor $AuNaCl_4$.	266
Botrytis fabae, (a fungus)	Unbuffered $TlNO_3$ solution. No spore germination stimulant. 50,000 spores/ml	8.6 x 10^{-5} M (~ 18 ppm)	The ED_{50} (median effective dose). Germination inhibited.	263
Mycoplasma laidlawii	Liquid medium containing cholesterol, Bacto-tryptose, penicillin G, etc.	TlOAc, 0.025 and 0.050% (wt/vol) (190 and 390 ppm Tl)	Redn. of cell membrane carotenoid content. Stage of carotenogenis affected unknown. Mycoplasmas similar to L-phase variants of bacteria. Both lack cell walls.	267
Mycoplasmas			Insusceptibility to TlOAc is a characteristic of mycoplasmas.	268
		~ 1,600 ppm Tl as TlOAc	Generally, growth uninhibited except that of T-mycoplasmas.	
		~ 390 ppm	T-mycoplasma growth inhibited (Edward, 1947; Morton and Leece, 1953; and Shepard, 1969, cited).	
Acholeplasmataceae (14 strains of A. laidlawii, two strains of A. granularum, one each of A. axanthum and A. sp. bovine group 6)	Solid Hayflick medium	TlOAc: 1,000 ppm	Reduced and prolonged growth of eight strains, completely inhibited one strain.	269
		8,000 ppm	Three remaining strains showed very poor growth.	

TABLE VII-12 (continued)

Organism	Medium	Concentration	Effect	References
Mycoplasmataceae (M. arthritidis, M. fermentans, M. hominis, M. salivarium, M. orale, and M. pulmonis)	Solid Hayflick medium	8,000 ppm (TlOAc)	Still grew very well.	269
Mycoplasma pneumoniae	-	4,000 ppm	Inhibited growth. Taylor-Robinson (1968) cited.	269
Mycoplasma strain "L5"	-	2,000 ppm	Inhibited growth. Edward (1947) cited.	269
Staphylococcus aureus S. faecalis	D.S.T. - agar containing TlOAc + sugar-bullion culture of the particular strain	5,333 ppm	All tested strains of the species still growing. (No growth at next higher concentration.)	269
Proteus mirabilis	D.S.T. - agar containing TlOAc + sugar-bullion culture of the particular strain	1,000 ppm (776 ppm Tl)	All tested strains of the species still growing. (No growth at next higher concentration.)	269
S. pyogenes A, List. monocytogenes, E. coli, S. typhimurium	D.S.T. - agar containing TlOAc + sugar-bullion culture of the particular strain	500 ppm	All tested strains of the species still growing. (No growth at next higher concentration.)	269
Mycoplasma lysodekticus, Aeromonas hydrophila	D.S.T. - agar containing TlOAc + sugar-bullion culture of the particular strain	250 ppm	All tested strains of the species still growing. (No growth at next higher concentration.)	269
Bacillus subtilis	D.S.T. - agar containing TlOAc + sugar-bullion culture of the particular strain	125 ppm	All tested strains of the species still growing. (No growth at next higher concentration.)	
Pseudomonas aeruginosa	D.S.T. - agar containing TlOAc + sugar-bullion culture of the particular strain	63 ppm	All tested strains of the species still growing. (No growth at next higher concentration.)	269

TABLE VII-12 (continued)

Organism	Medium	Concentration	Effect	References
Mycobacterium tuberculosis	Sauton's synthetic medium	Tl salt, 2.5-20 mg % (25-200 ppm)	Inhibition at $\geq$ 10 mg %. Reversible modification of culture characteristics.	270
Nitrobacter agilis	Liquid culture	0.82-16.3 ppm Tl as Tl_2SO_4	Nitrate formation retarded 48-360 hr with increasing Tl concentration.	271
Proteus mirabilis L-phase variant (strain L9)	Fluid	183 ppm	Growth unaffected.	268
		365 ppm	"Max. growth titre reduced by 1 in 10" compared with 183 ppm.	
	Solid (agar)	183 ppm	Slight inhibition.	
		365 ppm	Inhibition.	
Proteus mirabilis (Strain 9)	Fluid	183 ppm	Growth unaffected.	268
		365 ppm	Growth rate "delayed considerably" in medium with horse serum and yeast extract.	
	Solid	183 ppm	Not inhibited.	
		365 ppm	Inhibited swarming. Very small colonies. No morphol. alteration.	
Brucella abortus and its L forms (latter occur after intracellular exposure of B. abortus to antibiotics)	Hamster kidney cells	TlOAc, 1:5,000 added at end of growth period (155 ppm Tl).	All bacterial growth inhibited, but L-form arose in some cultures over a 10-day period.	272
Nitrosomonas (a nitrite-forming bacterium)	Omeliansky's medium + $MgCO_3$ + $(NH_4)_2SO_4$	~ 0.6-10 ppm Tl as Tl_2SO_4.	Increased growth 48-336 hr	214
		~ 1.3 ppm	Max. growth (max. nitrite formation).	
Cunninghamella elegans	Richard's medium, $TlNO_3$, 0-20 ppm colchicine	1 mequiv. Tl^+	Toxic. Fungus lost weight and showed least carbohydrate content. Respiration rate very low. In the presence of Tl, colchicine suppressed uptake of P and NO_3^-. Without colchicine, glucosan content high. Increased rate of sugar uptake favored transformation to noncarbohydrates.	273, 274

TABLE VII-12 (continued)

Organism	Medium	Concentration	Effect	References
Nitrobacter agilis, a nitrate-forming bacteria	Red soil suspension: 3.0 g soil, 0.05 g $CaCO_3$ Tl_2SO_4, q.s. 50 ml + $NaNO_2$	0.82-7.3 ppm Tl	Nitrate formation retarded.	216
		6.5 ppm	Very slow. After 360 hr, < 50% of the nitrite had been oxidized.	
		8.2 ppm	Nitrate formation stopped.	
Soil microorganisms isolated at the logarithmic growth phase from soil containing 100 ppm Tl (as $TlNO_3$)	- (in vitro)	10 ppm	Resistant cultures (no.): aerobic bacteria (10), aerobic spore-forming bacteria (13), fungi (12), anaerobic spore-forming bacteria (5).	215
		≤ 90 ppm	The 12 aerobic fungi and 2 aerobic bacteria still grew.	
		70 ppm	Two aerobic bacteria grew. Possibly, synthesis of a transport component necessary for Tl uptake was repressed or synthesis had been induced of enzymes that change Tl to a form not taken up.	
		50 ppm	One aerobic bacteria grew.	
		30 ppm	Two aerobic bacteria and two aerobic spore-forming bacteria grew.	
Fungi				
Candida albicans		$(C_6H_5)_2TlCN$ = 25.0 ppm	Minimum inhibitory concentration	275
Cryptococcus neoformans		All 4 diaryl derivs. > 100 ppm.	Minimum inhibitory concentration	
Trichophyton mentagrophytes (athlete's foot)		$(C_6H_5)_2TlSCN$ = 12.5 ppm	Minimum inhibitory concentration	
Microsporum canis (a common ringworm cause)		$(C_6H_5)_2TlSCN$, $(C_6H_5)_2TlCN$ = 12.5 ppm	Minimum inhibitory concentration	
Aspergillus niger		$(C_6H_5)_2TlCN$ = 25.0 ppm	Minimum inhibitory concentration	

TABLE VII-12 (continued)

Organism	Medium	Concentration	Effect	References
Bacteria				
Bacillus subtilis		All 4 diaryl derivs. = 3.12 ppm	Minimum inhibitory concentration	275
Staphylococcus aureus		$(C_6H_5)_2TlCN$ = 6.25 ppm	Minimum inhibitory concentration	
Salmonella typhi		$(C_6H_5)_2TlCN$ = 25.0 ppm	Minimum inhibitory concentration	
Escherichia coli		$(C_6H_5)_2TlSCN$ = 25.0 ppm	Minimum inhibitory concentration	
			$(C_6H_5)_2TlCl$ and its 1,10-phenanthroline complex showed the same toxicity. Neither inhibited fungal growth up to 100 ppm.	
Colletotrichum falcatum (causes red rot of sugar-cane)	Potato-dextrose agar	R_2TlX — R / X / ppm		275
		Phenyl / phenoxide / 15	Mycelial growth inhibition, ED_{50}	
		Phenyl / chloride / 13	(ED_{50} of the In and Ga analogs < 1 ppm)	
		Phenyl / thiocyanate / 9.2	ED_{50}	
		Phenyl / acetate / 7.1	ED_{50}	
		Phenyl / naphthoxide / 4.6	ED_{50}	
		m- or *p*-tolyl / acetate / 3.7	ED_{50}	
		o- or *m*-tolyl / phenoxide / 1	ED_{50}	
		o-,*m*-, or *p*-tolyl / chloride / < 1	ED_{50}	
		o-,*m*-, or *p*-tolyl / thiocyanate / < 1	ED_{50}	
		o-tolyl / acetate / < 1	ED_{50}	
		p-tolyl / phenoxide / < 1	ED_{50}	
		TlCl 80 ppm	16.6% inhibition of mycelial growth.	
		TlCl		
		$TlCl_3$ 80 ppm	6.99% inhibition of mycelial growth.	
		$C_6H_5TlCl_2$ 16 ppm	ED_{50}	
			Four diarylthallium(III) compounds sprayed twice at 80 ppm within 3 days did not adversely affect tomato seedlings.	
Aspergillus spp.	Agar	10^{-6} to 10^{-4} *M* (for Tl 0.2-20 ppm)	Ag^+, Cd^{2+}, Tl^+, and Hg^{2+} the most toxic	276

TABLE VII-12 (continued)

Organism	Medium	Concentration	Effect	References
Aspergillus niger	Agar	20 ppm Tl as $TlNO_3$ (5×10^{-5} *M*)	Minimum toxic concentration.	277
A. oryzae	Agar	20 ppm Tl	Minimum toxic concentration.	277
A. coespitosus	Agar	40 ppm Tl	Minimum toxic concentration.	277
A. terreus	Agar	20 ppm	Minimum toxic concentration.	277
A. awamori	Agar	20 ppm	Minimum toxic concentration.	277
Aspergillus flavus	Czapek's agar	0.50 m*N* Tl + (salt not specified) (~ 100 ppm)	Mycelial growth 50% that of controls (no Tl, extra K). With 2.5 m*M* KCl, mycelial growth is 331% compared with 191% without Tl.	278
		2.50 m*N* Tl + (~ 500 ppm)	No mycelial growth. With 2.5 m*M* KCl, mycelial growth was 172%.	
Aspergillus niger and *Penicillium italicum*	A standard nutrient solution 30°C	Tl_2SO_4 (1.6-2.4 ppm Tl)	Very good growth inhibition. The most toxic of > 120 compounds included those of Hg, Cr, Cu, and Zn. (Three weeks' duration)	279
Aspergillus niger	Nutrient solution	Tl_2SO_4, 1-2 ppm	Spore germination inhibited.	279
	Nutrient solution plus 50% wheat flour	500-800 ppm	Spore germination inhibited. Lower toxicity due to adsorption.	
	Nutrient solution plus 20% wheat flour	1,000-3,000 ppm	Spore germination inhibited.	
	Mineral solution plus 50% flour	1,000-5,000 ppm	Spore germination inhibited.	
	Mineral solution plus 20% flour	3,000-5,000 ppm	Spore germination inhibited.	

TABLE VII-12 (concluded)

Organism	Medium	Concentration	Effect	References
Aspergillus niger	Nutrient solution + 3% tragacanth	100-300 ppm	Spore germination inhibited.	
	Nutrient solution + 40% beer wort	1,000-3,000 ppm	Spore germination inhibited.	
Phycomyces	Coons medium	Tl acetate, nitrate, or sulfate		280
		0.2 γ/cm^3 (~ 0.2 ppm)	Progressive diminution of aerial mycelia.	
		2-10 γ/cm^3	Very weak development.	
		20 γ/cm^3	A few spores germinated.	
		40 γ/cm^3	Several spores are swollen. No germination.	

a/ The values given for this reference may be two times too high. Values are given as mM Tl, but the solutions were made up from Tl_2SO_4. In at least one place, the authors indicate that 0.5 mM Tl means (0.0005) (204) g/liter.

TABLE VII-13

ACCUMULATION OF ^{204}Tl BY THE ALGA *ULVA RIGIDA*[223, 224]

Conditions	Accumulation Coefficient, 48 hr			
	$^{204}Tl(I)$		$^{204}Tl(III)$[a]	
	14°C	24°C	14°C	24°C
In light	28.4	43.5	67.1	62.0
In darkness[b]		~ 20	Same as in light.	
In the presence of NaF, an inhibitor of glycolysis	~ 25			
In the presence of 2,4-dinitrophenol, an inhibitor of phosphorylation and oxidation	~ 2.5		About same as control.	

a/ The lack of temperature difference may be due to passive accumulation of Tl(III) complexes. Dead *Ulva* accumulated much less Tl(I) than the live alga.

b/ When placed in darkness, *Ulva* containing $^{204}Tl(I)$ loses thallium more slowly in pure seawater than when in the light. Thus, Tl(I) excretion depends on photosynthesis.

TABLE VII-14

EFFECTS OF THALLIUM ON TERRESTRIAL PLANTS

Plant	Culture	Concentration	Effect	Reference
Turkish tobacco seedlings	Water, sand, soil	≥ 0.04 ppm	Chlorosis. Least amount needed in water culture.	281
Tobacco	Water	≥ 0.016 ppm	Tl detected (spectrog.) in plant ash ≥ 1 ppm. Amount in tissue decreased from roots to youngest leaves. No Tl detected in frenched tobacco.	281
Tobacco	$TlNO_3$ in water	0.052 ppm Tl	Chlorotic phase within 5 days. Effect of Tl_2SO_4 similar.	281
	Sand	0.077 ppm		
	Sandy loam orchard soil	0.29 ppm		
	Clay loam field soil causing frenching	0.19 ppm		229
	Water	0.1 ppm Tl	Strap leaf phase within 12 days. Tl toxicity inhibited by steam sterilization of the soil, 30-60 ppm KI, 200 ppm $Al_2(SO_4)_3$ or $Ca(NO_3)_2$.	
Tobacco	Sandy loam, 2 moisture levels	35 and 75 ppm $TlNO_3$ (~ 27 and 58 ppm Tℓ)	More injurious at higher moisture content. Stems often killed at soil level.	281
		1 ppm Tl	Killed plants outright or slowed growth and produced chlorosis in earlier growth stages.	
Tobacco, <u>Nicotiana tabacum</u>	Sandy loam soil (pot)	Tl as Tl_2SO_4 1.7 ppm	At 30 days: 3.3 in. tall (compared with 4.2 in. controls). Normal root development.	217
		6.9 ppm	2.5 in. tall. No additional root growth.	
		14 ppm	1.0 in. tall. No additional root growth.	
		42 ppm	0.0 in. tall.	
	Silt loam soil (pot)		At 35 days:	217
		0.81 ppm	138 g (134 g controls) plants, fresh wt.	
		2.4 ppm	101 g	
		4.1 ppm	51 g	

TABLE VII-14 (continued)

Plant	Culture	Concentration	Effect	Reference
Tobacco, *Nicotiana tabacum* (cont.)	Sandy loam soil (pot)	1.7 ppm 6.9 ppm 14 ppm 42 ppm	Addition of 6 g 5-8-7 com. fertilizer/7 lb. soil had a negligible effect except at 1.7 ppm.	217
Tobacco	Soln. (introduced at 13-15 cm in ht., grown 2-3 weeks, $TlNO_3$ present for 7 weeks.)	1 ppm Tl as $TlNO_3$	Fresh wt. reduced 31.2% and nicotine content reduced 75%. Of 54 rare elements tested, Tl had the most deleterious effect on nicotine content followed by Zr ($ZrCl_4$ reduced the fresh wt. 96%.)	282
Lycopersicon esculentum (tomato)	Sandy loam soil (pot)	Tl as Tl_2SO_4	At 30 days:	217
		1.7 ppm	6.0 in. tall (controls 7.2 in.)	
		6.9 ppm	4.3 in. tall. Marked retardation of root development	
		14 ppm	3.0 in. tall. No additional root growth.	
		42 ppm	2.5 in. tall. No additional root growth.	
			In a similar expt., 6 g of a 5-8-7 com. fertilizer had a negligible effect on Tl toxicity except at 1.7 ppm Tl.	
Triticum aestivum (wheat)	Sandy loam soil (pot)	1.4 ppm	Slightly injurious	217
		6.9 ppm	Much more injurious	
		14 ppm	Very injurious. One-third as tall as controls. Lighter in color. Root development retarded.	
		28 ppm	Prevented lateral root development. Killed. Liming did not reduce injury.	
		56 ppm	Killed.	
		84 ppm	Killed soon after emergence. Leaf injury general type	
Fagopyrum esculentum (buckwheat)	Sandy loan soil (pot)	1.4 ppm	Slightly injurious	217
		6.9 ppm	Much more injurious	
		14 ppm	Very injurious. See wheat.	
		28 ppm	Prevented lateral root development. Great growth retardation.	
		56 ppm	Still alive.	
		84 ppm	Killed soon after emergence	

TABLE VII-14 (continued)

Plant	Culture	Concentration	Effect	Reference
Medicago sativa (alfalfa)	Sandy loam soil (pot)			217
		1.4 ppm	Slightly injurious	
		6.9 ppm	Much more injurious	
		14 ppm	Very injurious. See wheat.	
		28 ppm	Prevented lateral root development. Killed.	
		56 ppm	Killed.	
		84 ppm	Killed soon after emergence.	
Lolium perenne (rye grass)	Sandy loam soil (pot)			217
		1.4 ppm	Slightly injurious	
		6.9 ppm	Much more injurious	
		14 ppm	Very injurious. See wheat.	
		28 ppm	Prevented lateral root development.	
		56 ppm	Killed.	
		84 ppm	Killed soon after emergence. Leaf injury general type.	
Lolium perenne (plants established for several months)	Loam soil (pot)	Surface applications of Tl soln. (concn.?)		217
		2.65 lb/acre	No effect	
		5.30 lb/acre	Slight injury	
		7.95 lb/acre	Severe injury	
		12.25 lb/acre	Killed	

TABLE VII-14 (continued)

Plant	Medium	Concentration	Effect	Reference
Grass	Soil, Manoa Arboretum	Mongoose ground-meat baits contg. ~ 0.5 g/0.25 oz Tl_2SO_4 set a few in. above the soil in July, 1929, in Hawaii.	In July, 1931, bare patches 1-2 ft long and ~ 1/3 as wide below each stake. No plant life had been apparent in the intervening 2 yr.[a]	283
Lepidium sativum (peppergrass) seeds		N/ 600 Tl_2SO_4 (340 ppm Tl)	Complete and irreversible prevention of germination. Lower concns. not tested.	Dilling, 1926 Cited in Reference 283.
Zea mays	(Cut leaves in aerated 0.5% Tl_2SO_4 soln.)		Movements of Tl^+ as a response to illumination and CO_2 concn. are linked with those of K^+ in the stomata. Thus, Tl^+ cannot locate the terminal sites of transpiration as suggested by Maercker, 1965.[285]	284
Cucumis sativis (Cucumber) seedlings	Petri dish culture	10 ppm	Growth inhibition too severe to observe disturbances of orientation in roots and stems as shown by H_g compounds.	286
Vegetables and grasses	Soil (pot and field)	≥ 10 ppm	No injurious effects. Plant growth apparently stimulated. (Thallium-treated grain was eaten by ground squirrels so rapidly that 9 applications did not injure the vegetation. Several ranchers had estimated 10-25% increase in vegetative cover following ground squirrel eradication.)	287
Phleum pratense (a grass)	Soil	5-10 ppm 100 ppm 1000 ppm	Growth stimulated Growth retarded Growth inhibited or stopped	213
"General range of plants"	Soil	8 ppm 20-60 ppm 80 ppm	Growth normal Growth retarded Growth inhibited or stopped	213

TABLE VII-14 (continued)

Plant	Medium	Concentration	Effect	Reference
Oats, shoots (form mainly by cell division in the embryo)	Yolo clay loam (pot)	Tl_2SO_4 in moistening soln. 450-550 ppm	~ 17% reduction length	233
		800 ppm	50% reduction in length	
		1000 ppm	~ 73% reduction in length	
		2000 ppm	~ 83% reduction in length	
	Fresno sandy loam	150 ppm	~ 17% reduction in length	
		350 ppm	50% reduction in length	
		550 ppm	~ 73% reduction in length	
		800 ppm	~ 83% reduction in length	
		1500 ppm	No growth	
Oats, coleoptiles (develops mainly by enlargement of previously formed cells)	Yolo clay loam	800 ppm	10% reduction in length	233
		1000-2000 ppm	20% reduction in length	
	Fresno sandy loam	650 ppm	20% reduction in length	
		2000 ppm	50% reduction in length	
Oats (Kanota)	Yolo clay loam (pot, 500 g soil)	ppm Tl_2SO_4 15 30 60	Little or no effect on height compared with controls in 12 runs. Variable effects on wt. (light, temp., and humidity had been varied in these runs)	233
		120 ppm	Height and wt. significantly reduced in 6 of 12 runs.	
		180 ppm	Height and wt. always much reduced: ~ 50-75% for height, ~ 75-90% for wt.	
		240-600 ppm	Heights 3-17 in. and wts. 0.1 - 1.6 g compared with 12-33 in. and 3.4 - 9.8 g for the controls.	
Oats	Yolo clay loam Fresno sandy loam Stockton adobe clay Columbia fine sandy loam	Tl_2SO_4 in moistening solution corresponding to 7.5 - 600 ppm Tl_2SO_4 in the dry soil		233
		3.7-52.5 ppm	First deleterious effects on height and wt.	
		13.1-82.5 ppm	≥ 50% wt. reduction	
		36-151.8 ppm	≥ 50% height reduction	

TABLE VII-14 (continued)

Plant	Medium	Concentration	Effect	Reference
Avena sativa (oats)	Soil	22,000 ppm 40,000 ppm	Growth retarded Growth inhibited or stopped.	213
Trifolium sp.	Soil	5 ppm 100 ppm 1000 ppm	Growth stimulated Growth retarded Growth inhibited or stopped	213
Peas, sweet corn, field corn, clover, timothy	Loam soil in 6 x 6 x 3 in. trays	Tl_2SO_4 in soln. to provide 49 mg Tl on 61st day after germination. (amt. "represents the average quantity of thallium that might be applied in an ordinary bait spot of thalgrain placed in ground-squirrel control.")	No detrimental effects.	213
Peas, sweet corn, field corn, clover, timothy	Loam soil in 6 x 6 x 3 in. trays	49 mg Tl mixed with the soil of each tray before seeds were planted.	Delayed and reduced germination Death after slow growth	213
Peas, sweet corn, field corn, clover, timothy	Loam soil in 6 x 6 x 3 in. trays	After 12 days of growth, irrigation with 1100 ppm Tl (as Tl_2SO_4) solution on days 12, 20, 27, and 33. (400 mg Tl applied to each tray within ~ 1 month)	Growth completely inhibited	213
12 corn seeds, 1g timothy, or 1.5 g clover/jar	Loam soil in 1-gal jars	5 ppm 10 ppm	Corn growth stimulated. Clover less stimulated. Timothy slightly stimulated in later growth.	213
		100 ppm 1000 ppm 10000 ppm	Toxic effects in all seedlings Timothy germination stopped Corn and clover germination stopped	213
Vicia (vetch) *faba*, *Lupinus albus*, *Zea mays*	Distd. water	20 ppm	Growth retarded. Died in 2-4 days. Prat, 1932, cited.	213
Zea mays	Soil	10 ppm 5 ppm 100-1000 ppm 10,000 ppm	Growth stimulated Growth normal Growth retarded Growth inhibited or stopped	213

TABLE VII-14 (continued)

Plant	Medium	Concentration	Effect	Reference
Fagopyrum esculentum (buckwheat)	Silt loam soil (pot)	Tl as Tl_2SO_4	Plant weight:	217
		0 ppm	44 g at early-blooming stage	
		0.81 ppm	34 g; 23% retardation	
		1.6 ppm	30 g	
		2.4 ppm	29.2	
		3.2 ppm	22.5	
		4.1 ppm	14	
		8.1 ppm	2.1	
Zea mays (corn)	Sandy loam soil (pot)	Tl as Tl_2SO_4		217
		0.85 ppm	Slight retardation of tops and roots.	
		1.7 ppm	Striking retardation of tops and roots. Leaves yellowed soon after emergence.	
		6.9 ppm	Killed about 6 weeks after seeds were planted. Leaves yellowed soon after emergence. Leaf injury interveinal type.	
		28 ppm	Liming did not reduce injury.	
	Sandy loam, silt loam, or fibrous sedimentary peat soil (pot)	6.9 ppm, 14 ppm, or 42 ppm	Injury decreased in the order sandy loam > silt loam > peat.	
Maize Feed beans Carrots	Soil	Traces Tl	Stimulated germination and growth of leaves, stems, and roots in greenhouse and field.	230
		10 ppm Tl	Best results upon sowing seeds prewet with Tl_2SO_4 soln. Yield of maize from seeds increased 13%; carrot roots, 43%	
		500 ppm	Toxic	
Plants of economic importance	Soil	Suggested max. permissible level 0.25 ppm		288
Corn	?	?	Radiothallium accumulation decreased in the order stalk > sheaths > leaves > ears > tassel	289
Corn	Nutrient solution in subirrigated gravel cultures	Insufficient to cause toxicity.	Radiothallium was evenly distributed throughout the leaf	290

TABLE VII-14 (continued)

Plant	Medium	Concentration	Effect	Reference
Corn and Broad beans	$TlNO_3$ soln. ?	0.0001 M $TlNO_3$ (~ 20 ppm Tℓ)	Prat et al., 1932, cited. Plants died in 2-4 days	233
	Nutrient soln. (Shive, R_5C_2)	0.0001 M $TlNO_3$	Only 1-40% of the Tl was absorbed in 13 days. Finding agrees with the greater toxicity of Tl to plants grown in less fertile soils. Roots injured.	233
Pea seedlings, 3-day-old Beet seedlings, 3-day-old			Highest mutagenic activity of 17 elements with observable activity after 2-3 days.	291
Barley, excised roots			Tl is a competitive inhibitor of Rb uptake. Also inhibits Na uptake somewhat. Does not depress O_2 utilization.	288, 292
			Hawf and Schmid in unpublished results found Rb absorption also impeded in intact bean plants. In unpublished results of Schmid and Colclasure, Tl absorption found to be metabolically mediated.	
Barley, excised roots	Soln.	≤ 200 ppm	Tl competes with K for binding sites. Readily absorbed into protoplasm. Accumulation into vacuoles slow. Redn. of O_2 content in the presence of Tl did not retard the Tl absorption into vacuoles. Rb stimulated Tl transport into vacuoles. Na did not.	293
Barley, excised roots	Had been grown under sterile conditions	Tl_2SO_4 1.0 (408 ppm Tl) and 2.0 m M	O_2 uptake inhibited ~ 57%	232
		5.0 and 10.0 m M	O_2 uptake inhibited ~ 74%	
Barley and soybeans	soil		Battelle Pacific Northwest Labs. (BPNL): among 17 metals Cd, Ni, Cr, and Tl most readily taken up. However, Reference 295 (abstract) states that of 17 metals (radionuclides), only Ni, Zn, W, and Mo showed concn. factors > 1.	294
			Tl reduced yields, esp. barley	296

TABLE VII-14 (continued)

Plant	Medium	Concentration	Effect	Reference
Plants	Soil		BPNL, 1972 Increase of Tl in tissue nearly paralleled increase in soil concn. and yield redns. Accumulated in lower stem, does not reach photosynthetic areas.	296
Trifolium pratense seeds (clover)	Filter paper	Tl_2SO_4, ppm	% Germination	231
			Initial / Total	
		0	day 2:71%; day 10:95%	
		70	day 2:19% day 13:86%	
		100	day 2:10% day 13:83%	
		300	day 7:3% day 15:33%	
		600	day 10:5% day 16:5%	
		810	day 11:2% day 16:2%	
		1200	day 10:1% day 11:2%	
	Tl_2SO_4	70 ppm	Max. concn. allowing for chlorophyll development	
Medicago sativa (alfalfa)	Filter paper	0-300 ppm	day 2:15-69% Day 7-14:95%	
		600-2500 ppm	day 2-4:1-12% Day 9-14:81-83%	
	Tl_2SO_4	70 ppm	Max concn. for chlorophyll development	
Lupinus alba (lupine)	Filter paper	0-2500 ppm	day 3:10-39% day 5-11:48-49%	
	Tl_2SO_4	2500 ppm	Max. concn. allowing chlorophyll development	
Pisum sativum (peas)	Filter paper	0-810 ppm	48-50% germination	231
		1200 ppm	42% germination	
		2500 ppm	24% germination	
Phaseolus vulgaris (beans)	Filter paper	0-600 ppm	48-50% germination	
		810 ppm	37-40% germination	
		1200 ppm	37-40% germination	
		2500 ppm	27% germination	
	Tl_2SO_4	1200 ppm	Max. concn. allowing chlorophyll development	
Trigonella (fenugreek) *Foenum-graecum*	Filter paper	0-1200 ppm	98-100% germination	231
	Tl_2SO_4 soln?	810 ppm	Max. concn. allowing chlorophyll formation	

TABLE VII-14 (continued)

Plant	Medium	Concentration	Effect	Reference
Triticum aestivum (wheat)	Filter paper	0-70 ppm	94-97% germination	231
		100-810 ppm	86-89% germination	
		1200 ppm	76% germination	
		2500 ppm	56% germination	
	Tl_2SO_4	100 ppm	Max. concn. for chlorophyll	
Avena sativa (oats)	Filter paper	0-300 ppm	90-93% germination	231
		600 ppm	75% germination	
		810-1200 ppm	49-55% germination	
		2500 ppm	23% germination	
	Tl_2SO_4	100 ppm	Max. concn. for chlorophyll	
Senapis alba (white mustard)	Filter paper	0-100 ppm	97-98% germination	231
		300 ppm	92% germination	
		600 ppm	53% germination	
		810 ppm	41% germination	
		1200 ppm	26% germination	
	Tl_2SO_4	100 ppm	Max. concn. for chlorophyll	
Polygonum fagopirum (buckwheat)	Filter paper	2500 ppm	9% germination	
	Tl_2SO_4	70 ppm	Max. concn. for chlorophyll	
			These concns. also reduced the size of developing plants and inhibited development of the root systems.	
Glycine max (soybean)	Sandy loam (pot)	Tl as Tl_2SO_4		217
		1.4 ppm	Slightly injurious	
		6.9 ppm	Much more injurious. Above-ground portions about half as developed as controls. Portions of leaves yellow. Few lateral roots.	
		14 ppm	Very injurious	
		28 ppm	Prevented lateral root development. Killed.	
		56 ppm	Killed.	
		84 ppm	Killed soon after emergence. Leaf injury veinal type.	

TABLE VII-14 (continued)

Plant	Medium	Concentration	Effect	Reference
Phaseolus vulgaris (wax bean)	Sandy loam (pot)	1.4 ppm	Slightly injurious	217
		6.9 ppm	Slightly injurious	
		14 ppm	Only slight retardation. Leaf color normal	
		28 ppm	About half as large as controls, but considerable lateral root development. Liming did not reduce injury.	
		56 ppm	Still alive.	
		84 ppm	Killed soon after emergence. Leaf injury veinal type.	
Helianthus annuus (sunflower), detached leaves	Tl salt soln.	2 ppm	Rate of net photosynthesis not much lower than control.	297
		20 ppm	Rate much lower after 3 days	
		100 ppm	Rate much lower after 2 days Plants wilted by 1st or 2nd day	
		200 ppm	Rate much reduced immediately.	
			A 50% redn. of net photosynthesis rate occurred at 63 ppm Tl in the plant tissue.	
			Transpiration was reduced similarly. Stomatal function may be inhibited.	
Sunflower epidermal peels	Tl salt soln.	10 μM (~ 2 ppm)	> 90% redn. in stomatal opening	297
Helianthus annuus and Zea mays	Hoagland soln.	Tl as Tl_2SO_4 ~ 10-50 μmole/ℓ (2-10 ppm with the assumption of μmole of Tl^+ not the salt)	Tl reduced rate of net photosynthesis by ~ 50-75% (greater for corn). At a 50% redn, the leaf content was 82 ppm for both sunflower and corn. Cd, Ni, and Pb did not reduce net photosynthesis rate so rapidly nor so much. Photosynthesis was reduced more than transpiration.	298
		1 ppm	Leaf yellowing beginning along veins of plants. Extended into interveinal areas at higher concns.	
Sunflower, epidermal peel	Tl^+ soln. (4 hr in dark at ambient CO_2 + 4 hr in light at lowered CO_2 concns.)	0.2 ppm	Stomatal opening 70% of that of controls.	298
		2 ppm	Stomatal opening 13%	
		20 ppm	Stomatal opening 5%	

TABLE VII-14 (continued)

Plant	Medium	Concentration	Effect		Reference
Corn	Tl^+ soln. (4 hr in dark at ambient CO_2 + 4 hr in light at lowered CO_2 concns.)	0.2 ppm	Stomatal opening 75%		298
		2 ppm	Stomatal opening 52% (concn. of Tl about 0.1 that required for Cd, Ni, or Pb)		
		20 ppm	Stomatal opening 23%		
			Tl reduced corn top and root biomass 50-60%		
			Authors conclude that Tl not only inhibits the rate of net photosynthesis by stomatal closure but also by affecting steps in CO_2 assimilation.		
Vegetation in typical range area of Santa Clara County, Calif.	Altamount clay loam soil (field)	Applied as thalgrain (1-lb Tl_2SO_4/100 lb grain)	Reduction in Vegetation Density		213
			Originally	After 1 yr (and the rainy season)	
		10 lb Tl_2SO_4/acre (if upper 3 in. soil: 10 ppm Tl)	30%	20%	
		20 lb/acre	40%	80%	
		30 lb/acre	30%	95%	
		50 lb/acre	20%	97%	
		75 lb/acre	15%	97%	
		100 lb/acre	15%	99%	
		200 lb/acre	30%	99%	
		(Since baits were usually all taken within a week ~ 90% the first day -- Tl was usually removed before it had any effect on the vegetation)			
Ribes (wild gooseberry)	Soil (pots)	>1280 lb/acre	No injurious effects over 8 weeks.		213
Vegetation, pasture areas, Calif.	Soil	~ 100 ppm Tl_2SO_4 dry soil (27 lb/acre)	< 50% growth reduction		299
		~ 0.33 lb/acre (~ 1 ppm Tl in soil)	Not likely to sterilize soil		
		(Usual amt. initially used for ground squirrel control, with lighter applications following)			

TABLE VII-14 (concluded)

Plant	Medium	Concentration	Effect	Reference
Plantago maritima (300) seeds	Moist filter paper, 22°C	0.05mM Tl (10 ppm ?)	Germination (in 5 days) 82% compared with 81% for the controls. Addition of 0.50 mM K^+ increased germination percentage to 94-97% at 0.05 to 0.25 mM Tl. Addition of Na^+ without effect.	278
		0.10mM Tl	Germination 68%	
		0.25mM Tl (50 ppm ?)	Germination 36%	

a/ Horn et al.[213] report a personal communication from C. C. Barnum: Up to 15 g/liter Tl_2SO_4 in water solution applied to 1-sq yd plots of grass were not permanently toxic.

AUTHOR INDEX. SECTION VII.

The numbers in this list of authors refer to the reference numbers cited in the text. The full citations may be found in the bibliography of this section, which immediately follows this index.

BIBLIOGRAPHY. SECTION VII.

1. Il'in, N. A., "Identical Effects With Hypodermic, Oral, Local, or Intravenous Administration of a Pharmacon [Drug]: (Experiments With Thallium)," Arch. Intern. Pharmacodynamie, 60, 377-394 (1938); Chem. Abstr., 33, 4316 (1939).

2. Lie, R., G. Thomas, and J. K. Scott, "The Distribution and Excretion of Thallium-204 in the Rat, With Suggested Maximum Permissible Concentrations and a Bioassay Procedure," Health Phys., 2, 334-340 (1960).

3. Prick, J. J. G., W. G. S. Smitt, and L. Muller, Eds., Thallium Poisoning, Elsevier, Amsterdam, 1955.

4. Munch, J. C., and J. Silver, The Pharmacology of Thallium and Its Use in Rodent Control, U.S.D.A. Tech. Bull. No. 238, April 1931, 28 pp.

5. Heyroth, F. F., "Thallium. A Review and Summary of Medical Literature," Public Health Rep. (U.S.), Suppl., No. 197, 1-23 (1947).

6. Karsybekova, G. R., "Absorption and Excretion of Thallium," Izvest. Akad. Nauk Kazakh. SSR, Ser. Fisiol. i Med., No. 7, 116-121 (1956); Chem. Abstr., 51, 9943C (1957).

7. Sanotskii, I. V., "Issledovanie Toksichnosti Soedinenii Talliya" ["Toxicity of Thallium Compounds (Carbonate, Iodide, and Bromide)"], Toksikol. Novykh. Prom. Khim. Veshchestv, No. 2, 94-104 (1961).

8. Wadachi, Y., S. Tashiro, and M. Muramatsu, "Skin Contamination by Radioisotopes. IV. Pig Skin Contamination by ^{35}S, ^{125}Sb, ^{147}Pm, ^{204}Tl, and ^{210}Po," Nippon Genshiryoku Gakkaishi, 8, 642-644 (1966); Chem. Abstr., 67, 88127p (1967).

9. Koefoed, B. M., "The Cryptonephridial Systems in the Mealworm Tenebrio molitor: Transport of Radioactive Potassium, Thallium, and Sodium; A Functional and Structural Study," Cell Tissue Res., 165(1), 63-78 (1975); Chem. Abstr., 84, 133105u (1976).

10. Hagiwara, S., and K. Takahashi, "Anomalous Rectification and Cation Selectivity of the Membrane of a Starfish Egg Cell," J. Membr. Biol., 18(1), 61-80 (1974); Chem. Abstr., 81, 148849b (1974).

11. Ellory, J. C., J. Nibelle, and M. W. Smith, "The Effect of Salt Adaptation on the Permeability and Cation Selectivity of the Goldfish Intestinal Epithelium," J. Physiol. (London), 231, 105-115 (1973).

12. Kinsey, V. E., I. W. McLean, and J. Parker, "Crystalline Lens. XVIII. Kinetics of Thallium Tl^+ Transport in Relation to That of the Alkali Metal Cations," Invest. Ophthalmol., 10(12), 932-942 (1971); Chem. Abstr., 76, 97335t (1972).

13. Gehring, P. J., and P. B. Hammond, "The Uptake of Thallium by Rabbit Erythrocytes," J. Pharmacol. Exp. Therap., 145(2), 215-221 (1964).

14. Cavieres, J. D., and J. C. Ellory, "Thallium and the Sodium Pump in Human Red Cells," J. Physiol. (London), 243(1), 243-266 (1974); Chem. Abstr., 83, 53954j (1975).

15. Lishko, V. K., L. I. Kolchinskaya, and N. T. Parkhomenko, "Talii i Natrievii 'Nasos' Eritrotsitiv" ["Thallium and Sodium Pump of Erythrocytes"], Ukr. Biokhim. Zh., 45(1), 42-46 (1973); Chem. Abstr., 79, 864 (1973).

16. Rauws, A. G., "Thallium Pharmacokinetics and Its Modification by Prussian Blue," Naunyn-Schmiedeberg's Arch. Pharmacol., 284(3), 295-306 (1974); Chem. Abstr., 82, 26780x (1975).

17. Oehme, F. W., "Mechanisms of Heavy Metal Toxicities," Clin. Toxicol., 5(2), 151-167 (1972).

18. Gehring, P. J., and P. B. Hammond, "The Interrelation Between Thallium and Potassium in Animals," J. Pharmacol. Exp. Ther., 155(1), 187-201 (1967); Chem. Abstr., 66, 35985q (1967).

19. Reinhardt, G., and P. Zink, "Die Verteilung des Thalliums im tierischen Organismus bei akuten Vergiftungen" ["Distribution of Thallium in the Animal Organism in Acute Poisoning"], Beitr. Gerichtl. Med., 32, 226-229 (1974).

20. Truhaut, R., P. Blanquet, and L. Capot, "Distribution of Radiothallium in the Body of Rats and Rabbits During Experimental Intoxications," J. Med. Bordeaux et Sud-Ouest, 134, 725-734 (1957); Chem. Abstr., 52, 18898a (1958).

21. Truhaut, R., P. Blanquet, and L. Capot, "Distribution of Radioactive Thallium in the Rat," Compt. rend., 245, 116-119 (1957); Chem. Abstr., 52, 569d (1958).

22. Downs, W. L., J. K. Scott, L. T. Steadman, and E. A. Maynard, "Acute and Subacute Toxicity Studies of Thallium Compounds," Am. Ind. Hyg. Assoc. J., 21, 399-406 (1960).

23. Truhaut, R., "The Toxicology of Thallium," J. Occup. Med., 2, 334-336 (1960); a summary prepared by W. B. Deichmann of Truhaut's Recherches sur la Toxicologie du Thallium, Institut National Securite pour la Prevention des Accidents du Travail, Paris, 1959.

24. Frey, J., and M. Schlechter, "Experimentelle Untersuchungen über die Ausscheidungsgrösse des Thalliums in verschiedenen Körperflüssigkeiten" ["Experimental Investigations of the Extent of Elimination of Thallium in Various Body Fluids"], Arch. Exp. Pathol. Pharmakol., 193, 530-538 (1939).

25. Barclay, R. K., W. C. Peacock, and D. A. Karnofsky, "Distribution and Excretion of Radioactive Thallium in the Chick Embryo, Rat, and Man," J. Pharmacol. Exp. Therap., 107(2), 178-187 (1953).

26. Emara, M., and M. A. Soliman, "The Distribution of Ingested Thallium in the Tissues of Animals," J. Roy. Egyptian Med. Assoc., 33, 1-15 (1950); Chem. Abstr., 44, 5998i (1950).

27. Bradley-Moore, P. R., E. Lebowitz, M. W. Greene, H. L. Atkins, and A. N. Ansari, "Thallium-201 for Medical Use," J. Nucl. Med., 16(2), 156-160 (1975).

28. Potter, G. D., G. M. Vattuone, and D. R. McIntyre, "Fate and Implications of Ingested Thallium-204 in a Dairy Cow and a Calf," Health Physics, 20(6), 657-661 (1971); Chem. Abstr., 75, 105786n (1971).

29. Jenkins, C. E., N. A. Wogman, and H. G. Rieck, "Radionuclide Distribution in Olympic National Park, Washington," Water, Air, Soil Pollut., 1(2), 181-204 (1972).

30. Munch, J. C., "Development of Thallium as a Pesticide in the United States" in Pestic. Symp., Collect. Pap. Inter-Amer. Conf. Toxicol. Occup. Med., 6th, 7th, 1970, pp. 239-243.

31. Tewari, S. N., S. P. Harpalani, and S. S. Tripathi, "Detection and Determination of Thallium in Autopsy Material by Spectrophotometric Method," J. Indian Acad. Forensic Sci., 14(1), 10-16 (1975).

32. Malátova, I., and V. Dvořák, "Measurement of ^{208}Tl Distribution in vivo in Persons With Internal Contamination by Thorotrast," Strahlentherapie, 142(5), 586-589 (1971).

33. Ford, J. K., E. J. Eyring, and C. E. Anderson, "Thallium Chondrodystrophy in Chick Embryos. An Histological and Biochemical Investigation," J. Bone Joint. Surg. [Am], 50(4), 687-700 (1968).

34. Gibson, J. E., C. P. Sigdestad, and B. A. Becker, "Placental Transport and Distribution of Thallium-204 Sulfate in Newborn Rats and Mice," Toxicol. Appl. Pharmacol., 10(2), 408 (1967).

35. Gibson, J. E., and B. A. Becker, "Placental Transfer, Embryotoxicity and Teratogenicity of Thallium Sulfate in Normal and Potassium-Deficient Rats," Toxicol. Appl. Pharmacol., 16(1), 120-132 (1970).

36. Testoni, P., "Sul Passaggio dei Farmaci dalla Madre al Feto. Esperienze con Acetato Talloso" ["The Passage of Drugs From Mother to Fetus"], Arch. Intern. Pharmacodynamie, 53, 29-43 (1936); Chem. Abstr., 30, 7695[7] (1930).

37. Fitzek, A., and A. Henning, "Verteilung von Thallium im Organismus einer trächtigen Hauskatze" ["Distribution of Thallium in the Organism of a Pregnant Domestic Cat"], Deut. Tieraertzl. Wochschr., 83(2), 66-68 (1976).

38. Richeson, E. M., "Industrial Thallium Intoxication," Indust. Med. Surg., 27, 607-619 (1958).

39. Katayama, K., "Fecal and Urinary Excretion of Radioisotopes," Radioisotopes (Tokyo), 12(1), 81-90 (1963); Chem. Abstr., 60, 4583f (1964).

40. Lund, A., "Distribution of Thallium in the Organism and Its Elimination, Acta Pharmacol. Toxicol., 12, 251-259 (1956).

41. Thyresson, N., "Experimental Investigation of Thallium Poisoning in the Rat," Acta Derm. Venereol., 31, 3-27 (1951).

42. Polson, C. J., and R. N. Tattersall, Clinical Toxicology, 2nd ed., J. B. Lippincott Company, Philadelphia, Pa., 1969.

43. Paolaggi, F., B. Lancon, C. Dreux, and M. L. Girard, "Application de la Voltammétrie par Redissolution Anodique en Biochimie. III. Études sur le Thallium Urinaire et Tissulaire" ["Biochemical Application of Voltammetry by Anode Redissolution. III. Tissue and Urinary Thallium"], Ann. Biol. Clin. (Paris), 30(3), 279-287 (1972).

44. Shaw, P. A., "Toxicity and Deposition of Thallium in Certain Game Birds, J. Pharm. Exptl. Therap., 48, 478-487 (1933); Chem. Abstr., 28, 213[8] (1934).

45. Ehrhardt, K., "Influence on the Progeny of Thallium Poisoning in the Mother," Klin. Wochschr., 6, 1374-1375 (1927); Chem. Abstr., 21, 3962[2] (1927).

46. Mrozikiewicz, A., and W. Widy, "Pigmentary Changes in the Hair of Animals Fed on Milk From a Mother Poisoned With Thallium," Bull. Soc. Amis Sci. et Lettres Poznan C, 10, 35-39 (1960); Chem. Abstr., 55, 23811i (1961).

47. Arnold, W., J. J. Herzberg, E. Ludwig, and H. Sturde, "Dynamics of Epilation in Thallium Poisoning; Correlation of Clinical, Histological, and Toxicological Findings," Arch. Klin. Exp. Dermatol., 218(4), 396-414 (1964); Chem. Abstr., 61, 1164d (1964).

48. Stevens, W., C. Van Peteghem, A. Heyndrickx, and F. Barbier, "Eleven Cases of Thallium Intoxication Treated With Prussian Blue," Int. J. Clin. Pharmacol. Ther. Toxicol., 10(1), 1-22 (1974).

49. Prix, R., and I. Dvorackova, "Diagnosis of Acute Thallium Poisoning," Casopis Lekaru Ceskych, 102, 46-51 (1963); Chem. Abstr., 59, 6886d (1963).

50. Jacobs, M. B., "The Determination of Thallium in Urine," Am. Ind. Hyg. Assoc. J., 23, 411-413 (1962); Chem. Abstr., 58, 4802f (1963).

51. Goenechea, S., and K. Sellier, "Über den natürlichen Thalliumgehalt des menschlichen Körpers" ["The Normal Thallium Content of the Human Body"], Deut. Z. Ges. Gerichtl. Med., 60, 135-141 (1967).

52. Division of Radiological Health, Department of Health, Education, and Welfare, Ed., Radiologic Health Handbook, Revised ed., PB 121 784R, Department of Health, Education, and Welfare, Washington, D.C., 1960.

53. Schwartzman, R. M., and J. O. Kirschbaum, "The Cutaneous Histopathology of Thallium Poisoning," J. Invest. Dermatol., 39, 169-174 (1962); Chem. Abstr., 59, 12074f (1963).

54. Erich, C., and H. D. Waller, "Behavior of Sulfhydryl Compounds in Erythrocytes After in vivo Poisoning With Heavy Metals and Arsenic," Klin. Wochenschr., 45(19), 983-986 (1967); Chem. Abstr., 67, 115368b (1967).

55. McGregor, W. G., J. Phillips, and C. H. Suelter, "Purification and Kinetic Characterization of a Monovalent Cation-Activated Glycerol Dehydrogenase From Aerobacter Aerogenes," J. Biol. Chem., 249(10), 3132-3139 (1974); Chem. Abstr., 81, 46867p (1974).

56. Barden, R. E., and M. C. Scrutton, "Pyruvate Carboxylase From Chicken Liver. Effects of Univalent and Divalent Cations on Catalytic Activity," J. Biol. Chem., 249(15), 4829-4838 (1974); Chem. Abstr., 81, 132119c (1974).

57. Manners, J. P., K. G. Morallee, and R. J. P. Williams, "Thallium(I) as a Potassium Probe in Biological Systems," J. Chem. Soc. D, 16, 965-966 (1970).

58. Williams, R. J. P., "Biochemistry of Group IA and IIA Cations" in Bioinorganic Chemistry, Advances in Chemistry Series, No. 100, American Chemical Society, Washington, D.C., 1971, pp. 155-173.

59. Kayne, F. J., "Thallium(I) Activation of Pyruvate Kinase," Arch. Biochem. Biophys., 143(1), 232-239 (1971); Chem. Abstr., 74, 120498t (1971).

60. Bostian, K., G. F. Betts, W. K. Man, and M. N. Hughes, "Thallium Activation and Inhibition of Yeast Aldehyde Dehydrogenase," FEBS Lett., 59(1 88-91 (1975); Chem. Abstr., 84, 27359v (1976).

61. Antia, N. J., R. S. Kripps, and I. D. Desai, "L-Threonine Deaminase in Marine Planktonic Algae. III. Stimulation of Activity by Monovalent Inorganic Cations and Diverse Effects From Other Ions," Arch. Mikrobiol., 85(4), 341-354 (1972).

62. Britten, J. S., and M. Blank, "Thallium Activation of the (Na^+-K^+)-Activated ATPase of Rabbit Kidney," Biochim. Biophys. Acta, 159(1), 160-166 (1968); Chem. Abstr., 68, 111760w (1968).

63. Lee, A. G., "The Coordination Chemistry of Thallium(I)," Coord. Chem. Rev., 8, 289-349 (1972).

64. Robinson, J. D., "Interactions Between Monovalent Cations and the (Na^+ + K^+)-Dependent Adenosine Triphosphatases," Arch. Biochem. Biophys., 139(1), 17-27 (1970); Chem. Abstr., 73, 52532b (1970).

65. Natochin, Yu. V., and I. A. Skul'skii, "Difference in the Effect of Tl^+ and Ba^{2+} as Partial Analogs of K^+ on the Sodium Pump in Frog Skin," Dokl. Akad. Nauk S.S.S.R., 203(6), 1437-1440 (1972); Chem. Abstr., 77, 59147w (1972).

66. Maslova, M. N., Yu. V. Natochin, and I. A. Skul'skii, "Inhibition of Active Sodium Transport and Activation of Na^+, K^+-ATPase by Tl^+ in Frog Skin," Biokhimiya, 36(4), 867-869 (1971); Chem. Abstr., 138618n (1971).

67. Ivashchenko, A. T., and B. S. Balmukhanov, "Effect of Monovalent Ions on the Adenosine Triphosphatase Activity of Ascites Tumor Cells," Izv. Akad. Nauk Kaz. S.S.R., Ser. Biol., 12(6), 73-78 (1974); Chem. Abstr., 82, 106257x (1975).

68. Ivashchenko, A. T., B. S. Balmukhov, and Z. S. Toktamysova, "Effects of Monovalent Ions on the Glycolysis and Respiration of Ascites Tumor Cells," Tsitologiya, **15**(12), 1487-1491 (1973); Chem. Abstr., **80**, 104485e (1974).

69. Skul'skii, I. A., V. Manninen, and J. Jarnefelt, "Interaction of Thallous Ions With the Cation Transport Mechanism in Erythrocytes," Biochim. Biophys. Acta, **298**(3), 702-709 (1973); Chem. Abstr., **78**, 145767a (1973).

70. Skul'skii, I. A., V. Manninen, and J. Jarnefelt, "Thallium Inhibition of Ouabain-Sensitive Sodium Transport and of the (Sodium-Potassium Ion)-Dependent ATPase in Human Erythrocytes," Biochem. Biophys. Acta, **394**(4), 569-576 (1975); Chem. Abstr., **83**, 92528t (1975).

71. Ku, D., T. Akera, T. Tobin, and T. M. Brody, "Effect of Monovalent Cations on Cardiac (Sodium-Potassium Ion)-Dependent ATPase Activity and on Contractile Force," Naunyn-Schmiedeberg's Arch. Pharmacol., **290**(2-3), 113-131 (1975); Chem. Abstr., **84**, 795x (1976).

72. Barrera, H., and A. Gomez-Puyou, "Characteristics of the Movement of Potassium(+) Ion Across the Mitochondrial Membrane and the Inhibitory Action of Thallium(+) Ion, J. Biol. Chem., **250**(14), 5370-5374 (1975); Chem. Abstr., **83**, 92539x (1975).

73. Melnick. R. L., L. G. Monti, and S. M. Motzkin, "Uncoupling of Mitochondrial Oxidative Phosphorylation by Thallium," Biochem. Biophys. Res. Commun., **69**(1), 68-73 (1976); Chem. Abstr., **84**, 174405k (1976).

74. Inturrisi, C. E., "Thallium-Induced Dephosphorylation of a Phosphorylated Intermediate of the (Sodium + Thallium-Activated) ATPase," Biochim. Biophys. Acta, **178**(3), 630-633 (1969); Chem. Abstr., **71**, 56975e (1969).

75. Grisham, C. M., R. K. Gupta, R. E. Barnett, and A. S. Mildvan, "Thallium-205 Nuclear Relaxation and Kinetic Studies of (Sodium and Potassium Ion)-Activated Adenosinetriphosphatase," J. Biol. Chem., **249**(21), 6738-6744 (1974); Chem. Abstr., **82**, 53418s (1975).

76. Lindegren, C. C., "Mitochondria as Vehicles of Entry and of Transport," J. Biol. Psychol., **13**(2), 3-5 (1971).

77. Lindegren, C. C., and G. Lindegren, "Oxidative Detoxication of Thallium in the Yeast Mitochondria," Antonie van Leeuwenhoek, J. Microbiol. Serol., **39**(2), 351-353 (1973); Chem. Abstr., **79**, 62298 (1973).

78. Lindegren, C. C., and G. Lindegren, "Mitochondrial Modification and Respiratory Deficiency in Yeast-Cell Caused by Cadmium Poisoning," Mutation Res., 21(6), 315-322 (1973).

79. Hultin, T., and P. H. Näslund, "Ion Binding and Ribosomal Conformation and Function. Experiments With the Potassium Ion Analog, Thallium(+) Ion," Acta Biol. Med. Ger., 33(5-6), 753-760 (1974); Chem. Abstr., 83, 73663s (1975).

80. Hultin, T., and P. H. Näslund, "Effects of Thallium(I) on the Structure and Functions of Mammalian Ribosomes," Chem.-Biol. Interactions, 8(5) 315-328 (1974).

81. Burger, A., and K. Starke, "Effect of Thallium on ATPase of Amine-Storing Granules From Adrenal Medulla and Splenic Nerves," Experientia, 25(6), 578-579 (1969); Chem. Abstr., 71, 47787t (1969).

82. Bank, W. J., D. E. Pleasure, K. Suzuki, M. Nigro, and R. Katz, "Thallium Poisoning," Arch. Neurol., 26(5), 456-463 (1972).

83. Starke, K., A. Burger, and H. J. Schümann, "Thallium und Brenzcatechinminestoffwechsel" ["Thallium and Catecholamine Metabolism"], Naunyn-Schmiedebergs Arch. Pharmacol., 264(3), 310-311 (1969).

84. Truhaut, R., "Sur les Proprietes Antimitotiques des Sels de Thallium Monovalents et leur Signification Toxicologique" ["Antimitotic Properties of Monovalent Thallium Salts and Their Toxicological Significance"], Arzneimittel-Forsch., 14(7), 837-841 (1964); Chem. Abstr., 61, 12537f (1964).

85. Avanzi, S., "'Non-Congression,' 'Non-Disjunction,' and Other Mitotic Aberrations in Allium cepa Induced by Thallium Acetate," Caryologia, 9, 131-148 (1956); Chem. Abstr., 52, 9320a (1958).

86. Nannetti, L., and G. Marras, "Cytological Activity of Thallous Acetate on the Allium cepa Test," Arch. Ital. Sci. Farmacol., 7, 83-92 (1957) Chem. Abstr., 51, 16748d (1957).

87. Nanetti, L., and G. Marras, "Impiego del «Test» Allium cepa L. per Dimostrare L'Attivita' Citologica Esplicata Dall'Acetato Talloso" ["Use of the Allium cepa Test to Show the Cytological Activity of Thallous Acetate"], Boll. Soc. Ital. Biol. Sper., 32, 1488-1490 (1956

88. Beltrami, L., "Antimitotic Action of Thallium Chloride on Vegetative Apexes of Allium cepa," Arch. Ital. Sci. Farmacol., 8, 151-156 (1958) Chem. Abstr., 52, 20451a (1958).

89. Frizzi, G., and A. Lecis, "Effect of Thallium Acetate on Spermatogenesis of Anopheles maculipennis atroparvus," Riv. Biol. (Perugia), 58(4), 279-284 (1965); Chem. Abstr., 65, 4331a (1966).

90. von Rosen, G., "Breaking of Chromosomes by the Action of Elements of the Periodical System and by Some Other Principles," Hereditas, 40, 258-263 (1954).

91. von Rosen, G., "Mutations Induced by the Action of Metal Ions in Pisum," Hereditas, 43, 644-664 (1957).

92. Ganther, H. E., "Biochemistry of Selenium" in Selenium, R. A. Zingaro and W. C. Cooper, Eds., Van Nostrand Reinhold Company, New York, N.Y., 1974, pp. 546-614.

93. Rusiecki, W., and J. Brzezinski, "Effect of Sodium Selenate on Acute Thallium Poisoning," Acta Polon. Pharm., 23(1), 69-74 (1966); Chem. Abstr., 65, 4518e (1966).

94. Levander, O. A., and L. C. Argrett, "Effects of Arsenic, Mercury, Thallium, and Lead on Selenium Metabolism in Rats," Toxicol. Appl. Pharmacol., 14(2), 308-314 (1969).

95. Avanzi, S., "Inhibition of the Cytological Effect of Thallium Acetate by Cysteine," Caryologia, 10, 96-101 (1957); Chem. Abstr., 52, 10294i (1958).

96. Buschke, A., and W. Konheim, "AT10 und Thallium" ["Dihydrotachysterol and Thallium"], Schweiz. Med. Wochschr., 69, 702-703 (1939).

97. Selye, H., P. Jean, and R. Veilleux, "Sensitization by Thallium to Dihydrotachysterol Over-Dosage," Proc. Soc. Exptl. Biol. Med., 106, 408-409 (1961); Chem. Abstr., 55, 12641b (1961).

98. Selye, H., M. Cantin, and G. Gentile, "Effect of Various Electrolytes Upon Thallium Intoxication," Chemotherapia, 4, 1-7 (1962); Chem. Abstr., 57, 2534g (1962).

99. Rygh, O., "Scurvy, Calcification and Vitamin C," Research, 2, 443 (1949).

100. Rygh, O., "Causes of Dental Caries," Research, 3, 193-194 (1950).

101. Taylor, R., and B. Shore, "Biochemical Methylation of Selected Heavy Metals and Its Effects on Their Cellular Toxicity," Notice of Research Project, Smithson Science Information Exchange No. ZPE-11891, 1975.

102. Taylor, R., Lawrence Livermore Laboratory, personal communication (telephone), November 1976.

103. Aliev, A. M., and A. S. Gasanov, "Reaction of Some Salts of Rare and Rare Earth Elements With Group B Vitamins," Dokl. Akad. Nauk Azerb, S.S.R., 20(8), 79-84 (1964); Chem. Abstr., 62, 5142g (1965).

104. Sudmeier, J. L., and J. J. Pesek, "Chlorine 35 Nuclear Magnetic Resonance Studies of Metal Binding to Bovine Serum Albumin," Anal. Biochem., 41(1), 39-50 (1971).

105. Sundararajan, N. R., and R. McL. Whitney, "Binding of Thallous Ion by Casein," J. Dairy Sci., 52, 1445-1448 (1969); Chem. Abstr., 71, 98326z (1969).

106. Witschi, H. P., "Desorption of Some Toxic Heavy Metals From Human Erythrocytes in Vitro," Acta Haematol., 34(2), 101-105 (1965); Chem. Abstr., 64, 18160h (1966).

107. Hatem, S., "Etude Physico-Chemique de Sels Complexes Histamine-Argent et Histamine-Thallium(I)" ["Complex Silver-Histamine and Thallium(I)-Histamine Salts"], Helv. Chim. Acta, 43, 1431-1435 (1960); Chem. Abstr. 55, 16256c (1961).

108. Champy, C., and S. Hatem, "Sur le Caractère Neurotoxique du Thallium et son Affinité pour l'Histamine" ["Neurotoxic Property of Thallium and Its Affinity for Histamine"], Compt. Rend. Soc. Biol., 151, 520-521 (1957).

109. Sessions, H. K., and S. Goren, "Report of Investigation of Health Hazards in Connection With the Industrial Handling of Thallium," U.S. Naval. Med. Bull., 47, 545-550 (1947).

110. Masoud, A. N., J. T. Elder, and A. L. Czerwinski, "Chemistry and Pharmacology of Common Acute Poisoning in Children," Paediatrician, 2(1-2), 2-37 (1973).

111. Pritschow, A. L., A Study of the Distribution of Thallium in Tissues, Blood, Urine and Feces, Ph.D. Dissertation, University of Oklahoma, Oklahoma City, Oklahoma, 1959.

112. Curry, A. S., "Thallium" in Advances in Forensic and Clinical Toxicology, CRC Uniscience Series, CRC Press, Inc., Cleveland, Ohio, 1972.

113. Munch, J. C., H. M. Ginsburg, and C. E. Nixon, "The 1932 Thallotoxicosis Outbreak in California," J. Am. Med. Assoc., 100, 1315-1319 (1933).

114. Committee on Threshold Limit Values, Documentation of the Threshold Limit Values for Substances in Workroom Air, 3rd ed., American Conference of Governmental Industrial Hygienists, Cincinnati, Ohio, 1971.

115. Munch, J. C., "Human Thallotoxicosis," J. Am. Med. Assoc., 102(2), 1929-1934 (1934).

116. Cavanagh, J. B., N. H. Fuller, H. R. M. Johnson, and P. Rudge, "The Effects of Thallium Salts, With Particular Reference to the Nervous System Changes," Quart. J. Med., 43, 293-319 (1974).

117. Cavanagh, J. B., "Peripheral Neuropathy Caused by Chemical Agents," Crit. Rev. Toxicol., 2(3), 365-417 (1973).

118. Gefel, A., M. Liron, and W. Hirsch, "Chronic Thallium Poisoning," Israel J. Med. Sci., 6(3), 380-382 (1970).

119. Patterson, J. F., "Chronic Thallitoxicosis: Treatment of Choreiform Sequelae," Southern Med. J., 68(7), 923-925 (1975).

120. Rube and Hendricks, "Gewerbliche Thallium Vergiftung" ["Industrial Thallium Poisoning"], Med. Welt, 1, 733 (1927).

121. Teleky, "Gewerbliche Thallium Vergiftung" ["Industrial Thallium Poisoning"], Wiener Med. Wochschr., 78, 506-508 (1928); Chem. Abstr., 22, 2800 (1928).

122. Glomme, J., and B. Sjöström, "Yrkesbetingad Talliumförgiftning" ["Industrial Thallium Poisoning"], Svenska Laekartidningen, 52, 1436-1441 (1955).

123. Egen, B., "Gewerbliche Thallium Vergiftung" ["Industrial Thallium Poisoning"], Zentr. Arbeitsmed. Arbeitschutz, 5, 141-144 (1955).

124. Hill, W. H., and M. A. Murphy, "Appraisal of an Industrial Air Contaminant by Electrochromatography," Am. Ind. Hyg. J., 20, 387-391 (1959).

125. Budrin, Yu. P., and K. A. Meshcherskaya, "Otravleniya Soedineniyami Talliya" ["Poisonings With Thallium Compounds"], Sud.-Med. Ekspert., 18(4), 37-39 (1975).

126. Tikhova, T. S., "Voprosy Gigieny Truda v Proizvodstve Metallicheskogo Talliya i ego Solei" ["Problems of Labor Hygiene in the Production of Metallic Thallium and Its Salts"], Gigiena i Sanit., 29(2), 23-27 (1964); Chem. Abstr., 60, 13786c (1964).

127. Tikhova, T. S., "Tallii i Ego Soedineniya" ["Thallium and Its Compounds" in Novye Dannye po Toksikologii Redkikh Metallov i Ikh Soedinenii, Z. Izrael'son, Ed., Publishing House "Meditsina," Moscow, USSR, 1967, pp. 24-34.

128. Taylor, E. C., and A. McKillop, "Thallium in Organic Synthesis," Account of Chem. Res., **3**, 338-346 (1970).

129. Department of Labor Occupational Safety and Health Administration, "Occupational Safety and Health Standards. Rules and Regulations," Federal Register, **37**(202), Part II, 22140-22142 (1972).

130. American Conference of Governmental Industrial Hygienists, TLVs. Threshold Limit Values for Chemical Substances in Workroom Air Adopted by ACGIH for 1975, American Conference of Governmental Industrial Hygienists, Cincinnati, Ohio, 1975.

131. Korbakova, A. I., "Standard Levels of New Industrial Chemicals in the Air of Work Premises," Vestn. Akad. Med. Nauk S.S.S.R., **19**(7), 17-23 (1964); Chem. Abstr., **61**, 16694b (1964).

132. Winell, M., "An International Comparison of Hygienic Standards for Chemicals in the Work Environment," AMBIO, **4**(1), 34-36 (1975).

133. Ottinger, R. S., J. L. Blumenthal, D. F. Dal Porto, G. I. Gruber, M. J. Santy, and C. C. Shih, Recommended Methods of Reduction, Neutralization, Recovery, or Disposal of Hazardous Waste. Vol. I. Summary Report, TRW Systems Group for the U.S. Environmental Protection Agency, PB-224 580, National Technical Information Service, U.S. Department of Commerce, Springfield, Va., 1973.

134. Dawson, G. W., The Chemical Toxicity of Elements, BNWL-1815, UC-70, Battelle Pacific Northwest Laboratories, Richland, Washington, 1974.

135. U.S. Environmental Protection Agency, Comparison of NTAC, NAS, and Proposed EPA Numerical Criteria for Water Quality, U.S. Environmental Protection Agency, Washington, D.C. [undated, ~ 1974].

136. McKee, J. E., and H. W. Wolf, Eds., Water Quality Criteria, 2nd ed., Publication 3-A, California State Water Resources Control Board, 1963 (Reprint January 1973).

137. Williams, N., and A. L. Riegert, "Epidemic Alopecia Areata. An Outbreak in an Industrial Setting," J. Occup. Med., **13**(11), 535-542 (1971).

138. Heydlauf, H., "Ferric Cyanoferrate(II): An Effective Antidote in Thallium Poisoning," European J. Pharmacol., **6**(3), 340-344 (1969); abstr. in TOXLINE thallium search.

139. Kamerbeek, H. H., A. G. Rauws, M. ten Ham, and A. N. P. van Heijst, "Prussian Blue in Therapy of Thallotoxicosis. An Experimental and Clinical Investigation," Acta Med. Scand., 189, 321-324 (1971).

140. Manninen, V., M. Malkonen, and I. A. Skul'skii, "Elimination of Thallium in Rats as Influenced by Prussian Blue and Sodium Chloride," Acta Pharmacol. Toxicol., 39(2), 256-261 (1976); Chem. Abstr., 85, 41736t (1976).

141. Chusid, J. G., and L. M. Kopeloff, "Epileptogenic Effects of Pure Metals Implanted in Motor Cortex of Monkeys," J. Appl. Physiol., 17, 697-700 (1962); Chem. Abstr., 57, 15714f (1962).

142. Levkovich, L. I., "The Influence of External Temperature, Diet, and Emotion on the Pharmacological Action of Thallium," Bull. Biol. Med. Exptl. USSR, 5, 271-273 (1938); Chem. Abstr., 34, 2461[8] (1940).

143. Case, A. A., "Toxicity of Various Chemical Agents to Sheep," J. Am. Vet. Med. Assoc., 164(3), 277-283 (1974).

144. Munch, B., B. Clausen, and O. Karlog, "Thallium Poisoning in Red Foxes (Vulpes vulpes) and Badgers (Meles meles) in Denmark," Nord. Veterinarmed., 26, 323-338 (1974).

145. Clausen, B., and O. Karlog, "Loading With Thallium Among Wild Animals of the Marten Genus and Badgers in Denmark," Nord. Veterinarmed., 26, 339-350 (1974); abstr. in TOXLINE thallium search.

146. DeSloovere, J., M. Debackere, and J. Hoorens, "Poisoning of Domestic Animals," Vlaams Diergeneesk. Tijdschr., 40(1), 8-29 (1971); abstr. in TOXLINE thallium search.

147. Carpenter, J. L., "Feline Panleucopenia: Clinical Signs and Differential Diagnosis," J. Am. Vet. Med. Assoc., 158(6), 857-859 (1971).

148. Atkins, C. E., and R. K. Johnson, "Clinical Toxicities of Cats," Vet. Clin. N. Am., 5(4), 623-652 (1975); abstr. in TOXLINE thallium search.

149. Breukink, H. J., and J. R. Van Der Lee, "Thalliumintoxicatie bij de Kat" ["Thallium Poisoning in Cats"], Tijdschr. Diergeneesk., 101(11), 611-612 (1976).

150. Vismara, F., "Reperti Anatomo-Istopatologici dell'Avvelenamento Spontaneo da Tallio nel Cane" ["Lesions in Thallium Poisoning in 4 Dogs"], Atti. Soc. Ital. Sci. Vet., 15, 523-530 (1961).

151. Browning, E., Toxicity of Industrial Metals, Butterworths, London, 1961.

152. Mrozikiewicz, A., and W. Widy, "Toxicology of Thallium," Patol. Pol., 18(2), 239-244 (1967); Chem. Abstr., 67, 89308k (1967).

153. Selye, H., and I. Mécs, "Effect Upon Drug Toxicity of Surgical Interference With Hepatic or Renal Function," Acta Hepato-Gastroenterologic 21(3), 191-202 (1974) and 21(4), 266-273 (1974).

154. Malachovskis, A., "Thallium Intoxication and the Effect of Vitamin B_{12} on the Functional Condition of Liver," Liet. TSR Aukst. Mokyklu Mokslo Darb., Biol., 8, 23-25 (1968); Chem. Abstr., 70, 66369d (1969).

155. Toya, T., H. Harada, and Y. Tamura, "Systemic Studies on the Metabolism of Metals. IV. Distribution of Orally Administered Lead, Thallium, and Cadmium in Rats and Its Effect on the Endogenous Copper and Zinc," Tokyo-to Ritsu Eisei Kenkyusho Nempo, No. 13, 178-194 (1961); Chem. Abstr., 62, 3187h (1965).

156. Dzialek, S., "Depilatory Action of Thallium Acetate on Rats," Ann. Acad. Med. Lodz., 6, 74-81 (1965); Chem. Abstr., 65, 1283h (1966).

157. Sappino, G., "Il Tallio. Nota XI: Il Peso ed il Contenuto Idrico del Timo e della Tiroide nell Intossicazione Cronica da Tallio" ["Thallium. XI. The Weight and Water Content of the Thymus and Thyroid in Chronic Thallium Poisoning"], Arch. Intern. Pharmacodynamie, 53, 14-28 (1936); Chem. Abstr., 30, 7695^6 (1936).

158. Bensch, K. C., and H. M. Herman, "Histologic Changes Caused by Acute and Chronic Thallium Intoxication in Rats," Fed. Proc., 23(2), 199 (1964).

159. Herman, M. M., and K. G. Bensch, "Light and Electron Microscopic Studies of Acute and Chronic Thallium Intoxication in Rats," Toxicol. Appl. Pharmacol., 10(2), 199-222 (1967).

160. Ward, J. C., "Thallium Poisoning in Migratory Birds," J. Am. Pharm. Assoc., 20, 1272-1276 (1931); Chem. Abstr., 26, 1350^6 (1932).

161. Cromartie, E., W. L. Reichel, L. N. Locke, A. A. Belisle, T. E. Kaiser, T. G. Lamont, B. M. Mulhern, R. M. Prouty, and D. M. Swineford, "Residues of Organochlorine Pesticides and Polychlorinated Biphenyls and Autopsy Data for Bald Eagles," Pestic. Monit. J., 9(1), 11-14 (1975); Chem. Abstr., 84, 26612k (1976).

162. Bean, J. R., and R. H. Hudson, "Acute Oral Toxicity and Tissue Residues of Thallium Sulfate in Golden Eagles, Aquila chrysaetos," Bull. Environ. Contam. Toxicol., 15(1), 118-121 (1976).

163. Nehring, D., "Experiments on the Toxicological Effect of Thallium Ions on Fish and Fish-Food Organisms," Z. Fisch., 11, 557-562 (1962).

164. Zitko, V., W. V. Carson, and W. G. Carson, "Thallium. Occurrence in the Environment and Toxicity to Fish," Bull. Environ. Contam. Toxicol., 13, 23-30 (1975).

165. DiGaudio, M. R., and H. I. Hirshfield, "Respiratory Toxicity of Thallium in Paramecium bursaria" (meeting abstr.), J. Cell Biol., 70(2), A60 (1976).

166. DiGaudio, M. R., Thallium: Some Effects of an Environmental Toxicant, Ph.D. Dissertation, New York University, New York, N.Y., 1975.

167. Dilling, W. S., and C. W. Healy, "Influence of Lead and the Metallic Ions of Copper, Zinc, Thorium, Beryllium and Thallium on the Germination of Frogs Spawn and on the Growth of Tadpoles," Annals Appl. Biol., 13, 177-188 (1926).

168. Buschke, A., and B. Peiser, "Epithelwucherungen am Vormagen der Ratten durch experimentelle Thalliumwirkung" ["Epithelial Proliferation in the Forestomach of Rats by the Experimental Action of Thallium"], Zeitschrift fuer Krebsforschung (Berlin), 21, 11-18 (1923).

169. Oliver, H.-R., "L'Intoxication Chronique par l'Acetate de Thallium" ["Chronic Intoxication by Thallium Acetate"], Compt. Rend. Soc. Biol. [Paris], 96, 164-166 (1927).

170. Champy, C., S. Hatem, and J. Tharanne, "Végétations du Tractus Génital de la Souris après Intoxication Thallique Ménagée" ["Pathology of the Genital Tract of the Female Mouse After Controlled Thallium Intoxication"], Compt. Rend. Soc. Biol., 152, 906-907 (1958); Chem. Abstr., 53, 4512c (1959).

171. Hart, M. M., and R. H. Adamson, "Antitumor Activity and Toxicity of Salts of Inorganic Group IIIa Metals: Aluminum, Gallium, Indium, and Thallium," Proc. Natl. Acad. Sci. U.S.A., 68(7), 1623-1626 (1971).

172. Adamson, R. H., G. P. Canellos, and S. M. Sieber, "Studies on the Antitumor Activity of Gallium Nitrate (NSC-15200) and Other Group IIIa Metal Salts," Cancer Chemother. Rep., 59(3), 599-610 (1975).

173. Oertel, R. P., and E. A. Newmann, "Negative Skin Sensitization Test With KRS-5," Appl. Spectrosc., 26(5), 562 (1972).

174. Butcher, E. O., "The Effects of Thallium and Hair Growth in the Rat," J. Soc. Cosmetic Chemists, 15(3), 129-136 (1964); Chem. Abstr., 60, 15040b (1964).

175. Zaun, H., K. Neumann, and G. Werner, "The Fine Structure of the Hair in Vitamin A and Thallium Poisoning," Hautarzt, 23, 544-550 (1972); abstract in TOXLINE thallium search.

176. Gerdts, E., "Thallium Poisoning," Lancet, 2(7891), 1268 (1974).

177. Jindrichova, J., and I. Rockova, "Diagnosis of Thallium Intoxication by Examining Hair Roots," Vnitrni Lekar., 15(5), 471-474 (1969); abstract in TOXLINE thallium search.

178. LeJeune, J. R., "Les Oligo-Elements et Chelateurs. A. Les Oligo-Element ["Harmful Effects of Common Drugs on the Visual Apparatus. Trace Elements and Chelating Agents. A. Trace Elements"], Bull. Soc. Belge Ophtalmol., 160(1), 241-254 (1972).

179. Potts, A. M., and P. C. Au, "Thallous Ion and the Eye," Invest. Ophthalmol., 10(12), 925-931 (1971).

180. Tharanne, M., "Lesions of the Sympathetic Nervous System From Thallium Poisoning," Compt. Rend. Soc. Biol., 151, 542-543 (1957); Chem. Abstr. 52, 3133e (1958).

181. Hayashibara, Y., "Electroencephalographic (EEG) Study on Experimental Thallium Poisoning in Rabbits," Yonago Acta Med., 8(2), 85-88 (1964) Chem. Abstr., 62, 9677f (1965).

182. Pentschew, A., "Dysoria, a New Dimension of Pathology," Arch. Psychiatr Nervenkr, 213(2), 139-148 (1970).

183. Karkos, J., "The Neuropathological Findings in Thallium Encephalopathy Neurol. Neurochir. Pol., 21(6), 911-915 (1971); abstract in TOXLINE thallium search.

184. Tackmann, W., and H. J. Lehmann, "Refractory Period and Impulse Sequen Recorded From the Tibial Nerves of Guinea Pigs With Acute Thallium Polyneuropathy," Z. Neurol., 199(12), 105-115 (1971); abstract in TOXLINE thallium search.

185. Spencer, P. S., E. R. Peterson, A. R. Madrid, and C. S. Raine, "Effect of Thallium Salts on Neuronal Mitochondria in Organotypic Cord-Ganglia-Muscle Combination Cultures," J. Cell Biol., 58(1), 79-95 (1975); Chem. Abstr., 79, 101109d (1973).

186. Kennedy, P., and J. B. Cavanagh, "Spinal Changes in Neuropathy of Thallium Poisoning--Case of Neuropathological Studies," J. Neurol. Sci., 29(2-4), 295-301 (1976).

187. Ikeda, Y., "Toxicological Study on the Etiology of SMON," Nippon Rinsho, 29(2), 766-772 (1971); abstract in TOXLINE thallium search.

188. Utidjian, H., "Gallium Fluoride Poisoning," J. Occup. Med., 15(2), 134 (1973).

189. Hughes, M. N., W. K. Man, and B. C. Whaler, "The Effect of Thallium on Skeletal and Cardiac Muscle," J. Physiol., 256(2), 126P-127P (1976); Chem. Abstr., 84, 174533a (1976).

190. Rusznyak, I., T. Millner, L. Gyorgy, and S. Ormai, "Potassium-Like Properties of the Thallium Ion," Magy. Tud. Akad. V. Orv. Tud. Oszt. Kozlem., 19(1), 11-13 (1968); Chem. Abstr., 69, 94781e (1968).

191. Rusznyak, I., L. Gyorgy, S. Ormai, and T. Millner, "Potassium-Like Qualities of the Thallium Ion," Experientia, 24(8), 809-810 (1968); Chem. Abstr., 69, 65821n (1968).

192. Merguet, P., H. J. Schuemann, T. Murata, J. G. Rausch-Stroomann, E. Schroeder, D. Paar, and K. D. Bock, "Pathogenesis of Hypertension and Sinus Tachycardia in Human Thallium Poisoning," Arch. Klin. Med., 216(1), 1-20 (1969); Chem. Abstr., 70, 104713w (1969).

193. Bock, K. D., P. Merguet, G. Schley, H. J. Shuemann, J.-G. Rausch-Stroomann, V. Hocevar, E. Schroeder, and T. Murata, "Pathogenesis of Hypertonia and Tachycardia in Acute Thallium Poisoning and in Acute Intermittent Porphyria," Deut. Med. Wochenschr., 93(44), 2119-2124 (1968); abstract in TOXLINE thallium search.

194. Merguet, P., H. J. Schuemann, T. Murata, J. G. Rausch-Stroomann, K. D. Bock, and E. Schroeder, "Pathogenesis of Hypertension Due to Thallium Poisoning," Verh. Deut. Ges. Inn. Med., 74, 429-433 (1968); Chem. Abstr., 71, 99966v (1969).

195. Hantschmann, L., "The Development of High Blood Pressure in Thallium Poisoning," Hippokrates, 41(1), 150-151 (1970); abstract in TOXLINE thallium search.

196. Schütt, I., and U. Kersten, "Herzbefunde bei akuter Thalliumvergiftung des Hundes" ["Cardiac Conditions in Acute Thallium Poisoning in the Dog"], Kleintier-Praxis, 21(3), 94-100 (1976).

197. Lameijer, W., and P. A. van Zwieten, "Acute Cardiovascular Toxicity of Thallium(I) Ions," Arch. Toxicol., 35(1), 49-61 (1976).

198. Zondek, B., and S. Aschheim, "Ei und Hormon" ["Ovum and Hormone"], Kli Wochschr., 6, 1321-1322 (1927); Chem. Abstr., 21, 3942[2] (1927).

199. Arashima, T., "Influence of Chronic Thallium Poisoning on the Function of the Female Sex Organs of Rats," Japan. J. Med. Sci. IV. Pharmacol 7(2-3), Proc. Japan. Pharmacol. Soc., 135-136 (1933); Chem. Abstr., 29, 6952[8] (1935).

200. Di Gaudio, M. R., and H. I. Hirshfield, "Thallium Effects of Mating Types of Paramecium bursaria," J. Cell Biol., 59 (2 PT 2), 81 (1973)

201. Landauer, W., "Influence of Thallium Poisoning of the Father on His Descendants. Experiments on Roosters," Arch. Gewerbepathol. Gewerbehyg., 1, 791-792 (1931); Chimie & Industrie (Paris), 25, 1402 (1931); Chem. Abstr., 25, 5711[3] (1930).

202. Lallier, R., "The Toxicity of Thallium Ions to Eggs of Paracentrotus lividus," C. R. Acad. Sci., Paris, Ser. D., 267(10), 962-964 (1968); Chem. Abstr., 70, 9612d (1969).

203. Lallier, R., "Thallous and Thallic Ions and Their Effects on Sea Urchi (Paracentrotus lividus) Egg Development," C. R. Soc. Biol., 163(3), 598-601 (1969); Chem. Abstr., 72, 118875d (1970).

204. Hall, B. K., "Thallium-Induced Achondroplasia in the Embryonic Chick," Develop. Biol., 28(1), 47-60 (1972).

205. Bertran, E. C., "Hen Embryology. Morphological Modifications Caused by Trace Elements [Fluoride, Selenium, Arsenic, Thallium, and Copper]," Anales Inst. Invest. Vet. (Madrid), 13, 11-30 (1963); Chem. Abstr., 61, 6100b (1964).

206. Červeňanský, J., E. Kalman, D. Maar, V. Makovický, J. Micek, and B. Škrovina, "Experimentelle Teratogenese bei Kücken" ["Experimental Teratogenesis in Chicks"], Beitr. Orthop. Traumatol., 13(10), 597-599 (1966).

207. Škrovina, B., J. Miček, J. Hronská, and L. Spiššák, "Experimental and Human Achondroplasia," Teratology, 8(2), 237 (1973).

208. Roth, M., "Teratogenic Micromelia in the Chick Embryo: Osteo-Vasculo-Neural Macro-Relations," Folia Morphol. (Prague), 23(3), 236-246 (1975).

:09. Nogami, H., and Y. Terashima, "Thallium-Induced Achondroplasia in the Rat," Teratology, 8(1), 101-102 (1973); Japanese version: Congenital Anomalies Res. Assoc. Japan, 13(3), 199-200 (1973).

:10. Landauer, W., "The Teratogenic Nature of Thallium Polyhydroxy Compounds, Histidine, and Imidazole as Supplements," J. Exp. Zool., 143(1), 101-105 (1960); Chem. Abstr., 59, 4464d (1963).

:11. Hall, B. K., "Achondroplasia in the Embryonic Chick. Its Potentiation by Cortisone Acetate and Alleviation by Vitamin C," Can. J. Zool., 50(12), 1527-1536 (1972).

:12. Richards, O. W., "The Stimulation of Yeast Growth by Thallium, a 'Bios' Impurity of Asparagine," J. Biol. Chem., 96, 405-418 (1932).

:13. Horn, E. E., J. C. Ward, J. C. Munch, and F. E. Garlough, The Effect of Thallium on Plant Growth, U.S. Department of Agriculture, Circ. 409, 1936, 8 pp.

14. Khare, H. P., J. S. Chauhan, and S. P. Tandon, "Effect of Some Rare Elements on the Growth of Nitrite-Formers" [in English], Zentr. Bakteriol. Parasitenk., Abt. II, 120(2), 117-122 (1966); Chem. Abstr., 65, 9378g (1966).

.15. Schneiderman, G. S., T. R. Garland, R. E. Wildung, and H. Drucker, "Growth and Thallium Transport of Microorganisms Isolated From Thallium Enriched Soil," in Abstracts of the 74th Annual Meeting of the American Society of Microbiologists, Chicago, Illinois, May 1974, p. 2.

'16. Tandon, S. P., and M. M. Mishra, "Effect of Some Rare Elements on Nitrification by Nitrate-Forming Bacteria in Soil Suspension" [in English], Zentralbl. Bakteriol., Parsitenk., Infektionskr. Hyg., Abt. II, 122(2), 155-157 (1968).

:17. McCool, M. M., "Effect of Thallium Sulfate on the Growth of Several Plants and on the Nitrification in Soils," Contrib. Boyce Thompson Inst., 5(3), 289-296 (1933).

18. Magorian, T. R., K. G. Wood, J. G. Michalovic, S. L. Pek, and M. W. van Lier, Water Pollution by Thallium and Related Metals, Calspan Corporation for the U.S. Environmental Protection Agency, Cincinnati, Ohio, 1974. [This is the draft version of the final report issued under another title (see Reference 219, below) in 1977 and was obtained from the National Technical Information Service as PB-253 333.]

219. Calspan Corporation, Heavy Metal Pollution From Spillage at Ore Smelters and Mills, preliminary copy of the final report version provided in March 1977, Calspan Corporation, Buffalo, N.Y., for the U.S. Environmental Protection Agency, Cincinnati, Ohio (in press).

220. Lazorenko, G. E., and G. G. Polikarpov, "Al'ginovaya Kislota i Mekhanizm Fiksatsii Radionuklidov Burymi Vodoroslyami" ["Alginic Acid an Mechanism of Fixation of Radionuclides by Brown Algae"] in Radiatsinaya i Khimicheskaya Ekologiya Gidrobiontov [Radiation and Chemical Ecology of Hydrobionts], G. G. Polikarpov, Ed., Naukova Dumka, Kiev USSR, 1972, pp. 105-112.

221. Skul'skii, I. A., V. V. Glazunov, A. Ya. Zesenko, and A. A. Lyubimov, "Cell Selectivity of the Black Sea Alga Ulva rigida toward Tl+, Rb+, and Cs+ Ions," Tsitologiya, 14(7), 849-856 (1972); Chem. Abstr., 77, 111720u (1972).

222. Skul'skii, I. A., V. V. Glazunov, A. Ya. Zesenko, and A. A. Lyubimov, "Accumulation of Tl+ Ions in the Cells of the Black Sea Algae Ulva Rigida," Biofizika, 17(5), 824-831 (1972); Chem. Abstr., 78, 40509q (1973).

223. Polikarpov, G. G., A. Ya. Zesenko, and A. A. Lyubimov, "Dinamika Fizi Khimicheskogo Prevrashcheniya Radionuklidov Mnogovalentnykh Element v Srede i Nakoplenie ikh Gidrobiontami" ["Dynamics of Physicochemic Transformations of Radionuclides of Multivalent Elements in the Env ronment and Their Accumulation by Hydrobionts"] in Radiatsionnaya i Khimicheskaya Ekologiya Gidrobiontov, G. G. Polikarpov, Ed., Naukov Dumka (publisher), Kiev, USSR, 1972, pp. 5-42; Chem. Abstr., 78, 24 (1973).

224. Polikarpov, G. G., "Radioecology of Artificial Nuclides," Proc. Symp. Hydrogeochem. Biogeochem., 1, 356-371 (1970, publ. 1973).

225. Solt, J., H. Paschinger, and E. Broda, "Die energieabhängige Aufnahme von Thallium durch Chlorella" ["The Energy-Dependent Uptake of Thallium by Chlorella"], Planta, 101, 242 (1971); Chem. Abstr., 76(13), 175-176 (1972).

226. Broda, E., "Uptake of Heavy Cationic Trace Elements by Microorganisms Ann. Microbiol. Enzimol., 22(1-4), 93-108 (1972); Chem. Abstr., 81(22948q (1974).

227. Overnell, J., "Effect of Some Heavy Metal Ions on Photosynthesis in a Fresh Water Alga," Pestic. Biochem. Physiol., 5(1), 19-26 (1975).

28. Overnell, J., "Effect of Heavy Metals on Photosynthesis and Loss of Cell Potassium in Two Species of Marine Algae, Dunaliella tertiolecta and Phaeodactylum tricornutum," Marine Biol., 29(1), 99-103 (1975).

29. Tso, T. C., Physiology and Biochemistry of Tobacco Plants, Dowden, Hutchinson and Ross, Stroudsburg, Pa., 1972, pp. 42-81.

30. Kolotova, S. S., and N. I. Kharitonova, "Effect of Thallium on Plants," Uch. Zap., Perm. Gos. Univ., No. 206, 161-165 (1970); Chem. Abstr., 76, 2855a (1972).

31. Borzini, G., "Influenza degli Ioni Tallio sulla Germinazione di Semi Diversi e sullo Sviluppo Iniziale delle Piantine" ["The Influence of Thallium Ions on the Germination of Several Seeds, and on the Initial Growth of the Young Plants"], Boll. Staz. Patol. Vegetale (N.S.), 15, 200-231 (1935).

32. Barber, D. A., "Absorption of Ions by Microorganisms and Excised Roots," New Phytol., 73(1), 91-96 (1974).

33. Crafts, A. S., "Some Effects of Thallium Sulfate Upon Soils," Hilgardia, 10, 377-398 (1936).

34. Machata, G., "Uber den Thalliumgehalt in menschlichen Knochen" ["Thallium Content of Human Bones"], Deut. Z. Ges. Gerichtl. Med., 54(1), 95-98 (1963); Chem. Abstr., 60, 1029b (1964).

35. Weinig, E., and P. Zink, "Uber die Quantitative Massen spektrometrische Bestimmung des normalen Thallium-Gehalts in menschlichen Organismus," ["Quantitative Mass Spectrometric Determination of the Normal Thallium Content in the Human Body"], Arch. Toxikol., 22(4), 255-274 (1967); Chem. Abstr., 67(1), 723w (1967).

36. Dzialek, S., "Detoxicating Effect of Certain Preparations in Thallium Poisoning in Guinea Pigs," Med. Pracy, 16(2), 102-108 (1965); Chem. Abstr., 63, 18923h (1965).

37. Righetti, P., and S. Moeschlin, "The Therapeutic Effect of Dithiocarb (DTC) and Potassium Chloride on Experimental Thallium Poisoning in Guinea Pigs," Clin. Toxicol., 4(2), 165-171 (1971).

38. Schwetz, B. A., P. V. O'Neal, F. A. Voelker, and D. W. Jacobs, "Effects of Diphenylthiocarbazone and Diethyldithiocarbamate on the Excretion of Thallium by Rats," Toxicol. Appl. Pharmacol., 10, 79-88 (1967); Chem. Abstr., 66, 74770k (1967).

239. Kamerbeek, H. H., A. G. Rauws, M. ten Ham, and A. N. van Heijst, "Dangerous Redistribution of Thallium by Treatment With Sodium Diethyldithiocarbamate," Acta Med. Scand, 189(3), 149-154 (1971).

240. van Heijst, A. N. P., and S. A. Pikaar, "Experiences of the National Poison Information Center," Ned. Tijdschr. Geneeskunde, 114, 1611-1615 (1970); abstract in TOXLINE thallium search.

241. Kamerbeek, H. H., A. N. P. van Heijst, A. G. Rauws, and M. ten Ham, "Problems in the Treatment of Thallium Poisoning. A Clinical Trial With Deithyldithiocarbamate," Ned. Tijdschr. Geneeskunde, 114, 457-46 (1970); abstract in TOXLINE thallium search.

242. Rauws, A. G., M. ten Ham, and H. H. Kamerbeek, "Influence of the Antidote Dithiocarb on Distribution and Toxicity of Thallium in the Rat," Arch. Intern. Pharmacodyn., 182(2), 425-426 (1969); abstract in TOXLI thallium search.

243. Keller, R., W. Thimme, W. Dissmann, H. J. Bushmann, K. Dross, and J. Daugs, "Thallium Poisoning With Consumptive Coagulopathy," Schweiz. Med. Wochschr., 101(14), 511-515 (1971); abstract in TOXLINE thallium search.

244. Noirfalise, A., R. Versie, and C. Heuschem, "Tl Poisoning and Its Antidotes," Ann. Biol. Clin. (Paris), 24(5-6), 717-725 (1966); Chem. Abst 65, 9606c (1966).

245. Heyndrickx, A., F. Marchetti, G. Moreno, and M. H. Faes, "Absorption, Localization, and Elimination of Thallium by Rats Treated With Sodium EDTA," J. Pharm. Belg., 17(1-2), 31-39 (1962); Chem. Abstr., 61, 8791 (1964).

246. Bendl, B. J., "Thallium Poisoning," Arch. Dermatol., 100, 443 (1969).

247. Polakowski, P., "Pharmacological Properties of Dithizone and Its Influence on Excretion and Distribution of Various Metals in Tissues After Experimental Poisoning," Pol. Tyg. Lek., 29(5), 199-200 (1974).

248. Mrozikiewicz, A., and W. Widy, "Efficacy of Cytochrome C in Experimenta Animals Poisoned by Thallium," Bull. Soc. Amis Sci. Lettres Poznan Se C, No. 13, 77-78 (1964); Chem. Abstr., 61, 11233c (1964).

249. Janiak, T., "Action of Thioacetamide in Experimental Tl Poisoning in Dogs," Med. Weterynar., 16(1), 22-26 (1960); Chem. Abstr., 58, 10651b (1963).

50. Dvořák, P., "Removal by Rhyolitic Pumice of Internally Deposited Thallium From the Rat," Res. Exp. Med., 162(1), 63-66 (1974).

:51. Bokk, M. I., "Acceleration of the Removal of Thallium-204 From the Organism," Materialy 3-ei [Tret'ei] Nauchn.-Prakt. Konf. po Radiats. Gigiene, Leningrad, Sb., 56 (1963); Chem. Abstr., 61, 13612f (1964).

:52. Bokk, M. I., and L. A. Il'in, "Izyskanie Preparatov, Umenyshayushchikh Vsasyvanie Talliya v Zheludochno-Kishechnom Trakte" ["Preparations Reducing Thallium-204 Absorption in the Stomach and Intestine"], Radiobiologiya, 5(3), 434-439 (1965); Chem. Abstr., 63, 8706g (1965).

:53. Hart, M. M., C. F. Smith, S. T. Yancey, and R. H. Adamson, "Toxicity and Antitumor Activity of Gallium Nitrate and Periodically Related Metal Salts," J. Nat. Cancer Inst., 47(5), 1121-1127 (1971).

:54. Bienvenu, P., C. Nofre, and A. Cier, "Toxicité Générale Comparée des Ions Métalliques. Relation avec la Classification Périodique" ["General Comparative Toxicity of the Metallic Ions. Relation With the Periodic Classification"], Comp. Rend., 256, 1043-1044 (1963).

:55. Christensen, H. E , T. T. Luginbyhl, R. H. Hill, Jr., D. J. McGrath, M. Georgevich, F. L. Mitchell, and J. R. May, Eds., The Toxic Substances List 1973 Edition, U.S. Department of Health, Education, and Welfare, Public Health Service, U.S. Government Printing Office, Washington, D.C., 1973.

:56. Roschchin, A. V., "Toksikologiya Metallov i Profilaktika Professional'nykh Otravlenii" ["Toxicology of Metals and Prophylaxis of Occupational Poisonings"], Zh. Vses. Khim. Obshchest., 19(2), 186-192 (1974).

:57. Janiak, T., "Activity of Tl_2S in the Light of Some of the Methods of Tl_2SO_4 Poisoning Treatment," Med. Weternyar, 16(6), 336-338 (1960); Chem. Abstr., 58, 10651c (1963).

:58. Richards, O. W., "The Analysis of Growth as Illustrated by Yeast," Cold Spring Harbor Symposia, 2, 157-166 (1934).

:59. Norris, P., W. K. Man, M. N. Hughes, and D. P. Kelly, "Toxicity and Accumulation of Thallium in Bacteria and Yeast," Arch. Microbiol., 110(2-3), 279-286 (1976).

:60. McKenzie, D. A., "The Enrichment of Streptococci by Salts of Thallium," Proc. Soc. Agric. Bact., 45-46 (1941).

:61. Porter, J. R., Bacterial Chemistry and Physiology, Wiley, New York, N.Y., 1946.

262. Somers, E., "Plant Pathology: Fungitoxicity of Metal Ions," Nature, 184(4684), 475-476 (1959).

263. Somers, E., "The Fungitoxicity of Metal Ions," Ann. Appl. Biol., 49, 246-253 (1961).

264. Kunze, M., and I. Pramberger, "Sensitivity of Staphylococcus aureus and Staphylococcus epidermidis to Thallium," Zentralbl. Bakteriol., Parasitenls., Infektionskr. Hyg., Abt. 1: Orig., Reihe A, 222(4), 548-551 (1972); Chem. Abstr., 78, 40337g (1973).

265. Tuovinen, O. H., and D. P. Kelly, "Growth of Thiobacillus Ferrooxidans. IV. Influence of Monovalent Metal Cations on Ferrous Iron Oxidation and Uranium Toxicity in Growing Cultures," Arch. Microbiol., 98(2), 167-174 (1974).

266. DeJong, L. E. den D., and W. B. Roman, "Tolerance of Azotobacter for Metallic and Nonmetallic Ions," Antonie van Leeuwenhoek, J. Microbiol Serol., 37(1), 119-124 (1971).

267. Razin, S., and R. C. Cleverdon, "Carotenoids and Cholesterol in Membran of Mycoplasma laidlawii," J. Gen. Microbiol., 41(3), 409-415 (1965).

268. Watanabe, T., O. Fujita, and T. Horikawa, "Susceptibility of Proteus mirabilis and Its L-Phase Variant to Thallium Acetate," Bull. Tokyo Med. Dent. Univ., 18(4), 311-317 (1971).

269. Kunze, M., "Der Einfluss von Thalliumazetat auf das Wachstum von Achole plasmataceae, Mycoplasmataceae und einigen Bakterienspecies," ["Influ ence of Thallium Acetate Upon the Growth of Acholeplasmataceae, Mycoplasmataceae, and Several Bacterial Species"], Zentralbl. Bakteriol., Parasitenls., Infektionskr. Hyg., Abt. 1: Orig., Reihe A, 222, 535-5 (1972).

270. Ong, S. G., "Inhibiting Action of the Metals on the Growth of Mycobacterium tuberculosis. IV. Gallium, Indium, and Thallium," Proc. Konink Nederland. Akad. Wetenschap., 49, 1204-1210 (1946); Chem. Abstr., 41, 3170i (1947).

271. Tandon, S. P., and M. M. Mishra, "Effect of Some Rare Elements on Nitri fication by Nitrobacter agilis in Liquid Culture Medium," Proc. Nat. Acad. Sci., India, Sect. A, 39 (Pt. 2), 209-216 (1969); Chem. Abstr., 73, 111 (1970).

272. Hatten, B. A., and S. Sulkin, "Intracellular Production of Brucella L Forms. II. Induction and Survival of Brucella abortus L Forms in Tissue Culture," J. Bacteriol., 91(1), 14-20 (1966).

3. Naguib, M. I., and A. M. Salama, "Effect of Univalent Cations and Colchicine on Growth Respiration and Carbohydrate Metabolism of Cunninghamella elegans," Folia Microbiol. (Prague), **11**(6), 413-421 (1966); Chem. Abstr., **66**, 44471w (1967).

4. Salama, A. M., and M. I. Naguib, "Nitrogen and Phosphorus Metabolism of Cunninghamella elegans as Affected by Colchicine, in Presence of Various Monovalent Cations," Bull. Fac. Sci., Cairo Univ., **43**, 23-32 (1970); Chem. Abstr., **78**, 106433 (1973).

5. Srivastava, T. N., K. K. Bajpai, and K. Singh, "Antimicrobial Activities of Diaryl Gallium, Indium, and Thallium Compounds," Indian J. Agric. Sci., **43**(1), 88-93 (1973).

6. Tumanov, A. A., I. A. Filimonova, I. E. Postnov, and N. I. Osipova, "Fungitsidnoe Deistvie Neorganicheskikh Ionov na Vidy Gribov Roda Aspergillus Mich. ex Fr." ["Fungicidal Effect of Inorganic Ions on Some Species of Aspergillus Mich. ex Fr."], Mikol. Fitopatol., **10**(2), 141-143 (1976); Chem. Abstr., **85**(11), 73111b (1976).

7. Filimonova, I. A., A. A. Tumanov, I. E. Postnov, and N. I. Osipova, "Plesnevye Griby Roda Aspergillus kak Analiticheskie Indikatory" ["Molds of the Genus Aspergillus as Analytical Indicators"], Tr. Khim. Khim. Tekhnol., No. 3, 50-51 (1973); Chem. Abstr., **81**, 115391t (1974).

8. Siegel, B. Z., and S. M. Siegel, "Selective Role for Potassium in the Phytotoxicity of Thallium," Bioinorg. Chem., **4**, 93-97 (1975).

9. Haerdtl, H., "Grenzwerte der Fungiziditát Chemischer Substanzen unter verschiedenen Bedingungen" ["Threshold Values of Fungicides Under Different Conditions"], Zentr. Bakteriol. Parasitenk., Abt. II, **116**(5), 532-551 (1963); Chem. Abstr., **60**, 8374h (1964).

30. Schopfer, W.-H., "Recherches sur l'Action du Thallium sur un Champignon" ["Action of Thallium on a Fungus"], Compt. Rend. Soc. Phys. Geneve, **50**, 90-92 (1933).

31. Chilean Nitrate Educational Bureau, Inc., Bibliography of the Literature on the Minor Elements and Their Relation to Plant and Animal Nutrition, 4th ed., Vol. 1, Chilean Nitrate Educational Bureau, Inc., New York, N.Y., 1948.

32. Tso, T. C., T. P. Sorokin, and M. E. Engelhaupt, "Effects of Some Rare Elements on Nicotine Content of the Tobacco Plant," Plant Physiol., **51**, 805-806 (1973).

283. Brooks, S. C., "Thallium Poisoning and Soil Fertility," Science, 75(193[illegible] 105-106 (1932).

284. Pallaghy, C. K., "Localization of Thallium in Stomata is Independent of Transpiration," Aust. J. Biol. Sci., 25(2), 415-417 (1972).

285. Maercker, V., "The Transpiration of Guard Cells [of Higher Plants]," Protoplasma, 60(1), 71-78 (1965); Chem. Abstr., 63, 16788d (1965).

286. Puerner, N. J., and S. M. Siegel, "Effects of Mercury Compounds on the Growth and Orientation of Cucumber Seedlings," Physiol. Plant, 26(3), 310-312 (1972).

287. Horn, E. E., J. C. Ward, J. C. Munch, and F. E. Garlough, "The Effect of Thallium on Plant Growth," Science, 80 (2068), 167-168 (1934).

288. Yopp, J. H., W. E. Schmid, and R. W. Holst, Determination of Maximum Permissible Levels of Selected Chemicals That Exert Toxic Effects on Plants of Economic Importance in Illinois, PB 237 654/9GA, National Technical Information Service, U.S. Department of Commerce, Springfiel Va., 1974.

289. Yuan, T.-L., "The Accumulation and Distribution of Scandium, Chronium, Dodine, Cesium, Iodine, Cesium, and Thallium in the Corn Plant," J. Agric. Assoc. China [N. S.], 9, 67-72 (1955); Chem. Abstr., 50, 1987h (1956).

290. Sayre, J. D., "Some Hidden Secrets of Plant Nutrition. Accumulation of Radioisotopes in Corn Leaves," Fertilizer Rev., 28(1), 8-9 (1952); Chem. Abstr., 46, 4064b (1952).

291. Lück, H., and S. W. Souci, "Lebensmittel-Zusatzstoffe und mutagene Wirkung. II. Mutagene Stoffe in unserer Nahrung" ["Food Additives and Mutagenic Action. II. Mutagenic Materials in Our Foods"], Z. Lebensm.-Untersuch. u.-Forsch., 107, 236-256 (1958); Chem. Abstr., 52, 11302c (1958).

292. Schmid, W. E., "Influence of Thallium Ions on the Transport of Certain Cations by Excised Barley Roots," Trans. Ill. State Acad. Sci., 60(1), 61-67 (1967); Chem. Abstr., 66, 112992u (1967).

293. Bange, G. G. J., and F. Van Iren, "Absorption of Thallium Ions by Excised Barley Roots," Acta Bot. Neer., 19(5), 646-658 (1970); Chem. Abstr., 74, 85 (1971).

294. Piver, W. T., National Institute of Environmental Health Sciences, personal communication, March 10, 1975.

295. Battelle Pacific Northwest Laboratories, Fate of Heavy Metals and Heavy Metal Complexes in Soils and Plants, Progress Report for June 1 to September 30, 1973, Battelle Pacific Northwest Laboratories, Richland, Washington, 1973; abstract in TOXLINE thallium search.

296. Battelle Pacific Northwest Laboratories, Fate of Heavy Metals and Heavy Metal Complexes in Soils and Plants, Progress Report for October 1 to December 31, 1972, Battelle Pacific Northwest Laboratories, Richland, Washington, 1972; abstract in TOXLINE thallium search.

297. Bazzaz, F. A., R. W. Carlson, and G. L. Rolfe, "The Effect of Heavy Metals on Plants. I. Inhibition of Gas Exchange in Sunflower by Pb, Cd, Ni and Tl," Environ. Pollut., **7**(4), 241-246 (1974).

298. Carlson, R. W., F. A. Bazzaz, and G. L. Rolfe, "Effect of Heavy Metals on Plants. II. Net Photosynthesis and Transpiration of Whole Corn and Sunflower Plants Treated With Lead, Cadmium, Nickel, and Thallium," Environ. Res., **10**(1), 113-120 (1975).

299. Crafts, A. S., "The Effects of Thallium Sulfate on Soils," Science, **79**, 62 (1934).

VIII. ENVIRONMENTAL THALLIUM LOSSES AND ASSOCIATED HEALTH HAZARDS TO HUMANS AND OTHER LIFE FORMS

A. Magnitude, Chemical Forms, and Ultimate Fate of Environmental Losses and Human Exposure to Them

Table VIII-1 assesses the relative pollution potential of the industries examined in Section V. The magnitude of the losses, their probable chemical forms, their distribution to the environment, and some of the most likely human populations exposed are suggested. Except for coal-burning estimates, which are supported by fairly good U.S. materials-balance data, the other estimates have been largely based on very limited U.S. information or, even less satisfactorily, on European or Soviet processing information. For the most part, the assumptions made prior to extrapolating values to U.S. production volumes have been given in Section V.

1. Water and land pollution

a. Source and exposure: Most of the thallium-containing industrial solid wastes are confined on the industrial property (primarily copper tailings ponds). However, the environmental pollution potential by leaching, runoff, dam failure, seepage to ground water, etc., may be substantial. No estimates could be made for the magnitude of thallium losses by leaching from old dumps from milling, smelting, or refining activities or by acid mine drainage.

The water pollution potential of the copper industry is well established by the Calspan Corporation sampling and analysis program[1/] (see Table V-9), in which elevated thallium levels were detected in environmental samples and biota around copper smelters and around some of the other segments of the industry. Sites with the worst pollution potential are described in the Calspan Corporation report. Because of the physical state of copper milling wastes, thallium leaching from this fine material seems more likely than from smelter slags and other solid wastes. Collected coal fly ash and bottom ash disposed of on land also seem to be a very large collective source of fine, leachable thallium wastes. However, there are far fewer primary copper processors than coal-burning plants.

The Calspan Corporation report[1/] discusses the use of thallium-contaminated waters. Some waters were used to irrigate crops, e.g., alfalfa fed to beef cattle. Local food or drinking water contamination may be occurring in some of the geographical areas mentioned in that report, but it apparently would not be a constant menace in the arid Southwest. For example, it was the flood flow of Whitewater Creek (site of Kennecott Copper Corporation's Santa Rita pit in New Mexico) that was used for alfalfa irrigation. Low flow went to ground water recharge. The ion-exchange capacity of the soil was expected to remove toxic metals.

The water pollution potential of lead and zinc smelters is also indicated in Table V-9. The Calspan Corporation sampling program for the lead and zinc industries was not as extensive as for the copper industry, but some of the highest environmental thallium levels were found near a zinc smelter and electrolytic refinery complex. About 4 tons thallium sulfate solution may be discharged annually from U.S. electrolytic zinc and cadmium plants directly to water or after some treatment. Several zinc processors are located near populous regions (see Table V-3).

Dumps of dusts and slags from steelmaking and, possibly, ferroalloy production from manganese ore could attain a combined magnitude comparable with that of copper tailings and slags. Although distributed at more sites, the dumps would be closer to populous regions. The dust portion, especially, should be leachable. (Some dusts from sulfide ore smelting are also stored in waste banks subject to leaching and erosion.[1/]) No thallium values representative of iron ores and manganese ores processed in the United States were found. The loss estimates are based on assumed contents of 160 ppm in manganese ores and 1 ppm (about the crustal abundance) in iron ores. The estimate for thallium in iron smelting slags was based on an analogy with copper smelting. Leaching of thallium from gold tailings dumps and current mining and milling could be a problem because of the much higher (up to 100-fold) thallium concentrations than there are in solid wastes from other sources. Water pollution from the other industries considered is expected to be negligible because of the expected inert or insoluble chemical forms of thallium.

If 1 ppm thallium remains in the potassium salts formulated in U.S. fertilizers, 5 tons thallium would be spread on U.S. agricultural land. Five tons thallium is probably negligible compared with the amount of thallium present in the soil in which crop plants grow. However, 5 tons of plant-assimilable thallium (with the improbable assumption that the soil is already saturated with respect to thallium) is large compared with a calculated total annual U.S. dietary consumption of only 0.8 ton (0.6 ppb in the food, 1,600 g/day/person). The amounts of thallium in U.S. fertilizers and their uptake by plants grown in fertilized compared with nonfertilized soils should be determined as well as the possibility of runoff from fertilized fields. Although soil fixation probably renders most of the thallium immobile and unassimilable, fertilizer application represents a more direct route to man's food chain than most of the industrial processes assessed.

Another thallium source to the human food chain in years past might have been arsenical pesticides. In the 1940's, the Murray, Utah, lead smelter sold "plus 90% As_2O_3 for insecticide" containing 5,300 ppm thallium.[2/] In recent years, about 33,000 short tons arsenic (mostly organoarsenical herbicides) was sprayed on crops.[3/] A concentration of only 100 ppm thallium would have resulted in application of about 3 tons thallium annually, a number comparable to the amounts of thallium used as a pesticide at its peak. Where arsenic levels have built up in arable lands, thallium levels may also be high.

b. Chemical forms and fate: Smelter slags are predominantly iron silicate with heavy metals substituted for iron. The thallium present is at least in part in a water-soluble form; both the mud eroded from such slags and also the water runoff were found by Calspan Corporation[1/] to have relatively high thallium concentrations.

Thallium from discharges of electrolytes and leaching solutions or in leachate from sulfide ore mines and tailings is probably mostly in the form of thallium(I) sulfate solutions at low pH. If such solutions are basified prior to discharge, soluble TlOH is the most likely species released. The acidic effluents may contain ligands (e.g., chloride or organics) that would stabilize the thallic state and favor oxidation of thallous ions. However, in the reducing environment (rich in sulfides) encountered in mining and milling discharges, the thallous state is favored even in an acidic effluent. Thallium(I) is probably removed from solution at higher pH by ion exchange for K^+ in suspended clays, which accounts for the high levels found in sediments near polluted sites compared with much lower levels in the water (see Table V-9). High thallium in sediments may be due more to erosion of mud from tailings and slags than to sedimentation of the thallium dissolved in contaminated waters.

It has been proposed that chloride controls the speciation of dissolved thallium in natural waters,[4/] but adsorption by iron and manganese hydrated oxides seems to be an important mechanism for removing certain metal cations from solution under oxidizing conditions. A speciation study on thallium in fresh water such as that performed by Lu and Chen (1977)[5/] for several heavy metals in seawater under different redox conditions would be valuable for understanding the potential dangers of thallium water pollution, remobilization from sediments, and biomethylation in the environment.

The possibility of environmental biomethylation of thallium has been mentioned by a few workers.[6,7/] Agnes et al. (1971)[8/] reported that Tl(III), but not Tl(I), is methylated by methyl vitamin B_{12}. Wood[9/] expected that Tl(I) and Tl(III) could participate in an oxidation reduction couple such as that shown for tin:

$$Sn^{2+} + \text{Methyl-}B_{12}(Co^{III}) \longrightarrow MeSn^{4+} + \text{Reduced } B_{12}(Co^{I})$$

(Reactions with mercury, tin, selenium, and arsenic may occur in the environment to produce metal alkyls.) Thayer[10/] is currently exploring the possibilities of Tl(I) and Tl(III) methylation reactions under conditions more likely to be encountered in the environment. He suggests that popular organothallium(III) species, e.g., $(C_6H_5)_2Tl^+$, may be methylated and degraded to yield $(CH_3)_2Tl^+$.

2. Air pollution

a. Sources and exposure: Based on the magnitude of annual emissions, coal burning is the major source of thallium in urban air. We estimate that each of the approximately 415 coal-burning plants releases an average of about 0.4 ton of thallium per year. (Compare with our estimate of 2.5 tons per copper smelter.) Many cities have two coal-burning plants, and some have three or four (e.g., Detroit, Michigan; and Louisville, Kentucky). Thus, 0.8 to 1.6 tons thallium could be released annually in some cities, depending on the average thallium content of the coal used at each plant.

In Section V, we have estimated a ground-level thallium concentration near an actual coal-burning plant. If humans are exposed to this level of 0.0007 mg thallium/m^3/day (and inspire 20 m^3 air daily),[12] the daily intake would be about 11 to 14 μg if 75 to 100% absorption is assumed. Thus, coal-burning could be a greater source of thallium for some humans than is the diet (~ 2 μg/day). Clearly, separate estimates as to the "normal" human body burden should be made for people living in areas polluted and unpolluted by coal (or fuel oil**) burning.

The possibility of high thallium emissions from furnaces producing ferromanganese and silicomanganese in highly populated urban areas such as in Ohio should be explored. Representative manganese ores should be analyzed for thallium; and if the thallium concentrations exceed a few parts per million, an environmental sampling and analysis program would be advisable. These furnaces could be by far the greatest emitters of thallium per point source.

The heavy metal contents of typical iron ores should be as well documented as has been done for coal because of the sheer magnitude of raw material furnaced. We have calculated that about 6 tons thallium might be emitted from the numerous U.S. iron and steel works if the iron ore contains thallium at merely the crustal abundance--~ 1 ppm. Our assessment of steel production suggests that much of the thallium could end up in the steel.

---

* Maximum actual air concentrations of thallium measured at Chadron, Nebraska, would result in air intake of only 0.01 μg/day.

** Among cities with numerous power plants, nine New York City plants; five Baltimore, Maryland, plants; and three Portsmouth, New Hampshire, plants burn fuel oil amost exclusively.

The numbers we have developed for annual atmospheric emissions from copper smelters (38 tons) based on Soviet data appear out of line with estimates for zinc (1 ton) and iron and steel smelting operations (6 tons). Actual copper smelting emissions should be analyzed for thallium and other toxic heavy metals that are not economically interesting. Calspan Corporation[1/] states that thallium, lead, and cadmium in copper smelter* emissions "settle in soils and sediments but are not solubilized in water to any great extent." It would appear that water-soluble thallous oxide would be the predominant emitted form of thallium; but less soluble compounds such as Tl_2S, Tl_2SO_4, and TlCl could also be emitted. Except for the El Paso, Texas, and Tacoma, Washington, smelters, the other 13 copper smelters are not located near large urban areas.

The amount of thallium emitted from each lead (six) or zinc (four) smelter may be comparable with that from a typical coal-burning plant. However, the concentrations of thallium in sulfide ore smelter emissions could attain much higher values than the highest concentration reported for coal emissions (76 ppm in the finest fraction) due to recycling of materials. Ambient air levels in the areas of lead, zinc, and copper smelters should be compared with those to which people are exposed in coal-burning areas.

b. Chemical forms and fate: Major forms of thallium emitted from thermal processes should be Tl_2O, especially from coal-burning, Tl_2S; and Tl_2SO_4. The oxide (gives TlOH) and sulfate are water-soluble. Water and agricultural soil pollution or contamination, although worthy of consideration, do not seem as important as human intake from breathing high ambient levels of soluble thallium compounds. The air breathed could well be the major route of thallium intake by urban populations. Ultimately, the bulk of urban air pollutant particulates is probably fixed in soils or washed from paved areas into the local streams or sewage system via storm runoff.

B. Assessment of the Human Health Hazard

Waldbott (1973)[13/] classifies air pollutants according to their health effect. Thallium compounds would be classed as host-specific systemic pollutants, their most likely health effect (of those considered) being mutagenicity and teratogenicity. The host's individual susceptibility largely determines the target organ affected by host-specific pollutants. Other human health effects at chronic minute levels from inhalation and ingestion are almost impossible to predict and ascribe with any certainty to thallium alone. The similarity of Tl^+ to K^+ with regard to uptake, distribution, and enzyme effects indicate numerous bodily functions can be affected.

---

* The ASARCO Inc., smelter at El Paso, Texas, which operates on the Rio Grande River.

The reported human and mammalian studies lend themselves to few firm conclusions about the dangers of very low intakes of thallium. Rats given one-time s.c. doses of 0.24 μg Tl_2CO_3 per kilogram (comparable to 15 μg/70 kg man) showed disturbed hair regeneration but no other toxic manifestations.[14/] Men employed at a Canadian petroleum refinery, some of whom were suffering from alopecia areata (patchy baldness that did not especially resemble the diffuse alopecia recognized from thallium intoxication), excreted up to 12 to 23 μg thallium in their daily urine.[15/] If minor toxic manifestations of thallium can occur in people at chronic levels of about 10 to 20 μg/day, a possible daily air intake of about 11 to 14 μg near coal-burning power plants allows no safety factor. The low concentration factors and belief in little likelihood of bioconcentration may need to be reevaluated. A suggested limit of 5 ppb in drinking water, which would give 10 μg/day, should also be reevaluated. Up to 14 ppb has been reported in drinking waters.

Thallium sulfate is among the chemicals to be evaluated by Litton Bionetic, Inc., under an EPA contract for determination of potential for carcinogenesis, mutagenesis, teratogenesis, and alteration of reproductive function in rats.[16/] A modern reevaluation of thallium's oncological activity is certainly needed.

A fruitful first approach for determining if environmental thallium is having any effect on humans, when it is known that all major releases are associated with other known toxic elements, would be examination of birth-defect incidence (human or livestock) by geographical area. Other heavy metals produce birth defects, but those peculiar to thallium (such as leg-bone shortening) should be looked for. Birth defects data might be sough for the Alsar region in Yugoslavia. The thallium mineralization in this region is so high that livestock have been poisoned by eating plants containing very high thallium concentrations.

At this time, the best assessment of human health hazards can only be made by identifying direct routes to the food chain and by comparing the probable dietary intake with possible intake by inhalation in air-contaminated areas. Raising the body burden of an element as acutely toxic as thallium and with such deleterious effects on fetal development and mortality would seem inadvisable whether or not direct effects can be unequivocally ascribed to thallium releases. We have calculated above that inhaation of thallium in contaminated air could greatly exceed the intake from food and water. Air and water monitoring and food and tissue sampling are strongly urged to substantiate this hypothesis.

C. Hazard for Other Life Forms

1. Aquatic species: Calspan Corporation[1] reports that the highest background thallium level encountered in freshwater was 2.1 ppb, and 3.7 ppb was observed downstream from a copper mine. Where higher thallium levels were encountered (up to 2,400 ppb) concomitantly much higher lead, zinc, and/or copper levels and low pH militated against most life forms. Most of the experimental studies on fish and paramecia were done at thallium concentrations on the order of 10,000 ppb, which were lethal levels. The LD_{50} for Atlantic salmon, however, is 30 ppb.[17] Tadpoles were killed at 400 ppb.[18] Levels of 2,000 to 4,000 ppb were lethal within 3 days to Daphnia and Gammarus, small freshwater crustacean species.[19]

Levels as low as 2 to 3 ppm thallium in freshwater reduce photosynthetic activity of algae.[20,21] At 7 ppm, thallous compounds were toxic to algae.[22] Direct toxicity to most aquatic plants from thallium pollution alone appears to be remote. Where algae containing up to 0.85 ppm thallium[1] serve as an animal's primary food source, bioconcentration and direct toxicity to the primary or secondary consumer may occur.

2. Terrestrial animals: Calspan Corporation[1] found up to about 3 ppm thallium in biota near polluted areas. Such levels are several hundred times greater than normal levels. We estimate that herbivores might ingest about 6 ppb thallium in an uncontaminated diet, carnivores probably less (2 ppb).

"No-effect levels" for mammals have not been established experimentally, but levels of near 3 ppm thallium in an animal's diet are likely to produce toxic effects detrimental to the individual's survival. Effects on reproductive capacity, fetal development, and survival of the offspring are also probable, but chronic experiments have seldom been done at the parts per million level. The cumulative effects of all the heavy metals in a contaminated environment would have to be considered.

At levels corresponding to about 30 ppm thallium in the diet of rats for 15 weeks, both Tl(I) and Tl(III) compounds caused growth depression, alopecia, and high mortality in males. Levels of 18 ppm Tl(III) were not fatal but still produced alopecia and growth depression in males.[23] Tl(I) given at dietary levels of about 7 ppm to rats for 40 days produced blindness and affected movement in the oldest rats while the younger rats showed mainly depilatory and weight loss effects.[24] Rats given 8 ppm Tl(I) in their drinking water for 80 days showed reduced tissue copper levels.[25] Levels of about 4 ppm Tl(I) in the diet of rats for 15 weeks did not affect growth,[23] but levels corresponding to about 4 to 6 ppm Tl(I) in the diet of rabbits for 6 months produced behavioral and blood changes, mental retardation, and rear limb paralysis.[26/]*

3. Terrestrial plants: Environmental contamination by thallium up to about 20 ppm would probably affect few microorganisms unfavorably. Levels of 0.1 to 10 ppm thallium in the growth medium stimulate yeast and certain bacteria. The lower part of this range would correspond to background environmental levels. For bacteria, fungi, and mycoplasmas, the range of thallium concentrations having no effect on some species is 10 to ~ 1,600 ppm. Yeast and some bacteria still show some growth at up to 8,000 ppm thallium. Some soil organisms are unaffected by 10 to 90 ppm, while others show toxic effects at 30 to 90 ppm.[28/] Thallous ion at 0.8 to 8 ppm in soil suspensions retards or stops nitrate formation by *Nitrobacter agilis*; *Aspergillus niger* spore germination is inhibited by 1 to 2 ppm thallium.

Many crop plants are injured by concentrations of about 7 ppm thallium in the soil. Graminae and seeds are more resistant. Tobacco plants are especially sensitive, showing toxic effects at 1 ppm in the soil or 40 ppb in water. Toxic effects of thallium on higher terrestrial plants removed from the immediate sulfide smelter vicinity are not expected because thallium levels are not expected to be as high as those cited.

---

* In the studies cited, dose levels were given as unit weight per kilogram animal. For the purposes of our conversions to ppm thallium in the diet rats were assumed to eat 15 g food per day and rabbits 100 g.[27/]

TABLE VIII-1

ESTIMATED ANNUAL AMOUNTS OF THALLIUM RELEASED TO THE ENVIRONMENT FROM HUMAN ACTIVITIES

Initial Raw Material	Process	Thallium Content, Short Tons	Probable Initial Forms of Thallium	Probable Forms Released	Distribution to the Environment, short tons: Air	Water and/or Land	Land	Total	End Products, Short Tons Contained Thallium	Human Population Exposure and Other Remarks
Coal	Burning and coking	350	Tl_2S	Tl_2O	180		170	350		Highest in EPA region 3, 4, 5. Leaching of dumped ashes possible.
Copper ore	Mining, milling	530	Tl_2S	Tl_2S		410		410		Elevated levels in water close to sites. Most Tl^+ would be sedimented.
	Mine drainage, leaching old dumps	?		Tl_2SO_4		?		?		Possibly some crop irrigation waters have elevated levels.
	Roasting and/or smelting	120		Tl_2O, Tl_2S, Tl_2Se, Tl_2Te, Tl, Tl_2SO_4	24	48		72		Populations around thermal copper processing operations would be exposed to thallium fumes.
	Converting	48		Tl_2O	14	(20) recycled		14		Probably highest exposed populations in Tacoma, Washington, and El Paso, Texas.
	Refining	13		Tl_2O, Tl_2SO_4	?	(≤ 13)		(< 13)	Cu, (< 13) Ag, 0.01	High levels in sludge dumps. If thallium in gases rather than particulates, about 4 tons thallium in air emissions might be trapped in smelter sulfuric acid.
Manganese ore	Ferroalloy production	(360)[a/]	Tl_2O_3	Tl_2O	(140)[a/]		(220)[a/]	(36.0)		Exposure possible in highly populated urban areas. Ohio cities probably greatest potential exposure.
Crude oil	Burning residual fuel oils, leaching from asphalt	(16)[a/]	Porphyrin complex	Tl_2O	(8)[a/]	(8)		(16)		About 38% of residual fuel oil is burned in power plants.[29/]
Iron ore	Blast furnace smelting	300	Tl_2O_3, Tl_2S	Tl_2O, Tl_2S Tl_2SO_4	0.9		(recyled dust 83)	1	Pig iron, 216	
	Steelmaking	Pig iron 200		Tl_2O	~ 5		dust ~40 slag 160	~ 205		Pennsylvania, Ohio, and Indiana consume the most pig iron.[30/]
		Scrap 200							Steel, ~ 200	
	Iron founding	16			0.2					
Potash	Fertilizer use		TlCl, Tl_2SO_4 et al.	Tl^+	-	5		5		

TABLE VIII-1 (Continued)

Initial Raw Material	Process	Thallium Content Short Tons	Probable Initial Forms of Thallium	Probable Forms Released	Distribution to the Environment, short tons: Air	Water and/ or Land	Land	Total	End Products, Short Tons Contained Thallium	Human Population Exposure and Other Remarks
Gold ore	Mining and milling	30				30		30		Water contamination possible near tailings pond sites. Concentrations in solid wastes high.
	Leaching of old tailings and mines	?				?		?		Cyanide content probably a greater immediate problem than thallium. Wastes generally not chlorinated to destroy cyanides, although they might be in case of dam leaks or breaks.[31]
Zinc ore	Mining and milling, (Zn, Pb-Zn ores)	40 ↓	Tl_2S	Tl_2S		23		23		
	Roasting domestic and foreign concentrates	11		Tl_2O, Tl_2S, Tl_2Se, Tl_2Te, Tl_2SO_4	0.5			0.5	ZnO, 3	Human exposure from atmospheric emissions around thermal zinc and cadmium processing plants probably less than near a coal-burning power plant (~ 0.4 ton per coal plant.)
	Sintering, retorting, etc.	5+		Tl_2O, TlCl	0.1	0.1		0.2	Zn, 3	
	Electrolytic zinc production	5+ ↓	Tl_2SO_4	Tl_2SO_4		0.2	2.6	2.8	Zn, 2.1	
	Cadmium production	5+	Tl_2O, TlCl Tl	Tl_2SO_4, Tl_2O_3 Tl chromates		0.3	2	2.3	Cd, 0.1	
Sulfide ores	Thallium refining		Tl_2S	Tl_2SO_4, Tl, Tl_2O				nil	Tl, < 2	Stockpiled indoors. Even Cd releases are negligible. No exposure anticipated.
	Sulfuric acid manufacture		Tl_2S	Tl_2O_3 Tl^+, Tl^{3+}					H_2SO_4, 2-6	Little acid from smelter fumes and pyrites used for phosphate fertilizer production.
Lead ore	Mining and milling	See Zn ore above	Tl_2S	Tl_2S		9		9		
	Sintering and smelting	6-7		Tl_2S, Tl_2Se Tl_2Te, Tl, Tl_2O, Tl_2SO_4			(< 1)	(< 1)	Desilverizing Zn crust < 1 Fume to Cd Plant or Waelz kiln < 5 (most recirculated)	High concentrations of thallium could occur in emissions if recycled dusts constituted a large fraction of the furnace charge.

TABLE VIII-1 (Concluded)

Initial Raw Material	Process	Thallium Content Short Tons	Probable Initial Forms of Thallium	Probable Forms Released	Distribution to the Environment, short tons				End Products, Short Tons Contained Thallium	Human Population Exposure and Other Remarks
					Air	Water and/ or Land	Land	Total		
Feldspar	Mining	37	Tl aluminum silicate	Tl aluminum silicate						
	Glass and other product manufacture	37	ditto	Incorporated in glass structure					Glass, 30 Other products,	More accurate number would require thallium content in cullet used.
	Glass disposal	30					30	30		Recycling of waste glass not expected to release thallium nor is leaching in landfills or in container use.
Mica	Mining	15	Tl(I) silicate	Tl(I) silicate				nil		
	Manufacture	15	ditto	ditto				nil	Products, 15	No significant exposure expected
Totals		> 1572			> 225	> 525	> 405	> 1155	~ 500	

a/ The thallium content of representative U.S. materials was not found. The extrapolations are based on content in Soviet materials.

BIBLIOGRAPHY. SECTION VIII.

1. Calspan Corporation, Heavy Metal Pollution from Spillage at Ore Smelters and Mills, preliminary copy of the final report version provided in March 1977, Calspan Corporation, Buffalo, N.Y., for the U.S. Environmental Protection Agency, Cincinnati, Ohio (in press).

2. Zimmerley, S. R., Thallium, Salt Lake City Division, Metallurgical Branch, U.S. Bureau of Mines, Salt Lake City, Utah, 1947.

3. Bowen, H. J. M., Trace Elements in Biochemistry, Academic Press, Inc., New York, N.Y., 1966.

4. Batley, G. E., and T. M. Florence, "Determination of Thallium in Natural Waters by Anodic Stripping Voltammetry," J. Electronanal. Chem. Interfacial Electrochem., **61**(2), 205-211 (1975).

5. Lu, J. C. S., and K. Y. Chen, "Migration of Trace Metals in Interfaces of Seawater and Polluted Surficial Sediments," Environ. Sci. Technol., **11**(2), 174-182 (1977).

6. Thayer, J. S., "Organometallic Compounds and Living Organisms," J. Organometallic Chem., **76**, 265-295 (1974).

7. Wood, J. M., "Biological Cycles for Elements in the Environment, and the Neurotoxicity of Metal Alkyls" in Trace Metals in Water Supplies: Occurrence, Significance, and Control, Proceedings of the Sixteenth Water Quality Conference, Champaign, Illinois, February 12-13, 1974, pp. 27-38.

8. Agnes, G., S. Bendle, H. A. O. Hill, F. R. Williams, and R. J. P. Williams, "Methylation by Methyl Vitamin B_{12}." J. Chem. Soc. D. (Chem. Communications), No. 15, 850-851 (1971).

9. Wood, J. M., "Metabolic Cycles for Toxic Elements" in International Conference on Heavy Metals in the Environment, Toronto, Ont., Canada, October 27-31, 1975, pp. A-5 to A-6.

10. Thayer, J. S., Assistant Professor, University of Cincinnati, personal communication, March 1977.

11. Nielsen, G. F., Ed., 1976 Keystone Coal Industry Manual, Charles H. Daly, publisher, McGraw-Hill Mining Publication, New York, N.Y., 1976.

12. Division of Radiological Health, Department of Health, Education, and and Welfare, Ed., Radiologic Health Handbook, Revised ed., PB 121 784R, Department of Health, Education, and Welfare, Washington, D.C., 1960.

13. Waldbott, G. L., Health Effects of Environmental Pollutants, The C. V. Mosby Company, St. Louis, Mo., 1973.

14. Sanotskii, I. V., "Issledovanie Toksichnosti Soedinenii Talliya" ["Toxicity of Thallium Compounds (Carbonate, Iodide, and Bromide)"], Toksikol. Novykh. Prom. Khim. Veshchestv, No. 2, 94-104 (1961).

15. Williams, N., and A. L. Riegert, "Epidemic Alopecia Areata. An Outbreak in an Industrial Setting," J. Occup. Med., 13(11), 535-542 (1971).

16. Hart, E. R., R. P. Beliles, D. J. Brusick, D. W. Matheson, A. G. Manus, and H. R. Seibold, "Determination of Potential for Chronic Intoxicational Carcinogenesis, Mutagenesis, Teratogenesis, and Reproductive Potential of Five Compounds Generated from Coal Conversion and Utilization" (notice of research project), TOX-TIPS, March, pp. 10-31 (1977).

17. Zitko, V., W. V. Carson, and W. G. Carson, "Thallium. Occurrence in the Environment and Toxicity to Fish," Bull. Environ. Contam. Toxicol., 13, 23-30 (1975).

18. Dilling, W. S., and C. W. Healy, "Influence of Lead and the Metallic Ions of Copper, Zinc, Thorium, Beryllium and Thallium on the Germination of Frogs Spawn and on the Growth of Tadpoles," Annals Appl. Biol., 13, 177-188 (1926).

19. Nehring, D., "Experiments on the Toxicological Effect of Thallium Ions on Fish and Fish-food Organisms," Z. Fisch., 11, 557-562 (1962).

20. Overnell, J., "Effect of Heavy Metals on Photosynthesis and Loss of Cell Potassium in Two Species of Marine Algae, Duniella tertiolecta and Phaeodactylum tricornutum," Marine Biol., 29(1), 99-103 (1975).

21. Overnell, J., "Effect of Some Heavy Metal Ions on Photosynthesis in a Fresh Water Alga," Pestic. Biochem. Physiol., 5(1), 19-26 (1975).

22. DiGaudio, M. R., Thallium: Some Effects of an Environmental Toxicant, Ph.D. Dissertation, New York University, New York, N.Y., 1975.

23. Downs, W. L., J. K. Scott, L. T. Steadman, and E. A. Maynard, "Acute and Subacute Toxicity Studies of Thallium Compounds," Am. Ind. Hyg. Assoc. J., 21, 399-406 (1960).

24. Dzialek, S., "Depilatory Action of Thallium Acetate on Rats," Ann. Acad. Med. Lodz., 6, 74-81 (1965); Chem. Abstr., 65, 1283h (1966).

25. Toya, T., H. Harada, and Y. Tamura, "Systematic Studies on the Metabolism of Metals. IV. Distribution of Orally Administered Lead, Thallium, and Cadmium in Rats and Its Effect on the Endogenous Copper and Zinc," Tokyo-to Ritsu Eisei Kenkyusho Nempo, No. 13, 178-194 (1961); Chem. Abstr., 62, 3187h (1965).

26. Tikhova, T. S., "Voprosy Gigieny Truda v Proizvodstve Metallicheskogo Talliya i ego Solei" ["Problems of Labor Hygiene in the Production of Metallic Thallium and Its Salts"] Gigiena i Sanit., 29(2), 23-27 (1964); Chem. Abstr., 60, 13786c (1964).

27. Christensen, H. E., T. T. Luginbyhl, R. H. Hill, Jr., D. J. McGrath, M. Georgevich, F. L. Mitchell, and J. R. May, Eds., The Toxic Substances List 1973 Edition, U.S. Department of Health, Education, and Welfare, Public Health Service, U.S. Government Printing Office, Washington, D.C., 1973.

28. Schneiderman, G. S., T. R. Garland, R. E. Wildung, and H. Drucker, "Growth and Thallium Transport of Microorganisms Isolated from Thallium Enriched Soil" in Abstracts of the 74th Annual Meeting of the American Society of Microbiologists, Chicago, Illinois, May 1974, p. 2.

29. Smith, I. C., T. L. Ferguson, and B. L. Carson, "Metals in New and Used Petroleum Products and By-Products--Quantities and Consequences" in The Role of Trace Metals in Petroleum, T. F. Yen, Ed., Ann Arbor Science Publishers, Inc., Ann Arbor, Michigan, 1975, pp. 123-148.

30. Brantley, F. L., "Iron and Steel" in Minerals Yearbook 1972, Vol. 1 Metals, Minerals, and Fuels, U.S. Bureau of Mines, U.S. Government Printing Office, Washington, D.C., 1974, pp. 641-666.

31. Wilkinson, R. R., and G. R. Cooper, Study on Chemical Substances from Information Concerning the Manufacture, Distribution, Use, Disposal, Alternatives, and Magnitude of Exposure to the Environment and Man, Task III - The Manufacture and Use of Selected Inorganic Cyanides, Midwest Research Institute for the Environmental Protection Agency Office of Toxic Substances, Washington, D.C., 1976.

IX. CONCLUSIONS AND RECOMMENDATIONS

A. Conclusions

The largest collective source of thallium (180 tons) discharge into the human environment is air emissions from coal-burning power plants (over 400 in the United States). More people are exposed to thallium from coal burning than from any other source. An average coal burning power plant may emit 0.4 ton thallium; the maximum reported concentration for thallium in the particulates is 76 ppm. However, certain of the 15 U.S. copper smelters are probably the largest point sources of air emissions of thallium (estimated U.S. total 38 tons). In addition, certain copper, lead, and zinc smelters very likely emit much higher thallium concentrations in the particulates and gases than coal-burning power plants, especially when collected roasting and smelting dusts are recycled to concentrate cadmium. Ferroalloy production using manganese ores is conceivably the source of very high thallium emissions, but the thallium contents of the imported ores used in the United States have not been reported. The combustion of fuel oil and processing of iron ores may also be potential sources of large amounts of thallium, but again data in the thallium contents of U.S. supplies of these products are not available.

Copper mining, milling, smelting, and refining produce the greatest volume of wastes leading to water pollution by thallium. Zinc and cadmium plants could also discharge high-thallium solutions.

A direct route of thallium to the food chain could be the relatively small amounts of thallium in potash-derived fertilizers. Concentration factors from soil to plant ash, however, rarely exceed 10 for thallium.

The LD_{50}'s of thallium compounds for most species of birds and mammals as well as humans fall in the range 5 to 70 mg/kg, the toxic symptoms being similar in all species. Even tissue implants of thallium metal have been fatal to primates. Because of the physiological similarity of Tl^{+} to K^{+}, toxic effects can occur in most physiological systems.

The possibility of thallium(III) methylation in the environment or in the body cells is not nearly as important an issue as it is for mercury, because the difference in toxicities between the organic and inorganic forms is so much smaller for thallium. The difference in toxicities between the valence forms for thallium is not great.

The environmental speciation is unknown in freshwater at neutral pH, Tl(I) probably predominates unless species capable of strongly complexing Tl(III) are present. Dissolved thallium concentrations in water are probably kept low by efficient adsorption on suspended clay particulates by K^+ exchange or adsorption on iron and manganese oxide hydrates. Soils also probably fix most of the thallium coming in contact with them by the same mechanisms.

Very limited data indicate that inhalation of thallium-polluted air and sometimes drinking high thallium waters could contribute as much or more thallium than a normal human diet. Chronic intakes by people of 10 to 20 µg/day from these sources are possible at the levels presently found in the environment. Minor toxic effects (alopecia) from thallium at these levels in humans have possibly occurred. The systemic effects of thallium are so insidious that it is difficult to recognize it as the cause of the symptoms of acute lethal poisoning, let alone of milder symptoms. An epidemiological study of the human or livestock incidences of the peculiar types of birth defects caused by thallium in chicks and rats might be a first step in recognizing human ill effects from thallium in the environment.

Levels of thallium in water are usually high (up to 2.4 ppm) only where other toxic metals are in even higher concentrations and where the pH is very low. Background levels in natural waters are usually much below 2 ppb. By comparison, the LD_{50} for Atlantic salmon is 30 ppb, and lethality levels for the few other aquatic species that have been studied are in the ppm range. Accumulation of thallium by algae in contaminated waters can lead to moderate bioconcentration or, conceivably, direct toxicity in the primary or secondary consumer.

Chronic dietary consumption of biota in polluted areas (containing up to 3 ppm) by terrestrial species could produce effects serious enough to be detrimental to the individual's survival and, possibly, to that of its offspring.

Contaminated soil and plant surface levels of thallium are usually not high enough to produce the toxic effects seen in most microorganisms. However, 0.8 to 8 ppm thallium in soil suspension retards and/or inhibits nitrate formation by <u>Nitrobacter agilis</u>, which may affect soil fertility. Most crop plants are not seriously injured by thallium concentrations below 7 ppm, but other extremely sensitive plants like tobacco may be harmed.

B. Recommendations

Thallium concentrations should be determined in raw materials, products, and waste streams--especially air emissions--for the copper, lead, zinc, manganese ferroalloy, and iron and steel industries so that accurate assessments of their pollution potential can be made. Monitoring of some of these industries' discharges and emissions may be necessary. Controls or pretreatments prior to discharge should be devised and used where necessary and feasible. Potassium ferrocyanoferrate(II) (Prussian blue, an effective antidote for thallium poisoning, which exchanges K^+ for Tl^+) is a fairly cheap chemical that has been suggested for treating wastewater streams containing thallium in a percolation bed or as a sludge in basins.*

Ambient air levels should be determined in highly populated regions exposed to emissions from these sources and fossil fuel combustion. Market basket surveys like that of the FDA and drinking water analysis should also include thallium so that a better estimate of normal human thallium consumption can be made and compared with consumption in contaminated areas.

Long term animal studies with thallium at very low levels should be performed to determine the mutagenic, teratogenic, reproductive, carcinogenic, and systemic-toxicity potential for environmental thallium exposure in humans. A suggested drinking water limit of 5 ppb may be too high to prevent minor toxic symptoms and should be reevaluated.

Correlations can possibly be made between certain geographical areas expected to be high in thallium and the incidences of birth defects peculiar to thallium. This may be valuable in distinguishing toxicity due to thallium from the other toxic heavy metals released at the same time in higher amounts.

Thallium should be included in speciation studies. Thallium(III) compounds are often more water soluble and, thus, more mobile.

The amount of thallium in potash-derived fertilizers used in the United States should be determined as well as their potential for increasing plant uptake and local water pollution by runoff and erosion.

---

* Rauws, A. G., and J. H. Canton, "Adsorption of Thallium Ions by Prussian Blue," Bull. Environ. Contam. Toxicol., **15**(3), 335 (1976).

APPENDIX

APPENDIX

OCCURRENCE OF THALLIUM IN THE GEOCHEMICAL ENVIRONMENT

Source	Location	Thallium Concentration Mean	Thallium Concentration Range	No. of Samples	Method	Sensitivity (ppm)	Remarks	Reference
Freshwater Sediments								
Sediments, Wintergreen Lake	Kalamazoo County, Mich., 1973?	13.1 ppm	2.1-23.1 ppm	-	Atomic absorption (a.a.)	2.0	The lake was highly eutrophic. Tl was not found in 5 fish spp., 2 spp. of aquatic macrophytes, zooplankton, migratory goose feces, or water. The authors suggest the source may be particulate fallout from coal combustion.	1 2
Hot spring and drillhole discharge ppts.	New Zealand		320-5,000 ppm				High concns. of As, Sb, Hg, Tl, Ag, and Au may be due to the action of silica as an ion exchanger. SbS may act as a collecting agent by copptg. Au, Ag, and Tl. Weissberg, 1969, cited.	3
Argillaceous sediments		0.69 ppm		18	Spectrog.			4
Argillaceous sediments		0.36 ppm		323			Canney (1952) cited.	4
Scattered stream sediments	North and South Chitana River, McCarthy Quadrangle, Wrangell Mts. area		< 0.2-54.0 ppm	177/380	a.a.	0.2	USGS anal. The second highest value was 5.0 ppm. This area contains Cu deposits.	5
Stream sediments	Socorro and Sierra Counties, N.M.		< 0.2-1.0 ppm	8	a.a.		USGS anal. by Hubert and Crenshaw. Only one value more than < 0.2 ppm.	5
Marine Sediments								
Sediments, marine	At depths 720 to 5,408 m, 1969 Six Pacific Ocean stations, three Atlantic, one Bay of Bengal.	1.1 ppm	0.08-5.7 ppm	10	Neutron activation anal. (n.a.)		Tl is apparently adsorbed on the hydrous MnO_2 phase of ferro-manganese minerals and is incorporated into the lattice of illite and similar clay minerals. The Tl content of calcareous sediments was low: 0.08 to 0.35 ppm, varying inversely with the carbonate content.	6 7

Source	Location	Thallium Concentration		No. of Samples	Method	Sensitivity (ppm)	Remarks	Reference
		Mean	Range					
Sediments, marine (mostly sandy mud)	Minamata district, Japan		0.14-1.13 ppm		Rhodamine B			8
Mn nodules	Pacific	135 ppm	0-150 ppm	10				9 10
Mn nodules, average		100 ppm					Review of several papers.	11
Ocean mud	Depth 500 m, Antarctica	0.13 ppm					Brooks et al. (1960) cited.	11
Mn nodules	N.E. Pacific		135-205 ppm	9	a.a.			5
Manganese nodule	11°N, 113°48'W; depth 4,090 m		30-100 ppm		Spectrog.		Estimated.	12
Volcanic glass	11°N, 113°48'W; depth 4,090 m	0.24 ppm			Spectrog.			12
"Oozes" (clay materials "diluted" by org. remains of a siliceous or calcareous nature)			N.D. to 0.3 ppm		Spectrog.		High value in Fe-Mn-rich ooze.	12
Clays			0.16-1.5 ppm		Spectrog.			12
Deep-sea clays		0.8 ppm					Turekian and Wedepohl (1961) cited.	13
Deep-sea carbonate rocks		0.16 ppm					Turekian and Wedepohl (1961) cited.	13
Manganese nodules	Challenger Stations 248, 252, and 276				Spectrog.		Samples from the Challenger Expedition, 1873-1876.	14

APPENDIX (Continued)

Source	Location	Thallium Concentration Mean	Range	No. of Samples	Method	Sensitivity (ppm)	Remarks	Reference
	Acid-insol. fraction	N.D.					Detd. by Noddack and Noddack (1940)	14
	Acid-sol. fraction (excluding H2O and $CaCO_3$)	130 ppm	100-140 ppm					14
Manganese nodules	Pacific, Atlantic, and Indian Ocean	140 ppm	N.D. to 614 ppm		DC arc spectro-chem.		Avg. value given by Manheim (1965) 100 ppm.	15
Sedimentary Rocks								
Carbonate rocks		0.0X ppm					Turekian and Wedepohl (1961) cited.	13
Limestone, dolomite, Tennessee bauxite, pyritic slate, anthracite (Penn.)		N.D.			Spectrog.	~ 0.05		12
Limestones		1.7 ppm					"Generalized from Bowen" by Lisk	16
Cap rock	Hockley Mine Salt Dome, Texas	N.D.			Spectrog.	~ 0.05		12
Albian coquina	Southern Tadzhik deposit, Central Asia	10 n ppm (given as 0.00 n%)			Spectrog.		Contained coarse cryst. galena contg. 100 n ppm Tl. Values appearing in this table as an order of magnitude times n (an integer from 1 to 9) have been converted from the original given as a percent. Thus, 100 n ppm was given as 0.0 n%.	17

Source	Location	Thallium Concentration Mean	Range	No. Samples	Method	Sensitivity (ppm)	Remarks	Reference
Shungite, phosphorite-glauconite-sandstone, laterite, greywacke, loess, varved clay, clay, shale, carbonaceous schist			0.11-0.64 ppm		Spectrog.			12
Phyllite, slate, porcellanite			0.79-0.84 ppm		Spectrog.			12
Sedimentary rocks			0.1-3.0 ppm				Several papers tabulated.	11
Arkose	Fair Haven, New Haven, Conn.	1.0 ppm			Spectrog.			12
Sandstones, average		1.0 ppm					Greywacke excluded.	11
Sandstone	Finland	1.3 ppm			Spectrog.			12
Sandstones		0.82 ppm					Turekian and Wedepohl (1961) cited.	13
Argillite	Dickinson County, Mich.	1.4 ppm			Spectrog.			12
Anthraxolite	Sudbury, Ont.	2.8 ppm			Spectrog.			12
Shale (dictyonema)	Tallinn Estonia	3.1 ppm			Spectrog.			12
Shales		1.4 ppm					Turekian and Wedepohl (1961) cited.	13
Siltstone	Annie Creek Mine, Lead, S.D.	105 ppm			a.a.		USGS analysis.	5
Sedimentary rocks (clays and shales)		2 ppm 1 ppm					In sedimentary rocks of high oxidn. potential, Tl is probably present as Tl(III). (Vinogradov).	18 19

APPENDIX (Continued)

Source	Location	Thallium Concentration Mean	Range	No. of Samples	Method	Sensitivity (ppm)	Remarks	References
Sedimentary rocks of petroleum fields	USSR	2.4 ppm	0.0-6 ppm	95	Photo-metric			
Shales, average		0.9 ppm					Literature review	11
Clays			0.3-2.9 ppm	14/34	Spectro-chem. Spectral.		Most clays are from South Africa	21
Clays	Zirabulak Mtns., Central Asia		440-470 ppm				Clays contained very fine inclusions of plant fragments.	17
Pierre shale	Colorado	1.6 ppm			a.a			22
Pierre shale	South Dakota	25 ppm			a.a			22
Pierre shale (cretaceous)	South Dakota		0.3-25.8 ppm	89	a.a		USGS anal.	5
Sedimentary Minerals and Ores								
Sedimentary iron ores; siderite concretions and pseudo-oolitic ironstones from lower Jurassic sandstones, silt-stones, and claystones	Northern Caucasus	< 0.1 ppm			Colorimetric		When Tl is present in soln., hydrous iron oxides are always enriched in Tl. Iron oxides formed in the oxidized zone of sulfide deposits are also enriched in Tl. Presumably, little or no Tl was present when the sedimentary Fe oxides formed.	23
Coal (Jurassic Age)	Tadzhikistan, USSR		100 n to 1,000 n ppm		Spectrog.			24
Coal	Japan, Sakalin, and Manchuria	N.D.		70			Method not in abstr. 1949 Reference.	25

Source	Location	Thallium Concentration Mean	Range	No. of Samples	Method	Sensitivity (ppm)	Remarks	Reference
Coal (Jurassic Age)	Tadzhikistan, USSR	100 n to 1,000 n ppm			Spectrog.			24
Coal	Japan, Sakalin, and Manchuria	N.D.		70			Method not in abstr. 1949 Reference.	25
Coal	Zirabulak Mtns., Central Asia	370 ppm			Spectral		Coals interbedded with clays contg. 440 to 470 ppm Tl.	17
Brown coal	Sandy-carbonaceous sediments of the Middle Eocene		10 n to 100 n ppm		Not in abstr.		The anomalously high concns. suggest epigenetic accumulation by acid formation waters. High Tl concns. in accompanying iron disulfides.	26
Coal with sulfides	Middle Asia Basin	135 ppm		1	Quant. Spectrog. (two-arc technique)		Voskresenskaya (1968) established that pyrite and other sulfide inclusions in coals contain the thallium, not the organic coal mass. (Univalent thallium cannot be firmly held in the organic mass of the coal because it forms soluble humate and fulvate compounds.) The organic coal mass, however, must have served as the original concentrator of Tl.	23,27
Coals	Several Asian Basins		0-1.7 ppm		Colorimetric			23 27
Coal with thin coatings of pyrite		4.0 ppm		1				23 27
Western coals burned at power plants in the Southwest (1.76-29.60% ash)		< 0.25 ppm avg. of 19: 0.38 ppm	< 0.20 to 1.40 ppm	19/64	a.a.	0.20	The highest Tl value was found in a 3.85% ash coal (Late Cretaceous) from the Sorenson Mine, Wyoming.	28
Coal ashes of above		< 0.76 ppm avg. of 14: 0.83 ppm	< 0.20 to 1.70 ppm	14/16		0.20	Coal ash from above mine contained 1.10 and 1.20 ppm Tl. The average of the Tl content in these coals was < 0.2 ppm.	28
Anthracite coal ash	Abandoned Glen Alden power plant on Susquehanna R.	N.D.			a.a.	0.02	Calspan Report	29
Anthracite coal ash	Nesquehoning Creek (closed power plant)	N.D.						

APPENDIX (continued)

Source	Location	Thallium Concentration Mean	Thallium Concentration Range	No. of Samples	Method	Sensitivity (ppm)	Remarks	Reference
Bituminous coal	East Corning, N.Y., power plant		1.9 ± 0.9 ppm					29
Surface of fly ash of midwestern U.S. bituminous coals		28 ppm			Mass spec.		Tl concns. within the particle were much lower. A particle with aerodynamic diam. 1 μm may contain in the surface layer as much as 80% of the trace elemental mass. One plant was of small capacity using a chain grate stoker. The large modern plant used pulverized coal.	30
							Natusch et al. (1975) found that Tl (as well as As, Ca, Cd, Ga, K, Na, P, Pb, Rb, S, and Se) were preferentially associated with α-quartz and mullite ($3Al_2O_3 \cdot 2SiO_2$) rather than with the major matrix component, aluminosilicate glass.	31
Coal	Burned at Allen Steam Plant	<2 ppm			Spark source mass spec. (SSMS)	Uncertainty ± 50%	Mass flow <2.5 g/min.	32,33
Slag tank solids		2 ppm					Mass flow 0.2 g/min.	
Precipitator inlet		40 ppm 100 ppm					Mass flow 1.9 g/min. Mass flow 6.8 g/min.	
Precipitator outlet		30 ppm 30 ppm					Mass flow 0.059 g/min. Mass flow 0.056 g/min.	
							Schwitzgebel et al. (1975) also found by SSMS that Tl seems to be enriched in flue gas (gas plus fly ash).	34
Fly ash	Colorado power plant scrubber	0.6 ppm			a.a.		USGS anal.	5
Fly ash from southern Indiana coal, retained in coal-fired power plant	Illinois?	Unfractionated ~ 7.5 ppm			SSMS			35, 36
> 74 μm diam. (66.3% of the mass)		7 ppm						

Source	Location	Thallium Concentration Mean	Range	No. of Samples	Method	Sensitivity (ppm)	Remarks	Reference
44-74 µm (22.89%)		9 ppm						
20 - > 40 µm (9.29%)			5-9 ppm					
10-20 µm (1.11%)			12-15 ppm					
5-10 µm (0.33%)		20 ppm						
< 5 µm (0.08%)		45 ppm						
Airborne fly ash	Illinois?				SSMS		Avg. particle size for fly ash at the Allen Steam Plant is ~ 3 µm.[1231]	35,36
> 11.3 µm		29 ppm						
7.3-11.3 µm		40 ppm						
4.7-7.3 µm		62 ppm						
3.3-4.7 µm		67 ppm						
2.1-3.3 µm		65 ppm						
1.1-2.1 µm		76 ppm						
0.65-1.1 µm		-						
Coal ash, chiefly bituminous	U.S.		N.D. to 200 ppm	~ 900	Spectro-chem.	5		37
Coal	Illinois Basin	0.66 ppm	0.12-1.3 ppm	25	Opt. emis. spec.		Mostly face channel samples, 2-3/mine. Std. deviation 0.31. Possible bias on low side. Analyses by Ill. State Geol. Survey.	38
NBS Std. Coal Sample 1632		0.59 ± 0.03 ppm 0.37 ppm					Second value by method and analysts for samples above.	38
Herrin (No. 6) coal	Southwestern Ill.		1.3-1.9 ppm	4			One or two other coals of uncertain history ran up to 2.4 ppm Tl. Gluskoter states that "most and maybe all of the Tl in some samples are assocd. with the iron sulfide minerals in the coal."	38
Float coal dust	Pittsburgh Seam Mine	< 1 ppm			SSMS			39
Float coal ash	Pittsburgh Seam Mine	0.5 ppm						39
Upper Freeport coal fractions, < 1.0 µ to < 74 µ in diam.		< 0.5 ppm						39

APPENDIX (Continued)

Source	Location	Thallium Concentration Mean	Range	No. of Samples	Method	Sensitivity (ppm)	Remarks	Reference
Petroleum	Binagady fields, USSR	0.4 ppb 0.5 ppb		2	Colorimetric		Alk. water and hard water areas, resp. 0.39 to 1.17 ppm Tl in surrounding rocks.	40
Petroleum	Binagady fields, USSR	2.16 ppm (ash)					Katchenkov (1948)T617/ found Tl frequently in ashes of petroleums of all origins and ages by spectrographic analysis.	40
Igneous Rocks								
Basalt		0.10 ppm					Brooks and Ahrens (1961) cited.	41
Olivine rock, anorthosite, pyroxenite, horblendite, peridotite, olivine-basalt, basalt, leucite-basalt, tephrite, gabbro, hornblende-gabbro, olivine-diabase, andesite, diorite, tonalite, dacite, oligoclase-granite, lamprophyre			N.D. to 0.51 ppm		Spectrog.			12
Basalt, U.S. Geol. Survey Std. Rock BCR-1		0.59 ppm			n.a.		Wahler (1968) found 0.36 ppm.	7
Andesite, U.S. Geol. Survey, AGV-1		0.39 ppm			n.a.		Wahler found 0.27 ppm; Champ (1969) found 1.6 ppm.	7
Syenite	Marquis Township, Ontario	1.1 ppm			Spectrog.			12
Syenite	Marquis and Otto Townships, Ont.	1.3 ppm			Spectrog.			12
Syenites		1.4 ppm					Turekian and Wedepohl (1961) cited.	13
Ijolite	Finland	1.4 ppm			Spectrog.			12
Basalt	Keewanaw County, Michigan		< 0.2-0.7 ppm	13	a.a.		USGS anal.	5

Source	Location	Thallium Concentration Mean	Range	No. of Samples	Method	Sensitivity (ppm)	Remarks	Reference
Obsidian	Lake County, Oregon	1.4 ppm			Spectrog.			12
Obsidian	Millard County, Utah	5.6 ppm			Spectrog.			12
Nepheline-syenite	Methuen Township, Ontario	2.2 ppm						12
Arfvedsonite lujavrite	Ilimaussag intrusion, S.W. Greenland	6,500 ppm		1	Spectrog.		Lujavrite is a variety of nepheline syenite.	42
Augite syenite	Ilimaussag intrusion S.W. Greenland		4.8-7.0 ppm		Spectrog.			42
Agpaitic rocks	Ilimaussag intrusion, S.W. Greenland		33-340 ppm		Spectrog.			42
Granites		3.1 ppm	1.5-6.4 ppm		Spectrog.			12
Granite, U.S. Geol. Survey Std. Rock G-2		1.02 ppm			n.a.		Wahler (1968) found 0.85 ppm; Champ (1969) found 1.3 ppm.	7
Granites	Germany		0.3-3 ppm				Three papers cited, 1937-1941.	43
Granite		0.75 ppm					Brooks and Ahrens (1961) cited.	41
Granite, high-calcium		0.72 ppm					Turekian and Wedepohl (1961) cited.	13
Granite, low-calcium		2.3 ppm					Turekian and Wedepohl (1961).	13
Lepidolites and muscovites from granites and granite pegmatites		max. 221 g/ton and 58 g/ton, resp.					Ahrens (1945) cited.	43

APPENDIX (Continued)

Source	Location	Thallium Concentration Mean	Range	No. of Samples	Method	Sensitivity (ppm)	Remarks	Reference
Granite	Ilimaussag intrusion, S.W. Greenland		50-250 ppm		Spectrog.			42
Igneous rocks		1.3 ppm					Rankama (1954) cited.	13
Igneous rocks		0.3 ppm					Goldschmidt (1937) cited.	13
Igneous rocks		0.000X ppm					Clarke and Washington (1924) cited.	13
Acid (felsic) rocks: granites, liparites, rhyolites, etc.		2.5 ppm 1.5 ppm					Vinogradov's generalizations.	18,19
Basic (mafic) rocks: basalts, gabbros, norites, diabases, etc.		0.2 ppm					In crystalline rocks, the Tl content is related to that of Rb ($Rb_2O/Tl_2O \cong 100$).	18
Ultrabasic rocks: dunites, peridotites, pyroxenites.		0.06 ppm 0.01 ppm					Vinogradov (1959); Turekian and Wedepohl (1961); Vinogradov (1962).	13,18,19
Dunite, U.S. Geol. Survey Rock DTS-1		0.13 ppm			n.a.		Wahler (1968) found < 0.05 ppm.	7
Peridotite, U.S. Geol. Survey Rock PCC-1		0.35 ppm			n.a.		Wahler (1968) found < 0.05 ppm.	7
Intermediate rocks: diorites, andesites, etc.		0.15 ppm 0.5 ppm						18,19
Pegmatite	Finland	33.0 ppm			Spectrog.		Source of feldspar. Lepidolite (Li mica) is confined to granite pegmatites, whose source magma was rich in Li.	12
Ultramafic (ultrabasic) rocks			Mostly 0.07-0.30 ppm				The mineralogical compn. of the rock does not influence the Tl content. Wedepohl's generalizations.	11
Granodiorite, U.S. Geol. Survey Std. Rock GSP-1		1.87 ppm			n.a.		Wahler (1968) found 0.71 ppm; Champ (1969) found 1.6 ppm.	7

Source	Location	Thallium Concentration Mean	Range	No. of Samples	Method	Sensitivity (ppm)	Remarks	Reference
Mafic (basic) rocks			Mostly 0.05-0.7 ppm					11
Intermediate rocks			Mostly 0.15-1.4 ppm				Higher values are found in more acid types (syenite, monzonite, dacite).	11
Granitic-rhyolitic rocks:								
Granitic rocks			Mostly 0.6-3.5 ppm					11
Volcanic and hypabyssal rocks:								
Rhyolites and obsidians			Mostly 1.0-1.6 ppm					11
Rhyolite tuff	Near a beryllium deposit, Spor Mt., Utah		0.2-190 ppm	100			USGS anal.	5
Obsidian	Millard County, Utah	5.6 ppm					Shaw (1952)[12] cited.	11
Granodiorite, granite, granosyenite, quartz-syenite, and quartz porphyries		Range of means: 0.8-5.3 ppm						11
Quartz albitophyres		Range of means: 0.8-1.2 ppm						11
Calcium-alkali granites	9 from Malaya and 7 from Cornwall, Eng.		2.7-9.1 ppm	16			Method not given in the abstr.	44
Alkali granites	Northern Nigeria		0.7-5.0 ppm	47				44

APPENDIX (Continued)

Source	Location	Thallium Concentration Mean	Range	No. of Samples	Method	Sensitivity (ppm)	Remarks	Reference
Minerals from granitic rocks	Various		0.1-48 ppm				Numerous papers before 1972 reviewed. Highest value for zinnwaldite, lowest for quartz. Wedepohl summarizes that Tl is concd. in biotite. "Relatively high" amts. appear in K feldspar and amphibole.	11
Rocks (gold-producing area)	Cripple Creek, Colo.		0.3-11.1 ppm	110	a.a.		USGS anal. Rocks unidentified other than source. Almost all gold deposits occur near acidic igneous intrusions.	5
Rocks	Ely, White Pine County, Nev.	~ 10 ppm	< 0.2-1,075 ppm	1,089	a.a.		USGS anal. (G. L. Crenshaw and A. B. Hubert). Unidentified raw data. Ely is in Cu-producing area (Kennecott).	5
Perthite pegmatites	Glendale area Black Hills, S.D.	39 ppm	< 30-150 ppm	15	a.a.	30	Tl concn. in country rock < 30 ppm. Rb/Tl ratio (avg.) 60.	45
Rocks?	Badger Flats, Park County, Colo.		0.8-9.4 ppm	5	a.a.		USGS anal. by Hubert and Crenshaw. Park County minerals: peat, sand and gravel, stone, and Au.	5
Rocks?	Sierra County, N.M.		< 0.2-7.4 ppm	18	a.a.		USGS anal. by Hubert and Crenshaw. Sierra County com. minerals: sand and gravel.	5
Rocks?	Lake George and Tappan Mtn. Range, Colo.		< 0.2-13.4 ppm	72	a.a.		USGS anal. by A. E. Hubert and W. H. Ficklin.	5
Metamorphic Rocks								
Gneisses (norite, diorite, quartz diorite, quartz-monzonite, granodiorite, hornblende, enderbite, garnet, and granulite)			trace to 1.1 ppm		Spectrog.		Gneisses are silicate rocks.	12
Gneiss			<0.1-2.4 ppm	4	a.a.			46
Olive slate	Dutchess County, N.Y.	0.60 ppm		1			Balk and Barth (1936) cited.	
Black phyllite		0.27 ppm		several				
Black phyllite		0.44 ppm						
Black phyllitic slate		0.56 ppm						
Biotite phyllite		0.66 ppm						
Garnet schist		0.61 ppm						
Garnetiferous schist		1.1 ppm						

APPENDIX (Continued)

Source	Location	Mean	Range	Samples	Method	Sensitivity (ppm)	Remarks	Reference
Schists (garnetiferous, biotite-garnet, biotite-hornblende, garnet, sillimanite)			0.35-1.9 ppm		Spectrog.			12
Phyllites (black, biotite, biotite-muscovite, black phyllitic slate)			0.27-0.66 ppm		Spectrog.			12
Amphibolite	K.F.S.S.R.	0.40 ppm			Spectrog.			12
Amphibolite	Sayan Mtns., USSR		2-3 ppm		Chem.?		Slepnev (1961) cited.	12
Olive slate	Clove Quadr., N.Y.	0.60 ppm			Spectrog.			12
Schists in contact with copper ores	Copper-pyrrhotite ore belt, southern slopes of Caucasus Mtns., USSR	29.2 ppm					Tl content gradually decreased with increased distance from the ore. Method not in the abstr.	12
Alkalic Rocks								
Népheline syenites			≤ 2.7 ppm				These minerals are exceptions. Tl content is generally low.	12
Phonolites			≤ 7.2 ppm					
Leucitites			≤ 5.2 ppm					
Native Elements								
Graphite	Finland	0.49 ppm			Spectrog.			12
Graphite	Ontario	N.D.			Spectrog.	~ 0.05		12

APPENDIX (Continued)

Source	Location	Thallium Concentration Mean	Range	No. of Samples	Method	Sensitivity (ppm)	Remarks	Reference
Oxides and Hydrated Oxides								
Quartz, opal, magnetite, hematite, limonite, iron oxides, and pyrolusite			N.D. to ~ 1.0 ppm		Spectrog.			12
Silicates								
Neosilicates including olivine rock, almandite, garnet, epidote, staurolite, zircon, and titanite			N.D. to 0.36 ppm		Spectrog.			12
Inosilicates including hypersthene, diopside, augite, tremolite, and hornblende			N.D. to 0.65 ppm		Spectrog.			12
Phyllosilicates including Chlorite and kaolin			N.D. to $<$ 0.027 ppm		Spectrog.			12
Micas								12
Biotite	Not known	11.0 ppm			Spectrog.		Biotite is "black mica."	
Muscovite	Finland	16.0 ppm			Spectrog.		Muscovite is "white mica," $H_2KAl_3(SiO_4)_3$	
Muscovite	Custer Mt., Black Hills, S.D.	47.0 ppm			Spectrog.			12
Phlogopite	Finland	78.0 ppm			Spectrog.		Magnesium mica with decreasing amts. of Fe; biotite grades into phlogopite.	
Lepidolite	Custer Mt., S.D.	95.0 ppm			Spectrog.		Lithium mica.	
Tektosilicates including Quartz, calcic plagioclase, bytownite			N.D. to 0.029 ppm		Spectrog.			12
Albite	High Climb, Black Hills, S.D.	2.0 ppm			Spectrog.		$Na_2O \cdot Al_2O_3 \cdot 6SiO_2$, a principal feldspar species.	12

Source	Location	Thallium Concentration Mean	Thallium Concentration Range	No. of Samples	Method	Sensitivity (ppm)	Remarks	Reference
Nepheline	Not known	7.6 ppm			Spectrog.			12
Cleavelandite	Custer Mt., S.D.	6.8 ppm			Spectrog.			12
Microcline-perthite	Custer Mt., S.D.	84.0 ppm			Spectrog.		Microcline is a principal feldspar species. Most com. feldspar is perthite.	12
Microcline-perthite	High Climb, Black Hills, S.D.	21.0 ppm			Spectrog.			12
Amazonite	Ontario	33.0 ppm			Spectrog.			12
Amazonite		Nearly 100 g/ton (~ 110 ppm)			Spectrog.		Goldschmidt and Hörmann (1937) cited.	43
Leucite	Italy	65.0 ppm			Spectrog.		$KAl(SiO_3)_2$	43
Leucite		≤ 20 g/ton (~ 22 ppm)					Goldschmidt and Hörmann (1937) cited.	43
Cyclosilicates								
Beryl and tourmaline			< 0.031-0.40 ppm		Spectrog.		Ahrens found 38 ppm in a Madagascar beryl enriched in Cs.	12
Silicate rocks								
Mangerite, a variety of monzonite			< 0.1-0.19 ppm	4	a.a.			46
Granulite (banded)			< 0.1-0.22 ppm	3	a.a.			46
Anorthosite		< 0.1 ppm		1	a.a.	10^{-9} g in 1-20 mg samples	One of the 3 principal feldspar species. Limited com. importance.	46
Plagioclase feldspars		≤ 11 ppm			Spectrog.		Limited com. importance	43

APPENDIX (Continued)

Source	Location	Thallium Concentration Mean	Range	No. of Samples	Method	Sensitivity (ppm)	Remarks	Reference
Aluminosilicate veins assocd. with sulfide ores	Myl'nikov-Khorkirin deposit, Eastern Transbaikal, USSR		10 n to 100 n ppm					48
Jasperoid	Drum Mtns., Utah		4.1-5.9 ppm		a.a.		Various digestion procedures on same sample.	22
Jasperoid	Taylor District, Nev.		70-100 ppm	2	Emission spectrog.		USGS anal. Error ± 35% in this method.	5
Jasperoid	Dump of Franklin Lease Mine, Mercur District, Utah	150 ppm		1	Emission spectrog.		USGS anal. Error ± 35% in this method.	5
Jasperoid	Upper Terrace, Egan Range	100 ppm		1	Emission spectrog.		USGS anal. Error ± 35% in this method.	5
Jasperoid	Annie Creek Mine, W. Lead, S.D.	1,500 ppm		1	Emission spectrog.		USGS anal. Error ± 35% in this method.	5
Pegmatite minerals: Albite-microcline pegmatite:	Mongolian Altai							49
Microcline			10-61 ppm		Colorimetric	2 ppm		
Muscovite			15-21 ppm	2				
Gilbertite			22-28 ppm	2				
Lepidolite		72 ppm		1				
Beryl			4-7 ppm	4				
Albite		4 ppm		1				
Spodumene		4 ppm		1				
Biotites in general from granitic and alkalic rocks			3-15 ppm					11

Source	Location	Thallium Concentration Mean	Range	No. of Samples	Method	Sensitivity (ppm)	Remarks	Reference
Most K feldspars and possibly muscovites from granitic and alkalic rocks			1-6 ppm					11
K feldspars from granite pegmatites			Most 0.5-50 ppm				Occasionally ≤ 150 ppm found. Strongly differentiated pegmatites may contain up to 600 ppm.	11
Muscovites from granite pegmatites			≤ 280 ppm generally 2-60 ppm				Lepidolite, lithium-rich biotite, and pollucite of granite pegmatites also contain "appreciable amounts" of Tl.	11
Syenite pegmatites:								
K feldspars	Palabora pegmatites, E. Transvaal		0.4-1.0 ppm				These values are much lower than values from granitic pegmatites.	11
Phlogopite	Palabora pegmatites, E. Transvaal		< 1-1.9 ppm				These values are much lower than values from granitic pegmatites.	11
Minerals from alkalic pegmatites:	Lovozero Massif, USSR							11
K-Na feldspar		2.78 ppm						
Nepheline		1.3 ppm						
Aegirine		0.44 ppm						
Microcline pegmatite	Harding Mine, Dixon, N.M.	48 ppm			Spectrochem.		Ahrens (1948).	50
Muscovite	Harding Mine, Dixon, N.M.	145 ppm			Spectrochem.		Ahrens (1948).	50
Lepidolite	Harding Mine, Dixon, N.M.		105-210 ppm		Spectrochem.		Ahrens (1948).	50
Amazonite	Colorado		6-50 ppm		Spectrochem.		Ahrens (1948).	50
Lepidolite	Connecticut		70-95 ppm		Spectrochem.		Ahrens (1948).	50
Biotite	Strickland Quarry, Connecticut	80 ppm					Connecticut is one of the principal producers of mica.	11

APPENDIX (Continued)

Source	Location	Thallium Concentration Mean	Range	No. of Samples	Method	Sensitivity (ppm)	Remarks	Reference
Lepidolite	Pala, Calif.	185 ppm	175-200 ppm		Spectrochem.		Ahrens (1948).	50
Lepidolite	Maine	155 ppm			Spectrochem.		Ahrens (1948). Maine is a commercial producer of mica.	50
Lepidolite	Brown Derby, Colo.	250 ppm	230-270 ppm		Spectrochem.		Ahrens (1948). Colorado is a commercial producer of mica.	50
Lepidolite	Copper Mt., Wyoming	95 ppm			Spectrochem.		Ahrens (1948). Wyoming produces mica commercially.	50
Biotite	King's Mt., N. Carolina	385 ppm			Spectrochem.		Ahrens (1948). North Carolina is a principal mica producer.	50
K feldspars from metamorphic rocks (gneisses and granulites)	Norway	Range of means: 0.7-8.2 ppm	trace to 12 ppm				Heier (1960) and Heier and Taylor (1959) cited.	11
Potash feldspars (from pegmatites, chiefly microcline and perthite)	S. Africa		0.4-21 ppm Tl_2O	55	Spectrochem.		Ahrens (1948).	50
Pollucite ($CsAlSi_2O_6 \cdot H_2O$)	Tin Mt., Black Hills, S.D.	19 ppm Tl_2O			Spectrochem.		Ahrens (1948).	50
Pollucite	Greenwood, Maine	19 ppm			Spectrochem.		Ahrens (1948).	50
Pollucite			$\leq$ 106 ppm				Works of Ahrens cited.	12
Pollucite		100 ppm					Novokhatskii and Kalinin (1947) cited.	12
Biotite	Nevada		0.4-1.0 ppm	2	a.a.		USGS anal.	5
Muscovite	Nevada	0.7 ppm			a.a.		USGS anal.	5
Asbestos, chrysotile	Nine Canadian locations		0-71 ppm					51

APPENDIX (Continued)

Source	Location	Thallium Concentration Mean	Range	No. of Samples	Method	Sensitivity (ppm)	Remarks	Reference
Phosphates								
20% Superphosphate		0.2 ppm					Lisk generalizes data from several sources, mostly Schroeder and coworkers.	16
Natural Ca phosphates	In Nassau, Germany, sample	Detected		1/16	Spectrog. after chem. separation		Data of Jolibois and Hebert (1946)	5
Apatite	Custer Mt., S.D.	0.23 ppm			Spectrog.		Shaw (1952).	12
Sulfates, Carbonates, etc.								
Celestite-barite vein cutting limestones of the Bukhara series	Southern Tadzhik deposits, Central Asia		100 n to 7,000 ppm		Spectral			17
Thallium-jarosite	Daraiso Pb-Zn deposit, USSR		1.75-2.04%					11
Jarosite	USSR	200			Electro-dialysis			53
Jarosites in a Hg-Sb deposit	Dzhizhikrut deposit, USSR		≤ 5,500		Spectrog.		A main mineral of the yellow ochres.	54
Jarosites	Central Asia	1,344 g/ton	150-5,500 g/ton					55
Dolomite	Central Asia	3 g/ton		1/1				55
Black calcite	Western U.S. mining districts	350 ppm	0-2,000 ppm	20	Spectrog.		Black calcite is abundant and widespread in argentiferous manganese oxides in carbonate host rocks of the Lake Valley district, N.M., Tombstone district, Ariz.; White Pine (Hamilton) district, White Pine County, Nev.; Pioche district, Lincoln County, Nev.; Tybo district, Nye County, Nev.; and the Ophir district, Utah. Small tonnages of manganese ores from these districts were used to recover silver or make ferroalloys (1908-1934).	56

Source	Location	Thallium Concentration Mean	Range	No. of Samples	Method	Sensitivity (ppm)	Remarks	Reference
Plumbojarosites $PbFe_6(OH)_{12}(SO_4)_4$	Asia		3.7-44 ppm				Two papers cited.	11
Jarosite, (hydrous Fe K sulfate)	Lachin-Khana polymetallic deposit, USSR	19 ppm	1-40 ppm					11
Melanterite	Daraiso Pb-Zn deposit, USSR	100 ppm						11
Colorless cerussite, $PbCO_3$	Polymetallic deposits, E. Transbaikaliya, USSR		N.D. to trace					11
Black cerussite	Pb-Zn and polymetallic deposits, Asia		0-40 ppm					11
Manganese Oxide Minerals								
Psilomelane	Pinal County, Ariz.	70 or < 100 ppm		2	Colorimetric or X-ray fluorescence		All deposits from which Tl samples came believed to be hypogene in origin. Psilomelane is $H_4R_2Mn_8O_{20}$, where R = K, Ba, and Na oxides.	57
Psilomelane	Gila County, Ariz.	170 or 200 ppm						57
Hollandite-cryptomelane	New Mexico and Arizona		700-3,700 ppm	4			Cryptomelane is a potash-bearing oxide of variable composition.	57
Mn ores	Apache Mine, Gila County, Ariz.	Few thousand ppm					Herman (1943) cited.	57
Mn ores contg. psilomelane, braunite, and hollandite	Dzhezda deposit, Central Kazakhstan	800 and 1,100 ppm	mostly 100 to 300 ppm	2	D.T.A.		Voskresenskaya and Usevich (1957);58/ Maksimov (1960); and Voskresenskaya and Soboleva (1961)59/ cited.	57
Psilomelane	Kazakhstan, USSR	600 ppm						53

Source	Location	Thallium Concentration Mean	Range	No. of Samples	Method	Sensitivity (ppm)	Remarks	Reference
Braunite-psilomelane ore	Kazakhstan, USSR	500 ppm			Electro-dialysis			53
Mn ores	United States	1,900 ppm	200-3,400 ppm	5	X-ray fluorescent spectrog.		By comparison, chem. anal. by F. Cuttitta (1961) ranged from 170 to 3,800 ppm (average, 1,100 ppm).	60
Mn Oxides								
Rocks (including Mn ore samples)	Aguila District, Ariz.		< 0.2-230 ppm	51	a.a.		USGS anal. by G. [illegible]. Crenshaw and A. E. Hubert. Some samples were assocd. with Au; Tl values of these were usually low. Mn ore mining and sintering has occurred on a small scale in this area.61/	5
Rocks (including Mn ore samples)	Socorro County, N.M.		< 0.2-3,415 ppm	116	a.a.		USGS anal. by Crenshaw and Hubert. Eight samples had values ≥ 1,000 ppm; 19 samples, 500-999 ppm; 32 samples, 100-499 ppm. Thus, more than 50% of the samples have ≥ 100 ppm Tl.	5
Supergene Mn oxides	Arizona	7,000 ppm		1/6			In 1972 the production of U.S. Mn ores by state was (in tons): Minnesota, 119,000; New Mexico, 28,000; Montana, 600; and Utah, < 100.	9,10
	Arizona	2,300 ppm						
	Montana	1,750 ppm	0-4,000 ppm	4				
	New Mexico	275 ppm	0-1,500 ppm	2/6				
Hypogene vein Mn oxides	Arizona	1,100 ppm	0-7,000 ppm	6/8				9,10
	California	300 ppm	0-1,500 ppm	4/6				
	Nevada	1,000 ppm		3/3			1,500 ppm in two samples, detected in one.	
	Oregon, Texas	detected		2				
	New Mexico	1,300 ppm	0-7,000 ppm					
Hot spring apron Mn oxides	Ariz., Calif., Colo, Idaho, Nev., Utah	< 300 ppm	0-3,000 ppm	21				9,10
Stratified oxides	Ariz., Calif., Idaho, Nev.	120 ppm	0-300 ppm	9				9,10

APPENDIX (Continued)

Source	Location	Thallium Concentration Mean	Range	No. of Samples	Method	Sensitivity (ppm)	Remarks	Reference
Manganese Ores	Mysore, India		140-190 ppm	55	Emission Spectroscopy (performed in the U.S.)			62
Manganese Ores	Nagpur and Vizagpatam, India		10-100 ppm					
Hypogene Manganes Oxides	N.M., California	600 ppm	0-2,000 ppm	7	Spectrog.		The reference gives a map showing the sites of hypogene Mn oxides in the S.W. U.S., and Northern Mexico	63
Manganese Oxides	Canadian Maritime Provinces	210 ppm	50-1,500 ppm	10	Spectrog.			63
Hollandite	Luis Lopez district, N.M.	1,780 ppm	300-5,000 ppm	10				63

Source	Location	Thallium Concentration Mean	Range	No. of Samples	Method	Sensitivity (ppm)	Remarks	Reference
Iron Oxides								
Goethite, $Fe_2O_3 \cdot H_2O$	Daraiso Pb-Zn deposit, USSR	100 ppm					Mogarovskii (1961) cited.	11
Hydrogoethite, $3Fe_2O_3 \cdot 4H_2O$; geothite	Sarykan deposit, Middle Asia, USSR		5-14 ppm				Kulikova (1961) cited.	11
Hydrogoethite	Kumyshkan deposit, Middle Asia, USSR		2-4 ppm				Kulikova (1961) cited.	11
Hydrous iron oxides	Kurghashinkan deposit, Middle Asia, USSR		3-85 ppm		?		Kulikova (1961) cited. Among zones of Pb-Zn deposits in carbonate rocks, skarns, and shales, the highest Tl content was found in Fe oxides in carbonate rocks. Its contents in Fe oxides in skarns and shales were almost equal.	11 64
Limonite, $2Fe_2O_3 \cdot 3H_2O$	Polymetallic deposits, E. Transbaikaliya, USSR	6 ppm					Kulikova (1962) cited.	11
Mn-rich limonite	Polymetallic deposits, E. Transbaikaliya, USSR		10-50 ppm				Kulikova (1962) cited.	11
Iron oxides	Lachin Khana polymetallic deposit, USSR	?					The Tl contents of the Mn hydroxides of the same deposit were 40-95 ppm (mean 68 ppm). Dunin-Barkovskaya (1961) cited.	11
Magnetites	Alaska		0.2-1 ppm	23	Wet chem.			5
Iron oxides, primarily ochers (50) and gossans (10), chiefly in vein and bog deposits	Montezuma District, central Colo.		0.2-110 ppm	65	Chem.		USGS anal. by C. J. Neuerburg (1971). High value was a gossan vein sample. Second highest value was 19.4 ppm for same type sample.	5

APPENDIX (Continued)

Source	Location	Thallium Concentration Mean	Range	No. of Samples	Method	Sensitivity (ppm)	Remarks	Reference
Uranium deposit samples	Shirley Basin, Wyoming		0.3-1.2 ppm	60	a.a.		USGS anal.	5
Meteorites								
Stony meteorites		0.001 ppm					Vinogradov.	19
Iron meteorites							Reference reviews several papers.	11
Metal		1.35 ppm		5				
Troilite		113 ppm		4				
Achondrites		0.75 ppm		2				11
Chondrites			0.03-141 ppm				Reference reviews several papers.	11
Chondrites and achondrites		N.D.		3	Spectrog.	~ 0.05		12
Lunar Samples	Collected by Apollo 11 astronauts	0.63 ppb 2.44 ppb		2				11
Crust								
Continental crust		0.45 ppm					Taylor (1964).	13 41
Continental crust		0.55 ppm					Taylor and White (1967) cited.	11
Earth's crust		0.7 ppm					Brooks and Ahrens (1961) cited.	11
Earth's crust (see also igneous rocks)		1.0 ppm					Vinogradov (1962).	19 13
Earth's crust		1.7 ppm					Vinogradov (1959).	18

APPENDIX (Continued)

Source	Location	Thallium Concentration Mean	Range	No. of Samples	Method	Sensitivity (ppm)	Remarks	Reference
Earth's crust		1.0 ppm					Mason (1958).	13
Earth's crust		1.3 ppm					Shaw (1952).	12
Earth's crust		0.6 ppm					Rankama and Sahama (1950) cited.	11
Magmatic rocks in upper continental crust		0.830 ppm					Marowsky and Wedepohl (1971).	65
Earth's crust		0.8 ppm					Wedepohl (1972).	11
Earth's crust		3.0 ppm					Ahrens (1948) cited.	11
Earth's crust		0.3 ppm					Goldschmidt (1937) cited.	11
Earth's crust		0.1 ppm					Noddack and Noddack (1930) cited.	11
Earth's crust		0.00085 ppm					Clarke and Washington (1924) cited.	11

APPENDIX (Continued)

Source	Location	Thallium Concentration Mean	Range	No. of Samples	Method	Sensitivity (ppm)	Remarks	Reference
Sulfides								
Sphalerite	New Mexico: Blue Moon lode	20 ppm			Spectro-chemical		Numerous sphalerite samples were analyzed, but Tl was found in only these four samples. (There are only about 25 zinc deposits in the western U.S.) Tl was also sought in chalcopyrite but not found.	66
	Pawley mine, Uncompaghre district	40 ppm						66
	Mineral Hill mine	20 ppm					Tl was not found in two other samples from the same mine.	66
	Roberts district, Jameson, Fletcher, and Carletti property	100 ppm						66
Sphalerite			10-300 ppm		Spectroscopic flame source		Rusanov and associates cited.	12
Chalcopyrite	Finland	10 ppm				79	Joensuu (1948) cited.	43
Galena	Norway	< 100 ppm	< 10-1,000 ppm	39	Spectrog.	10	Oftedal (1940) scarcely ever found Tl in sphalerites. An upper limit for Tl in U.S. galena might be calcd. from Bi production. Bi almost always higher than Tl in Norwegian galena. Calc. 580 tons Bi in 1968 Pb processing.	67
Galena	Finland	0.27 ppm			Spectrog.			12
Lorandite, $TlAsS_2$	Barite veins within East ore body of the Carlin gold deposit, Nev.	59.5%		5	Chem. anal.		Mineral is commonly with realgar, orpiment, and an unknown Tl-Hg-As sulfide mineral.	68
Sphalerite	Low temp. deposits		0.01-1%	5/39	Spectrog.		Detected only in European low temperature Zn deposits of Mississippi Valley type. "...the absence of As and Tl is typical of Mississippi Valley sphalerite anal."	68

Source	Location	Thallium Concentration Mean	Thallium Concentration Range	No. of Samples	Method	Sensitivity (ppm)	Remarks	Reference
Copper ores	Central Kazakhstan	7 g/ton 9.4 g/ton						70
Cu concentrate	Central Kazakhstan	34 g/ton					Bulk of Tl found in tailing.	70
Polymetallic sulfide ores contg. chiefly the minerals sphalerite, pyrite, and arsenopyrite	USSR		N.D. to 55 g/ton		Spectral and Chem.		Tl usually secondary.	71
Sphalerite	USSR	Not given, presumably not present			Spectral and Chem.			71
Galena	USSR		N.D. to 71 g/ton	8	Spectral and Chem.			71
Pyrite	USSR		5-27 g/ton	3	Spectral and Chem.			71
Lead concentrate	USSR		10-210 g/ton	4	Spectral and Chem.		Up to 70-80% of the Tl and Ag in the ores appear in the Pb concentrate.	71
Zinc concentrate	USSR		N.D. to traces	3	Spectral and Chem.			71
Copper-pyrrhotite ore	Southern Slopes, Caucasus Mtns.		$\underline{n}$ to 100 $\underline{n}$ ppm					47
Pyrite-sphalerite ores	Southern Slopes, Caucasus Mtns.	10 $\underline{n}$ ppm					Higher concr. than in pyrite-pyrrhotite-chalcopyrite ores.	47
Bournonite	Southern Slopes Caucasus Mtns.	110 ppm						47
Galena	Southern Slopes Caucasus Mtns.	54.2 ppm						47
Sphalerite 1st generation 2nd generation	Southern Slopes Caucasus Mtns.	 3.9 ppm 6.9 ppm						47

APPENDIX (Continued)

Source	Location	Thallium Concentration Mean	Range	No. of Samples	Method	Sensitivity (ppm)	Remarks	Reference
Chalcopyrite	Southern Slopes Caucasus Mtns.	2.5 ppm						47
Pyrrhotite	Southern Slopes Caucasus Mtns.	1.5 ppm						47
Pyrite	Southern Slopes Caucasus Mtns.	1.4 ppm						47
Pyrite, 2nd and 3rd generations	Southern Slopes Caucasus Mtns.	5.3 ppm						47
Zinc, Pb, and Cu ores	Japan		≤ 100 and 800 ppm	125	Gravimetric ?		High values only in two marcasites.	72
Black ores	Hanoaka mine, Japan		≤ 120 ppm	31				72
Marcasites	Hanoaka mine, Japan		700-1,410 ppm	4				72
Pyrite	Hanoaka mine, Japan	110 ppm						72
Galena and sphalerite		traces						72
Colloform ore	Yunosawa Mine, Aomori Prefecture, Japan		≤ 4,800 ppm	12				72
Marcasite	Yunosawa Mine, Aomori Prefecture, Japan		500-2,000 ppm	4				72
Wurtzite	Yunosawa Mine, Aomori Prefecture, Japan	270 ppm		2				72
Pb-Zn deposits of pyrite, sphalerite, chalcopyrite, galena, and tetrahedrite	Kumyshkan Pb-Zn deposit, USSR	2 g/ton						73

Source	Location	Thallium Concentration Mean	Thallium Concentration Range	No. of Samples	Method	Sensitivity (ppm)	Remarks	Reference
Slightly crystallized varieties of iron disulfides	From unnamed deposit, USSR, also contg. cryst. and metacolloid varieties of chalcopyrite and sphalerite		≤ 72-83 ppm		Chem., photocolorimetric, or spectral		Tl probably adsorbed on surface of Pb, Zn, Cu, and Fe sulfides as they were deposited from colloidal solns.	74
Galenas	East Alps		≤ 100 g/ton	46	Spectrochem.		Tl is assocd. with As, Sb, and Bi.	75
Sphalerites	Southern Uzbekistan, USSR		-				These Pb-Zn deposits are generally poor in Ga and Tl, but the sphalerites of sedimentary ore deposits are rich in Ga.	76
"Ordinary lead-zinc ores"	USSR		1-10 ppm				Generalization	77
Copper pyrite deposits	Northern part of the Ural greenstone belt, USSR		3-50 ppm				Tl concentrates in the chalcopyrite-pyrite and sphalerite-pyrite.	78
Lead-zinc ores	Gorno, Italy	N.D.			Spectrog.		The presence of Tl had been previously reported.	79
Lead-zinc-copper ores	Tishinka Deposit, USSR		1-1.5 ppm				More Ga in sphalerite than in galena. Behavior of Tl probably same.	80
Sphalerite	Silesia-Cracow Zn and Pb deposits, Poland	430 ppm						81
Galena	Silesia-Cracow Zn and Pb deposits, Poland	36 ppm						81
Sphalerite concentrate	Gadore, Italy	95 ppm			Spectrog., polarog.			82
Sb-Hg deposit: Pyrites and marcasites	USSR		10 n to 1,000 n ppm					83
Colloform pyrite	USSR		2,000-2,600 ppm					83
Sb-Hg concentrate		200 ppm						83

APPENDIX (Continued)

Source	Location	Thallium Concentration Mean	Thallium Concentration Range	No. of Samples	Method	Sensitivity (ppm)	Remarks	Reference
A "thallium geochemical province":	Central Kyzylkum sulfide ores, Uzbek. SSR				?		Method not in abstr.	84
Pyrite			25-40 ppm					84
Marcasite			17-45 ppm				Goethite, FeO(OH), contained 375 ppm	84
								84
Country rock ferruginous schists			< 250 ppm					84
Galena	Zirabulak Mtns. Central Asia	100 n ppm			Spectral			17
Cinnabar	Sb-Hg deposits in the Gissar and Zaravshan Mtns., USSR	100 n ppm		48			Tl replaced Hg isomorphously. Somewhat higher Tl contents in the stibnite. Method not in abstr.	85
Pyrites	Hradiste near Kadan in Bohemia	140 ppm 170 ppm		2	Spectro-chem.			86
Marcasite	Hradiste near Kadan in Bohemia	330 ppm		1	Spectro-chem.			86
Mercury ores Pyrite	Nikitski Hg deposit, USSR		< 2.5-9 ppm		?		Tl is not reported in the cinnabar, stibnite, or kaolinite. Method not in abstr.	87
Polymetallic ores:	One polymetallic ore field in Uzbek. SSR							
Pyrite			3-16 g/ton		?		Method not in abstr.	88
Galena			1-8 g/ton					88

Source	Location	Thallium Concentration		No. of	Method	Sensitivity	Remarks	Reference
		Mean	Range	Samples		(ppm)		
Melnikovite-pyrite		87 g/ton						88
Melnikovite-marcasite		225 g/ton						88
Marcasite		78 g/ton						88
Sphalerite			2-14 g/ton					88
Iron sulfide	Pallieres Mine, Belgium		300-1,500 ppm					89
Pyrite finely dispersed in shales	Zirabulak Mtns., Central Asia	100 n ppm			Spectral.			17
Molybdenites (MoS_2)	Central Kazakhstan Mo-W deposits	< 1 g/ton traces		2 1				90
Wolframites (early)	Central Kazakhstan Mo-W deposits		< 1-161 g/ton					90
Wolframites (late)	Central Kazakhstan Mo-W deposits	3 g/ton		1				90
Cosalites (Pb bisulfide)	Central Kazakhstan Mo-W deposits		23-2,700 g/ton	4			Isomorphous substitution.	90
Bismuthines (native bismuth)	Central Kazakhstan Mo-W deposits		< 1-80 g/ton < 1-2 g/ton	5 2				90
Chalcopyrites	Central Kazakhstan Mo-W deposits		< 1-9 g/ton	6				90
Sphalerites	Central Kazakhstan Mo-W deposits		4.3-21 g/ton	4				90
Galenas	Central Kazakhstan Mo-W deposits		160-220 g/ton	5				90

APPENDIX (Continued)

Source	Location	Thallium Concentration Mean	Range	No. of Samples	Method	Sensitivity (ppm)	Remarks	Reference
Pyrites	Central Kazakhstan, Mo-W deposits		< 1-161 g/ton	31			Traces found in four other samples.	90
Pyrrhotite	Central Kazakhstan, Mo-W deposits		1.3-12.9 g/ton	2				90
Silicate-sulfide-iron deposits:	USSR							
Sphalerite		< 1 g/ton		2				91
Pyrrhotite		13 g/ton	10-20 g/ton	3				91
Franckeite ($Pb_5Sn_2Sb_2S_{12}$)		20 g/ton	5-50 g/ton	8				91
Galena		8 g/ton	< 1-47 g/ton	7				91
Boulangerite ($Pb_5Sb_4S_{11}$)		200 g/ton		2				91
Pyrite		3 g/ton	2-10 g/ton	5				91
Chalcopyrites	Central Urals		N.D. to 5 g/ton	9				92
Minerals of chalcopyrite-sphalerite association:								
Pyrite	Central Urals		N.D. to 530 g/ton	8				92
Chalcopyrite	Central Urals		N.D. to 158 g/ton	22			The high value sample was assocd. with colloform pyrite.	92
Sphalerite	Central Urals		N.D. to 23 g/ton	15				92

Source	Location	Thallium Concentration Mean	Range	No. of Samples	Method	Sensitivity (ppm)	Remarks	Reference
Tetrahedrite	Central Urals		N.D. to 2.5 g/ton	3				92
Galena	Central Urals	21 g/ton		1				92
Minerals of bornite-tennantite association:								
Bornite, chalcocite, tennantite, pyrite	Central Urals		N.D. to 7 g/ton	9				92
Chalcopyrite	Central Urals	30 g/ton		1				92
Minerals of chalcopyrite-pyrrhotite association:								
Ores	Central Urals	4 g/ton					Tl is found in the Cu conc. from beneficiation of the chalcopyrite ore. A large part of the Tl cannot be extracted and remains in the tailings in isomorphic admixture in sericite (a form of muscovite) and chlorite (a silicate of Al with Fe(II) and Mg)	92
Pyrite and pyrrhotite	Central Urals		≤ 62 g/ton					92
								92
Galenas from different Pb-Zn skarn-type deposits	Central Asia, Kazakhstan, Far East		N.D. to 100 g/ton	72			Review of several papers.	92
Galenas from high-temp. Pb-Zn deposits located in aluminosilicate rocks	Eastern Zabaikal		1-50 g/ton	20			Data of Meituv cited. Ten samples were < 8 g/ton.	93

APPENDIX (Continued)

Source	Location	Thallium Concentration Mean	Range	No. of Samples	Method	Sensitivity (ppm)	Remarks	Reference
Galenas from high-temp. Pb-Zn deposits located in aluminosilicate rocks	Central Asia	~ 12 g/ton	N.D. to 50 g/ton	15				93
Galenas from high-temp. Pb-Zn deposits located in aluminosilicate rocks	Mexico	6 g/ton		1				93
Galenas from high-temp. Pb-Zn deposits located in aluminosilicate rocks	Australia	5 g/ton		1				93
Galena, sphalerite, pyrite, chalcopyrite, geocronite (a Pb-Sb sulfide), boulangerite	In silicate rocks		N.D. to 50 g/ton				Data of numerous papers tabulated.	93
Galena, sphalerite, pyrite, chalcopyrite, geocronite (a Pb-Sb sulfide), boulangerite	In effusive sedimentary rocks		N.D. to 720 g/ton				Data of several papers tabulated.	93
Galena, sphalerite, pyrite, chalcopyrite, geocronite (a Pb-Sb sulfide), boulangerite	In carbonate rocks		N.D. to 3,120 g/ton				Highest value in geochronite. Highest value in galena, 300 g/ton. No data for sphalerite.	93
Low-temp. Pb-Zn deposits in carbonate rocks	South Kazakhstan SSR							93
Galena	Mirgalimsai deposit		N.D. to 37 g/ton					93

Source	Location	Thallium Concentration Mean	Thallium Concentration Range	No. of Samples	Method	Sensitivity (ppm)	Remarks	Reference
Pb concentrate	Mirgalimsai deposit	9 g/ton	2-19 g/ton	6				93
Pb concentrate	Mirgalimsai deposit	8 g/ton	2-15 g/ton	2				93
Pyrite	Mirgalimsai deposit	13 g/ton	2-15 g/ton					93
Galena, pyrite	Achisai deposit	N.D.						93
Pb concentrate	Achisai deposit	11 g/ton						93
Zn concentrate	Achisai deposit	2 g/ton						93
Pyrite conc.	Achisai deposit	3 g/ton						93
Galena	Baidzhansai deposit		4-5 g/ton					93
Pb concentrate	Baidzhansai deposit	60 g/ton						93
Pyrite conc.	Baidzhansai deposit	240 g/ton	180-300 g/ton					93
Galena	Aksuran deposit	50 g/ton	30-80 g/ton					93
Pyrite	Aksuran deposit	1,000 g/ton	300-1,700 g/ton					93
Sphalerite	Aksuran deposit	15 g/ton	6-30 g/ton					
Sphalerite	Bytom deposit	300 g/ton	40-700 g/ton				Data are given for several other deposits. Generally marcasite concns. > sphalerite > galena.	93
Galena	Bytom deposit	9 g/ton	trace to 35 g/ton					93
Marcasite	Bytom deposit	600 g/ton	70-1,000 g/ton					93
Pyrite	Bytom deposit	4,500 g/ton						93

APPENDIX (Continued)

Source	Location	Thallium Concentration Mean	Range	No. of Samples	Method	Sensitivity (ppm)	Remarks	Reference
Sphalerite	Southeastern and Central Europe		N.D. to 3,000 g/ton				Highest average: 750 g/ton.	93
Galena	Southeastern and Central Europe		N.D. to 300 g/ton				Highest average: 8 g/ton.	93
Sphalerite	North America							93
Sphalerite	Joplin, Missouri	17 g/ton	N.D. to 50 g/ton	3				93
Sphalerite	Galena, Kansas	15 g/ton	1-30 g/ton	3				93
Sphalerite	Jasper City, Carterville, LaSalle, Coil Camp	N.D.		1 each				93
Different types of zinc blende								93
Granular	Upper Silesia	18 g/ton	4-40 g/ton	6				93
Conchoidal	Upper Silesia	190 g/ton	20-700 g/ton	12				93
Ocherous (earthy)	Upper Silesia	160 g/ton	40-350 g/ton	5				93
Wurtzite	Upper Silesia	50 g/ton	40-70 g/ton	2				93
Sphalerites	Miess, Raibl, Blyaiberg-Kroit deposits		N.D. to 100 g/ton				Range of average values 5-50 g/ton. The concn. of Tl in the galena of these deposits ranged from N.D. to 30 g/ton.	93
Wurtzite	Miess, Raibl, Blyaiberg-Kroit deposits		50-300 g/ton					93

Source	Location	Thallium Concentration Mean	Range	No. of Samples	Method	Sensitivity (ppm)	Remarks	Reference
Wurtzite (conchoidal)	Miess, Raibl, Blyaiberg-Kroit deposits		50-5,000 g/ton					93
Chief sulfide minerals of low-temp. Pb-Zn deposits:	In silicate rocks, Caucasus						Review of several papers.	93
Galena			2-16 g/ton					
Sphalerite			1-45 g/ton					
Pyrite			53-64 g/ton					
Marcasite			140-3,400 g/ton					
Chief sulfide minerals of low-temp. Pb-Zn deposits:	In silicate rocks, Central Asia							93
Galena			N.D. to 200 g/ton					
Sphalerite			N.D. to 90 g/ton					
Pyrite			trace to 510 g/ton					
Marcasite		1,450 g/ton	11-3,060 g/ton	25				
Chief sulfide minerals of low-temp. Pb-Zn deposits:	In silicate rocks, Carpathia							93
Galena		4 g/ton		1				

APPENDIX (Continued)

Source	Location	Thallium Concentration Mean	Range	No. of Samples	Method	Sensitivity (ppm)	Remarks	Reference
Sphalerite	In silicate rocks, Carpathia		N.D. to 5 g/ton	6				93
Pyrite			N.D. to 700 g/ton	5				
Marcasite		250 g/ton		?				
Sphalerite	Silicate rocks, Northern region of the USSR		trace to 1 g/ton	3				93
Quartz-cinnabar-antimonite deposits:	Central Asia							55
Pyrite (cryst., early)		7.5 g/ton	7-8 g/ton	3				
Pyrite and marcasite (cryst., later)		389 g/ton	18-5,700 g/ton	29				
Other minerals including colloform pyrite and/or marcasite			≤ 5,700 g/ton					
Cryst. stibnite		5 g/ton	trace to 40 g/ton	7/13			Average based on those samples containing Tl.	
Colloform stibnite		15 g/ton	4-43 g/ton	9/11			Average based on those samples containing Tl.	
Cinnabar		5.5 g/ton	4-12 g/ton	3/4			Average based on those samples containing Tl.	
Realgar and orpiment		2.9 g/ton	2-17 g/ton	3/9			Average based on those samples containing Tl.	
Tetrahedrite		8 g/ton		1/2				

APPENDIX (Continued)

Source	Location	Thallium Concentration Mean	Range	Samples	Method	Sensitivity (ppm)	Remarks	Reference
Kaolinite-quartz-cinnabar deposits:	Central Asia							
Pyrite		52.3 g/ton	4-200 g/ton	12/12			Average based on those samples containing Tl.	
Sphalerite		40.0 g/ton	30-50 g/ton	2/2			Average based on those samples containing Tl.	
Cinnabar		5.0 g/ton	<1-8 g/ton	6/6			Average based on those samples containing Tl.	
Realgar		1 g/ton		1/1			Average based on those samples containing Tl.	
Vein gossan	Mystic Mine, Empire, Colo.		< 0.2-1.5 ppm		a.a.		Various digestion procedures.	22
Gossan	Montezuma, Colo.		4.9-5.5 ppm		a.a.		Various digestion procedures. Vlasov[91] states that the iron gossans of Tl-bearing deposits contain less Tl than the primary ore.	22
Orpiment (As_2S_3) and realgar (AsS)	Getchell, Nev.	400 ppm			a.a.			22
Galena, sphalerite, pyrite	Ophir, Utah	3.5 ppm			a.a.			22
Colloform sphalerite	New Prospect Mine, Eastern Tenn.	4.4 ppm			a.a.			22
Carbonaceous gold ore	Carlin, Nev.	8 ppm			a.a.			22
Carbonaceous gold ore	Carlin, Nev.		0.5-58 ppm		a.a.		USGS anal.	5

APPENDIX (Continued)

Source	Location	Thallium Concentration Mean	Range	No. of Samples	Method	Sensitivity (ppm)	Remarks	Reference
Pyrites assocd. with coal	Several Asian coal basins		2-360 ppm		Spectrog.		Higher concns. in pyrite from shale and granitoid source areas than from carbonate source areas.	25
Bedded pyrite	Adirondacks	1.9 ppm			a.a.		USGS anal.	5
Sulfides of the Paleogene and Oxfordian-Lusitanian claystones	Northern Caucasus		0.4-2.2 ppm				In coal deposits lacking in sulfides, Tl is absent from claystones and sandstones.	23
Massive sulfide	Milan mine, Milan, N.H.		6.6-6.7 ppm		a.a.		USGS anal.	5
Mesothermal vein (Ag-Au)	Maine ?		0.5-1.5 ppm				USGS anal.	5
Galena		Highest mean, 5.3 ppm; lowest, 0.27 ppm	0.0-5,000 ppm				Eight papers cited including Shaw (1952)	11
Sphalerite		Highest mean, 700 ppm; lowest, 5.8 ppm	0.0-1,000 ppm				Five papers cited.	11
Pyrite			0.0-210 ppm				The high value is colloform pyrite from the Sukhumi, Dusheti region, Caucasus.	11
Marcasite		Lowest mean, 0.24 ppm	≤ 9,000 ppm				Six papers cited.	11
Pyrrhotite			N.D. to 0.074 ppm				Two papers cited.	11
Chalcopyrite			N.D. to 0.8 ppm				Two papers cited.	11

Source	Location	Thallium Concentration Mean	Range	No. of Samples	Method	Sensitivity (ppm)	Remarks	Reference
Geocronite Pb_5SbAsS_8	Eastern Transbaikaliya, USSR		100-5,000 ppm		?		Kulikova (1962) cited.	11
Meneghinite	Eastern Transbaikaliya, USSR		100-1,000 ppm		?		Kulikova (1962) cited.	11
Boulangerite	Eastern Transbaikaliya, USSR		10-1,000 ppm		?		Kulikova (1962) cited.	11
Jamesonite	Eastern Transbaikaliya, USSR		10-100 ppm		?		Kulikova (1962) cited.	11
Miscellaneous								
Caliche ($NaNO_3$)	Chile		N.D.		Spectrog.	~ 0.05	Shaw (1952).	12
Ocean salts	3 locations, 3 depths		N.D.			~ 0.05		12
Crude potash salts and many com. samples			"extremely small amts."				Reissmann (1939) cited.	51
Sylvinite (rock contg. crude KCl)	Near Carlsbad, N.M.		N.D.		Spectrog.	~ 0.05	Shaw (1952).	12
Sylvinite	Uzbekistan salt deposits		0.23-0.60	4	Chem.			92
K and Na salt deposits	Poland		≤ 0.1 ppm	51			Trzebiatowski and Rozdzial (1957)	51
Halite (rock salt)	Grand Saline Salt Dome, Texas		≤ 0.05 ppm		Spectrog.	~ 0.05		12
Insoluble residue from rock salt			N.D.		Spectrog.	~ 0.05		12

APPENDIX (Continued)

Source	Location	Thallium Concentration Mean	Range	Samples	Method	Sensitivity (ppm)	Remarks	Reference
Geothermal brine	Salton Sea, Calif.	1.5 ppm						96
Scale deposited in pipes pumping geothermal brine	Salton Sea, Calif.	N.D.				Quant. spectrog.		96
Sylvinites	USSR	0.51 ppm				Not given		97
Carnallites	USSR	0.06 ppm				Not given		97
Langbeinite	USSR	N.D.				Not given		97
White sylvinite	Germany	0.09 ppm 0.13 ppm			Voltammetric		Geilman et al. (1960) found 0.31 ppm in technical KCl. Eschauer and Neeb (1960) found 0.15 ppm and 0.84 ppm in KCl. They also found 0.02 ppm in K_2SO_4	98,99
Colored sylvinite	Germany	0.003 ppm			Voltammetric			99
Sylvinite	USSR, 4 deposits	0.42 ppm	0.23-0.60 ppm	4	Colorimetric			95
Sylvinite	Ak-Bash deposit, USSR	0.44	0.20-1.09	24	Colorimetric			95
Kainite	Germany?	0.007 ppm			Voltammetric			99
Carnallite	Ak-Bash deposit, USSR	0.025 ppm			Voltammetric?			98
Fertilizer salts, 40-50%	Ak-Bash deposit, USSR		0.015-0.059 ppm	4	Voltammetric			98
Halite	Ak-Bash deposit, USSR, upper layers		0.0-0.17 ppm		Colorimetric		At depths 190.0-251.9 m, the average Tl content is 0.46 ppm (0-1.79 ppm). At 474.1-550.0 m, the average content is 2.68 ppm.	95

Source	Location	Thallium Concentration Mean	Range	Samples	Method	Sensitivity (ppm)	Remarks	Reference
Soil								
Soil	Over sphalerite vein, Montezuma, Colo.	4.9 ppm			a.a.			22
Humus ash	Montezuma, Colo.	5.6 ppm			a.a.		Exposed spruce tree root ash: 3.9 ppm	22
Soils		~ 0.1 ppm					Vinogradov.	18
Soils		5 ppm ?						16
Soils	Mercuryville, Calif.	2.7 ppm			Crystal violet spectrog.		Silica carbonate soil.	100
Soils	San Miguel County, Colo.		0.4-1.8 ppm	5	a.a.		USGS anal. by A. E. Hubert.	5
Soils	Coeur d'Alene District, Idaho		0.3-2.8 ppm	179	a.a.		USGS anal. by A. E. Hubert.	5
Soils	Egan Range, Nev.		0.3-4.5 ppm	14	a.a.		USGS anal. by A. E. Hubert.	5
Soils			0.02-5 ppm	30	a.a.		Although the table provided by the USGS is entitled "Soils," samples are variously identified by names of minerals, rocks, and ores. For example, samples of "copper ores" from Nevada, Utah, and California contained 0.10-0.90 ppm Tl. The highest values were for "sandstones."	5
Soils	Esaulovskii Pb-Zn deposit, USSR	0.03 g/ton			?			101
Soils	Bobrikovskii Au development, USSR	0.04 g/ton						101
Soils	Volnukhinskii Hg ore development, USSR	1.1 g/ton						101
Soils	Krasnodonskii region, USSR	0.36 g/ton						101
Soils	Golovinskii Hg ore development	0.045 g/ton						101

APPENDIX (Continued)

Source	Location	Thallium Concentration Mean	Range	No. of Samples	Method	Sensitivity (ppm)	Remarks	Reference
Water								
Pacific Ocean surface waters		13.0 ± 1.4 ng/ℓ (13×10^{-6} ppm)		6	Anodic stripping voltammetry		Authors state that thermodynamic arguments predict Tl(III) is the major species in both freshwater and seawater. $Tl(OH)_2^+$ and TlCl are the most likely forms of Tl(III) and Tl(I), respectively.	102
Seawater								
Bay of Biscay	2 to 2,500 m deep, 1969	10.1 ng/ℓ (avg. of 4 samples)	9.4-16.6 ng/ℓ	5	Neutron activation anal.		The high value at 500 m might have been due to decomposition of planktonic material that had fallen from the surface.	6
Irish Sea		18.7 ng/ℓ			Neutron activation anal.			7
Seawater		< 0.01 ppb					Ahrens et al. (1967) cited.	11
Brines, $CaCl_2$ types	Angara-Lena artesian basin, USSR	max. 2.76 ppm					Brines are enriched in K and Rb.	103
Waters from hot springs and drill holes	Thermal areas along the Taupo volcanic zone, New Zealand		≤ 7 ppb				Weissberg (1969) cited. At a depth of 1 km, the waters have pH 6-7. Amorphous sulfide ppts. from these waters contain 1 to 5,000 ppm Tl, representing only 0.4% of the Tl in the waters.	11
Freshwater	Catchment area having a sandstone base rock--Woronora Weir	3.7 ± 1.0 ng/ℓ (~ 4×10^{-6} ppm)			Anodic stripping voltammetry		Major species said to be Tl(III).	102
Freshwater	Mercuryville, Calif.	0.014 ppm			Crystal violet spectrophotometry			100
Freshwater	Unidentified mineral waters	40 µg/ℓ (0.040 ppm)			Spectrog.		Khitrov and Belousov cited.	102

APPENDIX (Continued)

Source	Location	Thallium Concentration Mean	Range	No. of Samples	Method	Sensitivity (ppm)	Remarks	Reference
Freshwater								
Spring water	Assocd. with sandstone C soil (0.45 ppm Tl)	0.04 ppb (winter) 1.6 ppb (summer)			a.a.		USGS anal.	5
Spring water	Assocd. with sandy loam A_1 (3 ppm)	14 ppb (summer)		17	a.a.			5
River water		0.02 ppb					Generalized (from Bowen or Tong and Lisk, 1972) by Lisk.	16
Potable water		(0.005 ppm)					Desired max. ambient environmental level.	104
Stratal waters of petroleum deposits	Azerbaidzhan, USSR	0.7 ppb 0.18 ppb	trace to 2.6 ppb	29	Photometric, colorimetric		Hard water.	20 40
Stratal waters of petroleum deposits	Azerbaidzhan, USSR	0.22 ppb			Colorimetric		Alkaline water.	40
Stratal waters of petroleum deposits	Azerbaidzhun, USSR	2.1 ppb	1.0-4.2 ppb	30	Photometric		Alkaline water.	20
River water 50 to 3,950 m from spraying locations	Hokkaido, Japan, 1972	N.D.					Forests are sprayed from helicopters with thallium sulfate in the fall to kill rats and other roderts. Method not in the abstr.	105
"Natural water"	Aldan region, USSR		0.01-1 ppb		Spectrog.			106
Mine waters	Lachin-Khana Pb-Zn deposit	2 ppb		1/7			Dunin-Barkovskaya (1961) cited.	11

Source	Location	Thallium Concentration Mean	Thallium Concentration Range	No. of Samples	Method	Sensitivity (ppm)	Remarks	Reference
Air								
Air	Chadron, Neb., weekly Jan. to to Dec. 1973	0.22 ng/m^3	0.07-0.48 ng/m^3	52	a.a.		Cu and Tl correlation coefficient indicates a common source during the summer months. Co and Tl correlation values were high and constant at all seasons. The enrichment factor for aerosol Tl relative to the Earth's crust with reference elements Al and Mn normalized to unity: 67 and 73, respectively. A value of 1 indicates soil derivation. By comparison, the enrichment factors for Ag, Cd, and Pb of ~ 300 to 500 indicate anthropogenic sources.	107
	June to Sept. 1973	0.22 ng/m^3						107
Air	Chadron, Neb., June to Sept. 1974	0.15 ng/m^3	0.04-0.23 ng/m^3	19				107
Manufactured and Refined Products								
USGS glass std. D		7.2 ppm			a.a.		15 ppm by spectrog. anal.	22
USGS glass std. B		0.5 ppm			a.a.			22
USGS glass std. E		13.2 ppm			a.a.			22
A silicate glass prepared by R&D dept. of Leeds and Northrup Co.		11.98%						22
Lead		< 1 and 32 ppm		2	Extn.-spectrophotometric		Tl is usually present in Pb, Cd, In, or Zn metal at < 100 ppm.	108
Indium		< 2 ppm		2	Extn.-spectrophotometric			108
Cadmium		1 and 56 ppm		2	Extn.-spectrophotometric			108
		11 and 30 ppm		2	?			109
Zinc		1 ppm		1	Extn.-spectrophotometric			108
Zinc Standard Reference Materials No. 683 No. 728		0.16 ppm 0.17 ppm			Isotope dilution SSMS			110

Bibliography. Appendix.

1. Mathis, B. J., and N. R. Kevern, "Distribution of Mercury, Cadmium, Lead and Thallium in a Eutrophic Lake," Hydrobiol., 46(2-3), 207-222 (1975).

2. Mathis, B. J., and N. R. Kevern, "Distribution of Mercury, Cadmium, Lead and Thallium in a Eutrophic Lake," PB 221993, Department of Fisheries and Wildlife, East Lansing, Mich., 1973.

3. Sabadell, J. E., and R. C. Axtmann, "Heavy Metal Contamination from Geothermal Sources," Environ. Health Perspect., 12, 1-7 (1975).

4. Shaw, D. M., "The Geochemistry of Gallium, Indium, and Thallium--A Review," Phys. Chem. Earth, 2, 164-211 (1957).

5. U.S. Geological Survey, Branch of Exploration Research, raw data, released by U.S.G.S. through Keith Robinson, Denver, Colo., November 1976. [Most thallium analyses were performed by A. E. Hubert, W. H. Ficklin, and G. Crenshaw]

6. Matthews, A. D., and J. P. Riley, "The Occurrence of Thallium in Sea Water and Marine Sediments," Chem. Geol., 6, 149-152 (1970).

7. Matthews, A. D., and J. P. Riley, "The Determination of Thallium in Silicate Rocks, Marine Sediments and Sea Water," Anal. Chim Acta, 48(1), 25-34 (1969).

8. Hamaguchi, H., N. Ohta, N. Onuma, and K. Kawasaki, "Inorganic Constituents in Biological Material. XIV. Contents of Thallium, Selenium, and Arsenic in Fish and Shells from the Minamata District, Kyushu," Nippon Kagaku Zasshi, 81, 920-927 (1960); Chem. Abstr., 54, 25337i (1960).

9. Hewett, D. F., M. Fleischer, and N. Conklin, "Depo its of the Manganese Oxides: Supplement," Econ. Geol., 58, 1-51 (1963).

10. Hewett, D. F., and M. Fleischer, "Deposits of the Manganese Oxides," Econ. Geol., 55, 1-55 (1960).

11. Wedepohl, K. H., Ed., Handbook of Geochemistry, Element 81, Vol. 2, No. 3, Springer-Verlag, Berlin, Heidelberg, 1972, pp. 81-B-1 to 81-O-1.

12. Shaw, D. M., "The Geochemistry of Thallium," Geochim. Cosmochim. Acta, 2, 118-154 (1952).

13. Parker, R. L., "Chapter D. Composition of the Earth's Crust" in Data of Geochemistry, 6th ed., M. Fleischer, Tech. Ed., Geol. Survey Professional Paper 440-D, U.S. Government Printing Office, Washington, D.C., 1967, pp. D1-D19.

14. Riley, J. P., and P. Sinhaseni, "Chemical Composition of Three Manganese Nodules from the Pacific Ocean," J. Marine Res. (Sears Found. Marine Res.), 17, 466-482 (1958).

15. Ahrens, L. H., J. P. Willis, and C. O. Oosthuizen, "The Composition of Manganese Nodules, with Particular Reference to Some of the Rarer Elements," Geochim. Cosmochim. Acta, 31(11), 2169-2180 (1967).

16. Lisk, D. J., "Trace Metals in Soils, Plants, and Animals" in Advances in Agronomy, Vol. 24, Academic Press, New York, N.Y., 1972, pp. 267-325.

17. Bokova, L. M., and E. A. Kon'kova, "Thallium Occurrences in the Central Asia Sedimentary Rocks and Deposits," Vopr. Mineralog. i Geokhim., Akad. Nauk Uz. SSR, Inst. Geol. i Geofiz., 116-119 (1964); Chem. Abstr., 62, 3842 (1965).

18. Vinogradov, A. P., The Geochemistry of Rare and Dispersed Chemical Elements in Soils, 2nd ed., translated from the Russian, Consultants Bureau, Inc., New York, N.Y., 1959.

19. Vinogradov, A. P., "Average Content of Chemical Elements in the Main Types of Igneous Rocks in the Earth's Crust," Geokhimiya, 1962b(7), 555-571 (1962); Geochemistry, 1962(7), 641-664 (1963).

20. Nuriev, A. N., and Z. A. Dzhabbarova, "O Raspredelenii Galliya, Talliya, Germaniya i Redkozemel'nykh Elementov v Osadochnykh Porodakh i Plastovykh Vodakh Neftyanykh Mestorozhdenii" ["Distribution of Gallium, Thallium, Germanium, and Rare-Earth Elements in Sedimentary Rocks and Stratal Waters of Petroleum Deposits"], Geokhimiya, 446-451 (1973).

21. Ahrens, L. H., "Trace Elements in Clays," S. African J. Sci., 41, 152-160 (1945).

22. Hubert, A. E., and H. W. Lakin, "Atomic Absorption Determination of Thallium and Indium in Geologic Materials" in Geochemical Exploration 1972.

23. Voskresenskaya, N. T., N. V. Timofeyeva, and M. Topkhana, "Thallium in Some Minerals and Ores of Sedimentary Origin," Geochemistry (USSR), 851-857 (1962).

24. Plesko, A. V., "The Accumulation of Thallium in Coals from Tadzhikistan," Izvest. Otdel. Estestven. Nauk. Akad. Nauk Tadzhik S.S.S.R., No. 1, 95-101 (1959); Chem. Abstr., 55, 8203 (1961).

25. Uoumaoa, Y., "Minor Inorganic Constituents of Coals," Chem. Researches (Japan), Inorg. and Anal. Chem., 5, 1-17 (1949); Chem. Abstr., 43, 8639 (1949).

26. Vinokurov, S. F., and N. T. Voskresenskaya, "Nature of Anomalous Thallium Concentration in Brown Coal Deposits," Geokhimiya, No. 12, 1468-1476 (1970); Chem. Abstr., 74(6), 154 (1971).

27. Voskresenskaya, N. T., "Thallium in Coal," Geochemistry (USSR), 158-168 (1968).

28. Swanson, V. E., Composition and Trace Element Content of Coal and Power Plant Ash. Part 2. Appendix J. Southwest Energy Study, Report of the Coal Resources Work Group, U.S. Department of the Interior, 1972.

29. Magorian, T. R., K. G. Wood, J. G. Michalovic, S. L. Pek, and M. W. Van Lier, Water Pollution by Thallium and Related Metals, Calspan Corporation for the U.S. Environmental Protection Agency, Cincinnati, Ohio, 1974. [This is the draft version of the final report issued under another title (see Reference 3, Bibliography VI.) in 1977 and was obtained from the National Technical Information Service as PB-253 333.]

30. Linton, R. W., A. Loh, D. F. S. Natusch, C. A. Evans, Jr., and P. Williams, "Surface Predominance of Trace Elements in Airborne Particles," Science, 19(4229), 852-854 (1976).

31. Natusch, D. F. S., C. A. Evans, P. K. Hopke, A. Loh, and R. Linton, "Characterization of Trace Elements in Coal Fly Ash" in International Conference on Heavy Metals in the Environment, Toronto, Canada, October 27-31, 1975, pp. C-73 to C-75.

32. Bolton, N. E., J. A. Carter, J. F. Emery, C. Feldman, W. Fulkerson, L. D. Hulett, and W. S. Lyon, "Trace Element Mass Balance Around a Coal-Fired Steam Plant" in Trace Elements in Fuel, Advances in Chemistry Series, No. 141, American Chemical Society, Washington, D.C., 1975, pp. 175-187.

33. Bolton, N. E., R. I. Van Hook, W. Fulkerson, W. S. Lyon, A. W. Andren, J. A. Carter, and J. F. Emery, Trace Element Measurements at the Coal-Fired Allen Steam Plant. Progress Report June 1971-January 1973, ORNL-NSF-EP-43, Oak Ridge National Laboratory, Oak Ridge, Tenn., March 1973.

34. Schwitzgebel, K., F. B. Meserole, R. G. Oldham, R. A. Magee, F. Mesich, and T. L. Thoem, "Trace Element Discharge from Coal-Fired Power Plants" in International Conference on Heavy Metals in the Environment, Toronto, Ontario, Canada, Oct. 27-31, 1975, pp. C-69 to C-70.

35. Davison, R. L., D. F. S. Natusch, J. R. Wallace, and C. A. Evans, Jr., "Trace Elements in Fly Ash. Dependence of Concentration on Particle Size," Environ. Sci. Technol., **8**(13), 1107-1113 (1974).

36. Natusch, D. F. S., J. R. Wallace, and C. A. Evans, Jr.," Toxic Trace Elements. Preferential Concentration in Respirable Particles," Science, **183**(4121), 202-204 (1974).

37. Peterson, M. J., and J. B. Zink, "A Semiquantitative Spectrochemical Method for Analysis of Coal Ash," Report of Investigations 6496, Bureau of Mines, U.S. Department of the Interior, Washington, D.C., 1964.

38. Gluskoter, H. J., Illinois State Geological Survey, personal communication to E. E. Angino, 1976.

39. Kessler, T., A. G. Sharkey, and R. A. Friedel, Spark-Source Mass Spectrometer Investigation of Coal Particles and Coal Ash, Bureau of Mines Progress Report No. 42, U.S. Department of the Interior, Washington, D.C., 1971.

40. Nuriev, A. I., and G. K. Efendiev, "Thallium and Gallium Contents in Stratal Waters, Crude Oils, and Surrounding Rocks," Azerb. Khim. Zh., No. 2, 112-116 (1966); Chem. Abstr., **65**, 15099 (1966).

41. Taylor, S. R., "Abundance of Chemical Elements in the Continental Crust: A New Table," Geochim. Cosmochim. Acta, **28**, 1273 (1964).

42. Hamilton, E. I., "The Geochemistry of the Northern Part of the Ilimaussa Intrusion, S.W. Greenland," Medd. Groenland, **162**(10), 1-104 (1964); Chem. Abstr., **61**, 13076 (1974).

43. Rankama, K., and T. G. Sahama, Geochemistry, Chicago University Press, Chicago, Ill., 1950.

44. Butler, J. R., "Thallium in Some Igneous Rocks," Geokhimiya, No. 6, 514-523 (1962); Chem. Abstr., 57, 8246 (1962).

45. Higazy, R. A., "The Distribution of Trace Elements in the Perthite Pegmatites of the Black Hills, South Dakota," Am. Mineralogist, 38, 172-190 (1953).

46. Langmyhr, F. J., J. R. Stubergh, Y. Thomassen, J. E. Hanssen, and J. Doleval, "Atomic Absorption Spectrometric Determination of Cadmium, Lead, Silver, Thallium and Zinc in Silicate Rocks by Direct Atomization from the Solid State," Anal. Chim. Acta, 71(1), 35-42 (1974).

47. Efendiev, G. Kh., N. A. Novruzov, and A. S. Geidarov, "Geochemistry of Thallium in Pyrite-Polymetallic Deposits," Izv. Akad. Nauk Azerb. S.S.R., Ser. Geol-Geogr. Nauk, No. 1 30-37 (1965); Chem. Abstr., 63, 6728e (1965).

48. Meituv, G. M., "Hydrothermally Altered Rocks and Rare Elements in the Myl'nikovo-Khorkirin Deposit (Eastern Transbaikal)," (Chetvertata) Konf. Molodykh Nauchn. Sotrudn. Inst. Mineralog. Geokhim. i Kristallokhim. Redkikh Elementov, Moscow, Sb., 56-60 (1962); Chem. Abstr., 63, 12894 (1965).

49. Solodov, N. A., "Distribution of Thallium among the Minerals of a Zoned Pegmatite," Geochemistry (USSR), 738-741 (1962).

50. Ahrens, L. H., "The Unique Association of Thallium and Rubidium in Minerals," J. Geol., 56, 578-590 (1948).

51. Lockwood, T. H., "The Analysis of Asbestos for Trace Metals," Hyg. Assoc. J., 35(5), 245-251 (1974).

52. Swaine, D. J., The Trace-Element Content of Fertilizers, Commonwealth Agricultural Bureau, Farnham Royal, Bucks, England, 1962.

53. Razenkova, N. I., and G. F. Galaktionova, "Eksperimental'noe Izuchenie Formy Nakhozhdeniya Talliya v Zone Okisleniya Sul'fidnykh Mestorozhdenii" ["Forms of Thallium Occurrence in the Oxidation Zone of Sulfide Deposits"], Tr. Inst. Mineralog., Geokhim. i Kristallokhim. Redkikh Elementov, Akad. Nauk S.S.S.R., No. 18, 5-19 (1963).

54. Novikova, T. I., "Mineralogy of the Yellow Ochers (Antimony) of the Dzhizhikrut Mercury-Antimony Deposit," Trudy. Inst. Geol. Akad. Nauk Tadzhik S.S.S.R., No. 2, 283-298 (1957); Chem. Abstr., 54, 15100 (1960).

55. Velikii, A. S., V. Yu. Volgin, and V. S. Ivanov, "Redkie Elementy v Syr'myano-Rtutnykh Mestorozhdeniyakh Srednei Azii" ["Rare Elements in Antimony-Mercury Deposits of Central Asia"] in Formy Nakhozhdeniya i Osobennosti Raspredeleniya Redkikh Elementov v Nekotorykh Tipakh Gidrotermal'nykh Mestorozhdenii [Forms of Occurrence and Characteristics of the Distribution of Rare Elements in Some Types of Hydrothermal Deposits], V. V. Ivanov, Ed., Publisher "Nauka," Moscow, 1967, pp. 180-212.

56. Hewett, D. F., and A. S. Radtke, "Silver-Bearing Black Calcite in Western Mining Districts," Econ. Geol., **62**(1), 1-21 (1967).

57. Crittenden, M. D., F. Cuttitta, H. J. Rose, Jr., and Fleischer, M., "Studies of Manganese Oxide Minerals. 6. Thallium in Some Manganese Oxides," Am. Mineralogist, **47**, 1461-1467 (1962).

58. Voskresenskaya, N. T., and T. D. Usevich, "The Occurrence of Thallium in Manganese Minerals," Geochemistry (USSR), 710-721 (1957).

59. Voskresenskaya, N. T., and L. T. Soboleva, "Once More on Thallium in Manganese Minerals," Geochemistry (USSR), 309-312 (1961).

60. Rose, H. J., Jr., and F. J. Flanagen, "X-Ray Fluorescence Determinations of Thallium in Manganese Ores," U.S. Geological Survey Professional Paper 450-B, Art. 32, pp. B-82 to B-83, 1962.

61. Fillo, P. V., Manganese Mining and Milling Methods and Costs, Mohave Mining and Milling Co., Maricopa County, Ariz., Information Circular 8144, Bureau of Mines, U.S. Department of the Interior, Washington, D.C., 1963.

62. Sreenivas, B. L., "Trace Element Content of Manganese Ores from Nagpur, Vizagpatam and Mysore," Indian Mineralogist, **7**(1 and 2), 79-80 (1966).

63. Hewett, D. F., Silver in Hypogene Manganese Oxides, Geological Survey Circular 553, U.S. Department of the Interior, Washington, D.C., 1968.

64. Kulikova, M. F., "O Kharaktere Raspredeleniya i Srednikh Soderzhaniyakh Nekotorykh Redkikh Elementov v Gidrookislakh Zheleza Zony Okisleniya Svintsovo-Tsinkovykh Mestorozhdenii Vostochnogo Zabaikal'ya i Srednei Azii" ["Distribution and Content of Certain Rare Elements in Hydrous Ferric Oxides within the Oxidation Zone of the Lead and Zinc Deposits of Eastern Transbaikal and Central Asia"], Dokl. Akad. Nauk S.S.S.R., **163**(5), 1248-1251 (1965); Chem. Abstr., **63**, 17708 (1965).

65. Marowsky, G., and K. H. Wedepohl, "General Trends in the Behavior of Cadmium, Mercury, Thallium, and Bismuth in Some Major Rock Forming Processes," Geochim. Cosmochim. Acta, 35(12), 1255-1267 (1971).

66. Burnham, C. W., "Metallogenic Provinces of the Southwestern United States and Northern Mexico," New Mexico, Bur. Mines and Mineral Resources, Bull. No. 65, 1-76 (1959).

67. Oftedal, I., "Untersuchungen über die Nebenbestandteile von Erzmineralien Norwegischer Zinkblendeführender Vorkommen," Skrifter utgitt av Det Norske Videnskaps-Akademi I Oslo, I. Matematisk-Naturvidenskapelig Klasse, No. 8, 1-103 (1941).

68. Radtke, A. S., C. M. Taylor, R. C. Erd, and F. W. Dixon, "Occurrence of Lorandite, $TlAsS_2$, at the Carlin Gold Deposit, Nevada," Economic Geol., 69, 121-124 (1974).

69. Stoiber, R. E., "Minor Elements in Sphalerite," Econ. Geol., 35, 501-519 (1940).

70. Lebedev, K. B., "Povedenie Redkikh Rasseyannykh Elementov pri Pererabotke Mednykh Rud Tsentral'novo Kazakhstana" ["Behavior of Rare, Dispersed Elements During the Processing of Copper Ores of Central Kazakhstan"], Vestnik. Akad. Nauk Kazakh. S.S.R., 16(8), 58-63 (1960); Chem. Abstr., 55, 2391g (1961).

71. Li, A. F., and N. N. Burlutskaga, "Redkie i Rasseyannye Elementy v Polimetallicheskikh Rudakh Zabaikal'ya" ["Rare and Disperse Elements in Polymetallic Ores of Transbaikal'ya"], Tr. Vostochno-Sibirsk. Filiala, Akad. Nauk S.S.S.R., Sibirsk. Otdel., No. 41, 63-66 (1962).

72. Kagaya, B., "Thallium in Japanese Ores and Its Significance in Metallogenesis," J. Mining Coll., Akita Univ. Ser. A, 2(2), 60-88 (1962); Chem. Abstr., 60, 5209b (1964).

73. Badalov, S. T., and A. A. Zemlyanov, "Rare and Dispersed Elements in the Kumyshkan Lead-Zinc Deposit," Dokl. Akad. Nauk Uz. S.S.R., 20(5), 39-42 (1963); Chem. Abstr., 59, 15044 (1953).

74. Fillippova, N. A., B. L. Dobrotsvetov, and V. A. Korosteleva, "Determination of the Nature of Associations of Thallium in Ore of a Pyrite Deposit," Sb. Nauchn. Tr. Gos. Nauchn. Issled. Inst. Tsvetn. Metal., No. 19, 785-794 (1962); Chem. Abstr., 60, 3882 (1964).

75. Schroll, E., "Trace-Element Paragenesis (Microparagenesis) of the Galenas of the East Alps," Anz. Math.-Naturw. Kl., Osterr. Akad. Wiss., 6-12 (1951); Chem Abstr., 47, 5313 (1953).

76. Badalov, S. T., "Geochemistry of Rare and Dispersed Elements in Pb-Zn Ore Deposits in the Southern Uzbekistan," Vopr. Mineralog. i Geokhim. Akad. Nauk Uzb.S.S.R., Inst. Geol. i Geofiz., 93-115 (1964); abstract in TOXLINE thallium search.

77. Garmash, A. A., K. F. Kuznetsov, and G. M. Meituv, "Osobennosti Rasprostraneniya Redkikh Elementov v Svintsovo-Tsinkovykh Mestorozhdeniyakh i Metodika ikh Izucheniya dlya Podscheta Zapasov" ["Features of the Distribution of Rare Elements in Lead-Zinc Deposits and Method of Study of Resources Estimation"], Tr. Inst. Mineralog., Geokhim. i Kristallokhim Redkikh Elementov, Akad. Nauk S.S.S.R., 6, 56-71 (1961); Chem. Abstr., 57, 4363 (1962).

78. Mutalov, M. G., "Distribution of Dispersed Elements in Ores of Copper Pyrite Deposits," Geol.-Mineralog. Osobennosti Mednorudn. Mestorozhd. Yuzhn. Urala, Ufa, Sb., 49-57 (1962); Chem Abstr., 60, 9030 (1964).

79. Fruth, I., and A. Maucher, "Trace Elements and Sulfur Isotopes in Sphalerites from Lead-Zinc Deposits of Gorno," Miner. Deposita, 1(3), 238-250 (1966); Chem. Abstr., 67(10), 8749 (1967).

80. Litvinovich, A. N., K. A. Bespaev, B. V. Man'kov, and K. P. Sitnikov, "Distribution of Rare Trace Elements in Ores of the Tishinka Deposit," Vestn. Akad. Nauk Kaz. S.S.R., 20(10), 56-63 (1964); Chem. Abstr., 62, 6272 (1965).

81. Haranczyk, C., "Geochemistry of the Ore Minerals from Silesia-Crawcow Zinc and Lead Deposits," Polska Akad. Nauk, Geol. Trans., No. 30 3-101 (1965); abstract in TOXLINE thallium search.

82. Canneri, G., and D. Cozzi, "The Determination of Rare Metals in Zinc Blende by Spectrographic Methods," Chimica e Industria (Milan), 36 354-357 (1954); Chem. Abstr., 48, 13570 (1954).

83. Tikhomirova, V. V., "Thallium in Antimony-Mercury Deposits," Mineral'n Syr'e. Moscow, Sbornik., No. 1, 93-97 (1961); Chem. Abstr., 55, 26881 (1961).

84. Ruzmatov, S. R., "Thallium Content in the Central Kyzylkum Ore Occurrences," Mineral. Geokhim. Sul'fidnykh Mestorozhd. Uzb.; Akad. Nauk Uzb. S.S.R., Inst. Geol. Geofiz., 126-130 (1966); Chem. Abstr., 66, 9149 (1967).

85. Plesko, A. V., "Thallium-Containing Stibnites and Cinnabar," Tr. Voronezhsk. Gos. Univ., 62, 174-181 (1963); Chem. Abstr., 62, 10225 (1965).

86. Litomisky, J., and O. Paukner, "Quantitative Determination of Thallium in Pyrite by a Spectro-Chemical Method," Casopis Mineral. Geol., 8, 167-174 (1963); Chem. Abstr., 60, 20 (1964).

87. Bol'shakov, A. P., "Accessory Rare Earth and Dispersed Elements in Ores and Minerals of the Nikitski Mercury Deposit," Dopovidi Akad. Nauk Ukr. R.S.R., No. 8, 1096-1098 (1963); Chem.Abstr., 60, 2681 (1964).

88. Chebotarev, G. M., "Distribution of Thallium in Minerals of One Polymetallic Ore Field in the Uzbekistan," Vopr. Mineralog., i Geokhim., Akad. Nauk Uzb.S.S.R., Inst. Geol. i Geofiz., 165-171 (1964); Chem. Abstr., 62, 3817 (1965).

89. Duchesne, J. C., "Presence of Thallium in Iron Sulfides of the Pallieres Mine," Ann. Soc. Geol. Belg., Bull., 87(7), 225-231 (1964); Chem. Abstr., 62, 8881 (1965).

90. Popova, N. I., T. N. Nechemostov, and I. S. Razina, "Redkie Elementy v Molibdeno-Vol'framovykh Mestorozhdeniyakh Tsentral'nogo Kazakhstana" ["Rare Elements in Molybdenum-Tungsten Deposits in Central Kazakhstan"] in Formy Nakhozhdeniya i Osobennosti Raspredeleniya Redkikh Elementov v Nekotorykh Tipakh Gidrotermal'nykh Mestorozhdenii [Forms of Occurrence and Characteristics of the Distribution of Rare Elements in Some Types of Hydrothermal Deposits], V. V. Ivanov, Ed., Publisher "Nauka," Moscow, 1967, pp. 5-51.

91. Ivanov, V. V., Yu. A. Tarkhov, and I. E. Maksimyuk, "Redkie Elementy v Nekotorykh Olovorudnykh Mestorozhdeniyakh S.S.S.R." ["Rare Elements in Some Tin-Ore Deposits of the USSR"] in Formy Nakhozhdeniya i Osobennosti Raspredeleniya Redkikh Elementov v Nekotorykh Tipakh Gidrotermal'nykh Mestorozhdenii [Forms of Occurrence and Characteristics of the Distribution of Rare Elements in Some Types of Hydrothermal Deposits], V. V. Ivanov, Ed., Publisher "Nauka," Moscow, 1967, pp. 52-75.

92. Vorob'eeva, M. S., and N. D. Sindeeva, "Redkie Elementy v Serno- i Mednokolchedannykh Mestorozhdeniyakh Srednego Urala" ["Rare Elements in the Pyrite and Chalcopyrite Deposits of the Central Urals"] in Formy Nakhozhdeniya i Osobennosti Raspredeleniya Redkikh Elementov v Nekotorykh Tipakh Gidrotermal'nykh Mestorozhdenii [Forms of Occurrence and Characteristics of the Distribution of Rare Elements in Some Types of Hydrothermal Deposits], V. V. Ivanov, Ed., Publisher "Nauka," Moscow, 1967, pp. 76-110.

93. Ivanov, V. V., A. A. Garmash, G. M. Meituv, and N. V. Nechelyustov, "Kadmii, Tallii, i Gallii v Svintsovo-Tsinkovykh Mestorozhdeniyakh" ["Cadmium, Thallium, and Gallium in Lead-Zinc Deposits"] in Formy Nakhozhdeniya i Osobennosti Raspredeleniya Redkikh Elementov v Nekotorykh Tipakh Gidrotermal'nykh Mestorozhdenii [Forms of Occurrence and Characteristics of the Distribution of Rare Elements in Some Types of Hydrothermal Deposits], V. V. Ivanov, Ed., Publisher "Nauka," Moscow, 1967, pp. 5-51.

94. Anonymous [probably V. V. Ivanov], "Thallium" in Geochemistry of Rare Elements, Vol. 1, of Geochemistry and Mineralogy of Rare Elements and Genetic Types of Their Deposits, K. A. Vlasov, Ed., Russian ed., Publishing House "Nauka," Moscow, 1964; Z. Lerman, translator, Israel Program for Scientific Translations, Jerusalem, 1966 [distributed by Daniel Davey and Co., Inc., New York, N.Y.], pp. 491-524.

95. Kasymkhodzhaeva, U. S., and R. G. Osichkina, "Aktsessornye Rubidii i Tallii v Solyanykh Otlozhdeniyakh Uzbekistana" ["Accessory Rubidium and Thallium in Uzbekistan Salt Deposits"], Khim. Tekhnol. Miner. Udobr., 177-183 (1971); Chem. Abstr., **77**, 154,975y (1972).

96. Skinner, B. J., D. E. White, H. J. Rose, and R. E. Mays, "Sulfides Associated with the Salton Sea Geothermal Brine," Econ. Geol., **62**, 316-330 (1967).

97. Ladynina, I. N., and G. N. Anoshin, "Neketorye Zakonomernosti Raspredeleniya Rubidiya, Talliya, i Broma v Protsesse Fromirovaniya Mestorozhdenii Kaliinykh Solei" ["Some Principles of the Distribution of Rubidium, Thallium, and Bromine in the Formation of Potassium-Salt Deposits Geologiya i Geofizika, No. 3, 64-74 (1962).

98. Geilmann, W. K. Beyermann, K. H. Neeb, and R. Neeb, "Thallium ein regelmässig vorhandenes Spurenelement im tierischen und pflanzlichen Organismus" ["Thallium as a Trace Element for Animals and Plants"], Biochem Z., **333**, 62-70 (1960); Chem. Abstr., **55**, 14528f (1961).

99. Eschnauer, H., and R. Neeb, "Beiträge zur analytischen Chemie des Weines. IX. Voltammetrische Bestimmung geringster Thallium-Gehalte im Wein" ["Analytical Chemistry of Wines. IX. Voltammetric Determination of Trace Amounts of Thallium in Wine"], Z. Lebensm.-Untersuch. u.-Forsch., 112, 275-280 (1960).

100. Kothny, E. L., "Trace Determination of Mercury, Thallium and Gold with Crystal Violet," Analyst, 94(1116), 198-203 (1969).

101. Dvornikov, A. G., L. B. Ovsyannikova, and O. G. Sidenko, "Nekotorye Osobennosti Koeffitsientov Biologicheskogo Pogloshcheniya i Biogeo-khimicheskikh Koeffitsientov na Gidrotermal'nykh Mestorozhdeniyakh Donbassa v Sbyazi s Prognozirovaniem Skrytogo Rtutnogo Orudeneniya" ["Some Characteristics of the Coefficients of Biological Absorption and the Biogeochemical Coefficients at Hydrothermal Deposits of the Donets Basin in Relation to Predicting Hidden Mercury Mineralization"], Geokhimiya, No. 4, 626-633 (1976); Chem. Abstr., 85(10), 65891r (1976).

102. Batley, G. E., and T. M. Florence, "Determination of Thallium in Natural Waters by Anodic Stripping Voltammetry," J. Electroanal. Chem. Interfacial Electrochem., 61(2), 205-211 (1975).

103. Shvartsev, S. L., and L. N. Gomonova, "Thallium in Brines of the Angara-Lena Artesian Basin," Dokl. Akad. Nauk S.S.S.R., 220(5), 1183-1185 (1975); Chem. Abstr., 83, 46117k (1975).

104. Dawson, G. W., The Chemical Toxicity of Elements, BNWL-1915 UC-70, Battelle Pacific Northwest Laboratories, Richland, Wash., 1974.

105. Kitayama, M., and M. Saito, "Studies on the Prevention of Poisoning by Agricultural Chemicals (Part 17). Water Pollution by Rodenticide (Thallium Sulfate) Strewn on Forest Region From Helicopter," Hokkaidoritsu Eisei Kenkyusho-ho (Rep. Hok), 22, 92-95 (1972); abstract in TOXLINE thallium search.

106. Borovitskii, V. P., A. D. Miller, and V. N. Shemyakin, "Determination of Small Amounts of Gold in Natural Waters of the Aldan Region," Geokhimiya, No. 4, 483-488 (1966); Chem. Abstr., 65, 1958b (1966).

107. Struempler, A. W., "Trace Element Composition in Atmospheric Particulates During 1973 and the Summer of 1974 at Chadron, Neb.," Environ. Sci. Technol., 9(13), 1164-1168 (1975).

108. Tsukahar, I., M. Sakakiba, and T. Yamamoto, "Extraction-Spectrophotometric Determination of Traces of Thallium in Lead, Cadmium, Indium and Zinc Metals with Tri-n-Octylamine," Anal. Chim. Acta, 83, 251-25 (1976).

109. ASTM, 1974 Annual Book of ASTM Standards, Part 12. Chemical Analysis of Metals; Sampling and Analysis of Metal Bearing Ores, American Society for Testing and Materials, Philadelphia, Pa., 1974.

110. Paulsen, P. J., R. Alvarez, and D. E. Kelleher, "Determination of Trac Elements in Zinc by Isotope Dilution Spark Source Mass Spectrometry, Spectrochim. Acta, Part B, 24(10), 535-544 (1964).

SUBJECT INDEX